THE ROUTLEDGE COMPANION TO MILITARY RESEARCH METHODS

This new handbook is about the practices of conducting research on military issues.

As an edited collection, it brings together an extensive group of authors from a range of disciplinary perspectives whose chapters engage with the conceptual, practical and political questions raised when doing military research. The book considers a wide range of questions around research about, on, and with, military organisations, personnel and activities, from diverse starting-points across the social sciences, arts and humanities.

Each chapter in this volume:

- Describes the nature of the military research topic under scrutiny and explains what research practices were undertaken and why.
- Discusses the authors' research activities, addressing the nature of their engagement with their subjects and explaining how the method or approach under scrutiny was distinctive because of the military context or subject of the research.
- Reflects on the author's research experiences, and the specific, often unique, negotiations with the politics and practices of military institutions and military personnel before, during and after their research fieldwork.

The book provides a focussed overview of methodological approaches to critical studies of military personnel and institutions, and processes and practices of militarisation and militarism. In particular, it engages with the growth in qualitative approaches to military research, particularly research carried out on military topics outside military research institutions. The handbook provides the reader with a comprehensive guide to how critical military research is being undertaken by social scientists and humanities scholars today, and sets out suggestions for future approaches to military research.

This book will be of much interest to students of military studies, war and conflict studies, and research methods in general.

Alison J. Williams is Senior Lecturer in Political Geography in the School of Geography, Politics and Sociology at Newcastle University, UK. She is co-editor of *From Above: war, violence and verticality* (2013) and co-author of *The Value of the University Armed Service Units* (2015).

K. Neil Jenkings is Senior Research Associate in the School of Geography, Politics and Sociology at Newcastle University, UK. He is the author/editor of numerous titles, including most recently *The Value of the University Armed Service Units* (co-author, 2015).

Matthew F. Rech is Lecturer in Human Geography in the School of Geography, Earth and Environmental sciences at Plymouth University, UK. His research engages with everyday militarism and popular culture, particularly in the British context.

Rachel Woodward is Professor of Human Geography in the School of Geography, Politics and Sociology at Newcastle University, UK. She is the author of *Military Geographies* (2004) and co-author of *Sexing the Soldier* (Routledge, 2007).

THE ROUTLEDGE COMPANION TO MILITARY RESEARCH METHODS

Editors: Alison J. Williams, Newcastle University;
K. Neil Jenkings, Newcastle University;
Matthew F. Rech, Plymouth University;
Rachel Woodward, Newcastle University

Routledge
Taylor & Francis Group

LONDON AND NEW YORK

First published 2016
by Routledge
2 Park Square, Milton Park, Abingdon, Oxon OX14 4RN

and by Routledge
52 Vanderbilt Avenue, New York, NY 10017

First issued in paperback 2020

Routledge is an imprint of the Taylor & Francis Group, an informa business

© 2016 selection and editorial material, Alison J. Williams, K. Neil Jenkings, Rachel Woodward and Matthew F. Rech; individual chapters, the contributors

British Library Cataloguing in Publication Data
A catalogue record for this book is available from the British Library

Library of Congress Cataloging-in-Publication Data
CIP data has been applied for.

ISBN 13: 978-0-367-58162-6 (pbk)
ISBN 13: 978-1-4724-4275-8 (hbk)

Typeset in Perpetua
by Apex CoVantage, LLC

CONTENTS

Contents

FIGURES

CONTRIBUTORS

Stephen Atherton is a Lecturer in Education at Aberystwyth University, having previously completed a PhD in Human Geography also at Aberystwyth University. Stephen's research has focused on transitions from military to civilian life and the complex negotiations of masculine identities that are prevalent during this period. Stephen has also conducted research on the impact of military service on families and on the education of service children.

John Beck is Professor of Modern Literature and Director of the Institute for Modern and Contemporary Culture at the University of Westminster. John is the author of *Dirty Wars: Landscape, Power and Waste in Western American Literature* (University of Nebraska Press, 2009) and has also published on landscape photography and militarisation in *Cultural Politics*, *Theory, Culture & Society*, and *Tate Papers*, among other publications.

Aaron Belkin is a scholar, author, and advocate who has written and edited more than twenty-five scholarly articles, chapters and books. Since 1999, Aaron has served as founding director of the Michael Palm Center, which the *Advocate* named as one of the most effective gay rights organisations in the United States. He designed and implemented much of the public education campaign that eroded popular support for military anti-gay discrimination, and when 'don't ask, don't tell' was repealed, the president of the Evelyn and Walter Haas Jr. Fund observed that 'this day never would have arrived (or it would have been a much longer wait) without the persistent, grinding work of the Michael Palm Center.' Harvard law professor Janet Halley said of Aaron: 'Probably no single person deserves more credit for the repeal of "don't ask, don't tell."' Aaron's most recent book, *Bring Me Men*, is a study of contradictions in American warrior masculinity and the ways in which smoothing over those contradictions makes US empire seem unproblematic. It was first published by Columbia University Press in 2012 and then picked up by Oxford University Press in 2013.

Daniel Bos is a Teaching Associate in Historical and Cultural Geography at the University of Nottingham. Daniel's main research interest is on the relationship between popular culture and world politics. His PhD research, based at Newcastle University, explored the popular geopolitics of military-themed videogames, investigating their production and how their geopolitical and militaristic content is understood, experienced and internalised by their players.

Amanda Chisholm is a Lecturer in International Relations at Newcastle University. Having recently finished an ethnographic research project on Gurkha security contractors in Afghanistan, Amanda continues to be motivated by the ways in which everyday on-the-ground security practices are shaped by and are shaping global security markets. Informed by feminist security and political economy scholarships, Amanda remains interested by the ways in which race, class and gender intersect and make certain security subjectivities and security supply chains possible while foreclosing on others. By applying a postcolonial feminist analysis on security, she has, and continues to, actively contribute to the growing field of critical gender studies on private military and security companies (PMSCs). Amanda's current research plans continue to involve ethnographic methods to examine how states, markets, companies and contractors draw upon racial and gendered norms to establish security labour supply chains in remote and hostile environments within the Gulf and Asian regions.

Wayne Cocroft is a Senior Investigator with Historic England. For over twenty years Wayne has specialised in the archaeological recording and assessment of redundant defence sites and military production facilities. He is the author of *Dangerous Energy: The Archaeology of Gunpowder and Military Explosives Manufacture* (English Heritage, 2000); co-author of *Cold War Building for Nuclear Confrontation 1946–1989* (with Roger J. C. Thomas and P. S. Barnwell, English Heritage, 2003) and *War Art. Murals and Graffiti – Military Life, Power and Subversion* (with Danielle Devlin, John Schofield and Roger J. C. Thomas, Council for British Archaeology, 2006); and co-editor of *A Fearsome Heritage: Diverse Legacies of the Cold War* and *The Home Front in Britain 1914–18* (with John Schofield and Catrina Appleby, Council for British Archaeology, 2015). Wayne has also published many articles on the manufacture of munitions and the archaeology of rocketry. He is a Fellow of the Society of Antiquaries.

Gair Dunlop researches, writes and makes artworks exploring the interconnections of people, places and technologies. Gair has been working on military airfields, heritage cryptanalysis sites, new towns and atomic research centres. The core interest is in dreams of progress, their corruption and their transmission. Project materials include archive, contemporary and absurdist visions of technology and entropy. Finished works include single-screen films, video installations, and site-specific and interactive works. He teaches at Duncan of Jordanstone College of Art and Design, Dundee.

Christopher Elsey is currently a Research Associate based in the School of Social Sciences at Loughborough University. Primarily, Chris's work adopts a video ethnographic approach to study interactional practices and activities within various worksite settings rooted in an ethnomethodological analytic mentality and informed by conversation analytic studies. To date the worksites that Chris has researched include an educational setting for adults with a range of learning difficulties and disabilities; bedside teaching encounters and interactions between doctors, medical students and patients in primary and secondary care; 'Memory Clinic' consultations between neurologists, patients and family members (to improve diagnosis of dementia and other memory disorders); and military communication during an incident of friendly fire.

Matthew Farish is an Associate Professor in the Department of Geography and Planning at the University of Toronto. Author of *The Contours of America's Cold War* (University of Minnesota Press, 2010), his current research concerns the historical geographies of militarism and militarisation in the United States during the twentieth century. He is writing a history of the Distant Early Warning (DEW) Line.

Neil Ferguson is Professor of Political Psychology at Liverpool Hope University. He has been the Director of the Desmond Tutu Centre for War and Peace Studies, a Visiting Lecturer to Lock Haven University of Pennsylvania and the University of York and a Research Fellow at University of St Andrews; he has also previously lectured at the University of Ulster. His research and writings deal with moral development and a number of topics located within political psychology. Neil is currently the President of the MOSAIC – Moral and Social Action Interdisciplinary Colloquium and is a member of the Governing Council for the International Society of Political Psychology (ISPP). He also serves on the editorial boards of the *Journal of Moral Education* and the *Journal of Social and Political Psychology* and is a trustee of the *Journal of Moral Education Trust*.

Matthew Flintham is an artist and writer specialising in the hidden geographies of militarisation, security and surveillance. Matthew has a BA (Hons) in Fine Art from Central Saint Martins, an MA in Humanities and Cultural Studies from the London Consortium, and a PhD in Visual Communications from the Royal College of Art. His work intersects academic and arts practices, exploring speculative relationships between architecture, power and place, and the possibilities for arts methods to reveal hidden or immaterial relations in the landscape. Matthew has been Leverhulme Artist-in-Residence in the School of Geography, Politics and Sociology (GPS) at Newcastle University, and is currently Research Associate at the Centre for Architecture and the Visual Arts (CAVA) at Liverpool University. His research is featured in the edited volume *Militarized Landscapes: From Gettysburg to Salisbury Plain* (Continuum, 2010), *Tate Papers* (Issue 17) and *Emerging Landscapes: Between Production and Representation* (Ashgate 2014).

Isla Forsyth is an Assistant Professor in the School of Geography, University of Nottingham. Isla's work is positioned in cultural and historical geography with specific interest in critical military geographies, history of science and technology, and biography. Her previous research has focused on the history of British military camouflage, and she is currently studying covert military activities in the Desert War in WWII and the use of animals in warfare.

Emily Gilbert is an Associate Professor cross-appointed between the Canadian Studies Program at University College and the Department of Geography at the University of Toronto. Emily's current research deals with questions relating to contemporary changes to the military, especially its monetisation; victim compensation; and citizenship, mobility, borders and security. She is the co-editor of *War, Citizenship, Territory* (with Deborah Cowen, Routledge, 2008) and *Nation-States and Money: The Past, Present and Future of National Currencies* (with Eric Helleiner, Routledge, 2014).

John Hockey teaches a Master's Degree in Research Methodology at the University of Gloucestershire. John's current research interests include the application of sociological phenomenology to sport and occupations as well as the practice of ethnography. He has published extensively across the sociologies of sport, occupations and education, research which has been ethnographically grounded and theoretically propelled. At the British Sociological Association Conference in 2010 John was awarded a Sage prize for sociological innovation following a published paper on sensory phenomenology. He is also author of *Squaddies: Portrait of a Subculture* (University of Exeter Press, 1986).

K. Neil Jenkings is Senior Research Associate in the School of Geography, Politics and Sociology at Newcastle University. Neil is currently a co-investigator on the ESRC-funded 'Keeping Enough in Reserve: The Employment of Hybrid Citizen-Soldiers and the Future Reserves 2020 Programme'. Previous to this he was co-investigator on the ESRC-funded projects 'The Value of the University Armed Service Units' and 'The Social Production of the Contemporary British Military Memoir'. Neil was also the researcher on the ESRC-funded study 'Negotiating Identity and Representation in the Mediated Armed Forces'. In addition to publishing widely on various aspects of military and society, Neil has researched and published on the sociology of health and social care, rock climbing and aesthetics. He has a PhD in Sociology from a study of socio-legal decision making.

Sue Jervis, a former civil servant, is a trained psychodynamic counsellor who has worked as an independent practitioner, clinical supervisor and counselling service manager. Sue holds a MA in Psychoanalytic Studies and a PhD based on psycho-social research with service spouses. Being married to a (now retired) serviceman herself, Sue has personal experience of military life, including accompanying her husband to several overseas postings. She has presented papers and workshops both on the emotional experiences of service spouses and on the use of psychoanalytically informed reflexive research. More recently, she has supervised a number of undergraduate psycho-social students' dissertations. She has published a monograph, *Relocation, Gender and Emotion: A Psycho-Social Perspective on the Experiences of Military Wives* (Karnac, 2011), and has chapters in a number of edited books.

Kenneth MacLeish is Assistant Professor of Anthropology and Medicine, Health and Society at Vanderbilt University. Ken studies how war, broadly considered, takes shape in the everyday lives of people whose job it is to produce it: US military service members and their families and communities. He is the author of *Making War at Fort Hood: Life and Uncertainty in a Military Community* (Princeton University Press, 2013) and more recently 'The Ethnography of Good Machines' in *Critical Military Studies*. Ken's current research examines relationships between the harm, healing and enhancement of soldiers; ideas about military and veteran suicide, risk and resilience; and the production of morality in military medical interventions.

Michael Mair is Senior Lecturer in Sociology at the University of Liverpool. In the course of research which has been largely sociological, ethnographic and practice-oriented in character, Michael has explored a range of topics from an ethnomethodological perspective. Michael's most recent research falls into two main areas: politics, government and the state; and the methodology and philosophy of research. The focus of that work includes the politics of accountability in different settings, including military investigations into friendly fire deaths, and work on methodological practice in the social sciences, incorporating research on qualitative, quantitative and 'digital' methods. The different strands in Michael's research are brought together by a focus on practical action and reasoning, and the role of language, evidence and inquiry within them, in and across a range of sites and settings.

Jocelyn Mawdsley is a Senior Lecturer in European Politics at Newcastle University. Jocelyn's research has looked at both comparative defence policies in Europe and the growing role of the European Union in security policy. She has published widely on European armaments issues, most recently on the A400M project, Franco-British defence relations and the growth of the homeland security industrial sector. Her current research concentrates on security technologies and export controls, unmanned aerial vehicles (UAVs), large states and the Common Security and Defence Policy (CSDP) and the potential of interpretivism in security studies. Jocelyn is co-editor of the journal *European Security*.

Ross McGarry is a Senior Lecturer in Criminology within the Department of Sociology, Social Policy and Criminology at the University of Liverpool. Ross has written widely in international journals on criminology, victimology and military sociology, including the *British Journal of Criminology*, *Armed Forces and Society* and *Critical Criminology: An International Journal*. He is the co-editor (with Sandra Walklate) of *Criminology and War: Transgressing the Borders* (Routledge, 2015) and the forthcoming *Palgrave Handbook of Criminology and War* (Palgrave, 2016). He is co-author (also with Sandra Walklate) of *Victims: Trauma, Testimony and Justice* (Routledge, 2015) and the author of the forthcoming book *Criminology and the Military* from Routledge's New Directions in Critical Criminology series. Prior to becoming an academic Ross served as a military policeman in the British Armed Forces (RAF), and he previously worked for the civilian police service as a Projects Officer.

Ann Murphy has a PhD from Newcastle University, a Master's Degree in Sociology from Newcastle University, and a BSc Degree with First Class Honours from the University of Northumbria in Sociology and Social Research. Her PhD research examined army wives and their engagements with military and militarised landscapes, using an innovative photo-elicitation methodology which enabled military wives to document through photographs and reflective text their engagements with the landscapes in which they lived as spouses to military personnel. Her research interests include militarism in its varied social expressions, phenomenological approaches to landscape, the politics of gender and gender equalities, and ethics in sociological research.

Matthew F. Rech is a Lecturer in Human Geography in the School of Geography, Earth and Environmental Sciences at Plymouth University. Matthew's work engages with everyday forms of militarism and militarisation and their manifestation as representations, visualities and materialities. He recently completed a research associateship on the ESRC-funded project, 'Ludic Geopolitics: Children's Play, War Toys and Re-enchantment With the British Military' (led by Tara Woodyer, Portsmouth), which explored how contemporary geopolitics are expressed and enacted through play. In 2012 he was awarded a PhD from Newcastle University, where he previously completed Bachelor's and Master's Degrees in Human Geography. Recent publications appear in *Political Geography*, *Transactions of the Institute of British Geographers* and *Critical Military Studies*.

Ian Roderick is an Associate Professor of Communication Studies at Wilfrid Laurier University in Canada and the Special Issues Editor for the journal *Critical Discourse Studies*. Ian's research interests include critical military studies, studies in technology and society, and multimodal critical discourse analysis. He has previously published articles on the representation of soldiers by defence contractors and on military robotics and technological fetishism. His book, *Critical Discourse Studies and Technology: A Multimodal Approach to Analysing Technoculture* (Bloomsbury, 2016), is part of the new Bloomsbury Advances in Critical Discourse Studies series.

Steve Rowell is a research-based artist who works with still and moving images, sound, installation, maps and spatial concepts. Currently based in Los Angeles, Steve has lived in Berlin, Chicago, and Washington, DC, over the past twenty years. His transdisciplinary practice focuses on overlapping aspects of technology, perception and culture as related to ontology and landscape. Steve contextualises the built environment with the surrounding medium of nature, appropriating the methods and tools of the geographer and archaeologist. Since 2001, he has collaborated with the Center for Land Use Interpretation (Los Angeles), SIMPARCH (Chicago) and the Office of Experiments (London).

John Schofield is Professor and Head of Archaeology at the University of York. An archaeologist by training, John completed a PhD at Southampton University in 1989, after which he worked for English Heritage until 2010, prior to taking up his post. During his time with English Heritage, and subsequently at York, John specialised in cultural heritage and the archaeology of the contemporary past, with particular interests that include military architecture and landscape. He served English Heritage as Head of Military Programmes from 2002 until 2010, and has been Vice President of the International Council on Monuments and Sites (ICOMOS) International Scientific Committee on Military Fortifications and Landscape. John's fieldwork has included research in the UK, the US and across Europe, notably in Malta. John is a member of the Chartered Institute for Archaeologists and a Fellow of the Society of Antiquaries of London.

Justin Sikora is a Historic Resource Specialist with OC Parks in Orange County, California. Justin co-manages the interpretation, exhibit development, event planning, volunteer management, and internship coordination for the county's historic sites. His current research explores the post–World War II transformation of Southern California from agrarian ranch farming to rapid suburbanisation, and the tensions this brought within a profoundly militarised landscape. This investigation will be the subject of an exhibit which features the photographs of a photojournalist from Orange County who documented these changes from the late 1940s to the mid-1950s. Justin completed his doctoral research at Newcastle University, England, at the International Centre for Cultural and Heritage Studies, where his research explored on-site interpretation at historic battlefields and how sites' interpretive presentations influence visitors' valuations of these spaces.

Paul V. Smith is the Head of Student Support Services in the School of Social Sciences, University of Manchester. Paul's doctoral thesis, awarded in July 2013, was an ethnomethodological critique of the idea of practice in the New Literacy Studies. His research currently addresses practical reasoning and other epistemological themes – justification, evidence, discipline, recognisability – within the broad topic of academic writing. He has published on methodology, in particular the implications of an ethnographic methodology for the study of academic writing, and most recently has been developing an interest in the implications of the postsecular society for universities. This will be developed most prominently in studies of students' religious identity as they pertain to their academic work, drawing from ethnomethodology and from Foucault. His disciplinary background is in sociology, social anthropology and education. His involvement in the joint work on friendly fire extends his focus on practical accountability, multimodal analyses and plausible reasoning.

Jane Tynan is a design historian and Senior Lecturer at Central Saint Martins, University of the Arts London. Much of Jane's research concerns visual and material aspects of war and conflict, with a focus on uniforms and the body. She has published on militarized bodies in popular culture and the adoption of military codes by police and other civilian groups. Her book *British Army Uniform and the First World War: Men in Khaki* (Palgrave, 2013) explores the social and cultural meaning of military uniforms worn by combatants on the Western Front. Her current research project deals with images of insurgency.

David Walker is a Research Fellow in the Jubilee Centre for Character and Virtues at the University of Birmingham. David's interest in character and identity began during a first full career in the British Army. In 2010 he completed a PhD in Sociology and Social Policy at Durham University before moving to the United States to work at Purdue University, Indiana, in the Human Development and Family Studies Department (HDFS), where he researched National Guard members and their families. From September 2012 to February 2015 David was Principal Investigator for the Character and Virtue Education in British Schools project, and is now Principal Investigator for the Soldiers of Character research project, a three-year study of ethical climate and character among British soldiers and officers.

Vron Ware is a Professor of Sociology and Gender Studies at Kingston University London. Vron has published widely on issues of racism and gender, national identity, Britishness and the legacy of colonial history. Her study of Commonwealth soldiers in the contemporary British Army (*Military Migrants: Fighting for YOUR Country*) was first published by Palgrave Macmillan in 2012 and reissued in 2014.

Patrick G. Watson is a Lecturer in the Sociology Department and Social Psychology programme at McMaster University, Canada. Patrick's research focuses on questions of perception, evidence and knowledge in official settings such as legal proceedings or in government meetings. In particular, he is interested in the interaction between those consulted for their 'expertise' and those who decide how to proceed in light of expert advice. Alongside his journal and conference papers, he is currently working on two books: *Counting Cars, Counting Time, Counting Votes: The Gardiner Decision and the Prospect of 'Evidence Based Governance'* and *Scientific Governance: An Ethnography of Experts and Elected Officials*.

Neal White is Director of Emerge, the Experimental Media Research Group in the Media Production Department and Professor at the Bournemouth University. Neal's research and work as an artist, both of which draw on the relationships between art, design and new technologies, overlap and are driven by a long-standing concern with the impact of art and artists on society. Having cofounded the BAFTA award–winning, London-based art and technology group Soda (1997–2002), his interest in creative computing and critical media art led him into teaching and then research. In 2004, Neil founded the Office of Experiments (2004), an international network of artists, academics and architects based in London that has since developed research and projects with support from both within and beyond academia.

Alison J. Williams is Senior Lecturer in Political Geography in the School of Geography, Politics and Sociology at Newcastle University. Alison's research interests focus on the geographies of military airspaces and the geopolitics of aerial power projection. To date, this has included ESRC-funded research on the aerial geographies of the interwar Pacific, the contemporary geographies of UK military airspaces and, most recently, analysis of the value of the UK's university armed service units. In particular, her work has investigated the ways in which airspaces are performed, how geopolitics can critique the aerial dimension, the spatialities of the pilot-drone assemblage, and the ways in which international boundaries can be secured and maintained through the use of air power. Recent publications include *From Above: War, Violence and Verticality* (Hurst, 2013, co-edited with Peter Adey and Mark Whitehead) and *The Value of the University Armed Service Units* (Ubiquity 2015, co-authored with Rachel Woodward and K. Neil Jenkings).

Louise K. Wilson is a visual artist and Lecturer in Art and Design at the University of Leeds. Louise's work primarily uses sound to ask philosophical and material questions about the spatio-temporal physicality of certain sites and our perceptions of them. Published writing includes an interview with Paul Virilio (CTHEORY); an essay for *Private Views: Artists Working Today* (Serpents Tail); artist pages for *Zero Gravity – A Cultural Users Guide* (Arts Catalyst); and book chapters for *A Fearsome Heritage: Diverse Legacies of the Cold War* (Left Coast Press) and *Contemporary Archaeologies: Excavating Now* (Peter Lang). She has a PhD from the University of Derby.

Rachel Woodward is Professor of Human Geography at Newcastle University. Rachel's research interests include military geographies, landscapes and the politics of military land use; and the sociology of the armed forces and civil-military relations. She has recently completed a major study of contemporary military memoirs, to be published as *Bringing War to Book* with K. Neil Jenkings (Palgrave, forthcoming), and a study of student military units published as *The Value of the University Armed Service Units* with Alison J. Williams and K. Neil Jenkings (Ubiquity, 2015). She is currently part of a research team investigating reservist identities and employment issues in the British Armed Forces.

ACKNOWLEDGEMENTS

This book has developed out of, and would not have been possible without, a range of fruitful discussions with a host of people. These have continued over a number of years, including informal conversations at a range of Geography, Sociology, Military Studies, and other academic conferences, workshops and gatherings. Central to the emergence of this book were the contributions from those who attended the Military Methods workshop we ran at Newcastle University in June 2011. We are most grateful to those people who spent two days with us discussing the need for, and possible format of, what has become this Military Methods Research companion.

We are grateful to colleagues in the Power, Space, Politics cluster in Geography at Newcastle University, and to members of the university's Military, War and Security Research Group for ongoing intellectual stimulation about critical military scholarship.

We are indebted to the authors who have contributed to this volume, for their support of this endeavour and commitment to its publication. We would also like to thank Brenda Sharp and the team at Ashgate for their support throughout this process, the anonymous reviewer for their helpful feedback, and Hannah Lyons for her work in helping to prepare the manuscript for submission.

Finally, we would like to thank Katherine Nicholson, Joe Painter, Billy Williams, and Sophie Yarker for their personal support and encouragement that has enabled each of us, in our own ways, to see this book through to publication.

Alison J. Williams, K. Neil Jenkings,
Matthew F. Rech and Rachel Woodward

1

AN INTRODUCTION
TO MILITARY RESEARCH
METHODS

Matthew F. Rech, K. Neil Jenkings,
Alison J. Williams and Rachel Woodward

What's so special about military research methods? And why a whole book dedicated to them? This edited collection aims to comment on and give testament to the specificity of military research and the variety of methods deployed to address it. Military research poses a unique set of practical challenges for researchers working in civilian research contexts – challenges which are seldom found in other spheres of social-scientific research. These might relate to issues of access (to certain spaces, to research participants, to classified or redacted documents), to gatekeeper relations amid a convoluted and often gendered military hierarchical culture, or to the sensitivities of remembrance and the violation of bodies. Our original starting point with this book, informed by our own experiences and those of colleagues investigating a range of military-related topics in the social sciences and humanities, was to explore these sorts of issues in quite practical terms. We feel there is an urgent need for this since, although social science and humanities research into the military and the militarisation of Western democracies has developed and expanded in recent years, there is much more work to do. We argue that this lack of research is due, at least in part, to the unique challenges of developing military research methodologies, and hope that this collection may facilitate new and empirically rich scholarship from critical military perspectives.

This book is also warranted because, in our experience, military research almost inevitably requires some sort of personal engagement with questions about the politics of research, and with positionality. This might entail an explicit statement by the researcher on their attitude towards questions of military power and its consequences, or a more personal, internal negotiation of one's relationship to the military establishment. Reflexive awareness of researcher position means different things to the contributors to this volume. For some, doing military research entails the development or utilisation of a critical distance from the object of critique (i.e. militaries), one which involves an exploration of the myriad social, political and cultural consequences of military forces, militarisation and war-making. For others, it entails a more proximate inspection of the internal dynamics of military institutions and life-worlds through research facilitated by, or perhaps produced for, military organisations themselves.

For still others, military forces or institutions may constitute part of the context for research which, while not explicitly directed at the military, is irrevocably shaped by it. What we hope to show through this collection is that, above all else, there is a specificity to military research which suggests the need for attentiveness to the practicalities of research in military contexts, a reflexivity about that context, and a sensitivity to the ramifications of methods employed whatever the researcher's position.

The outcomes of military research are in part orientated towards the concepts and disciplinary debates which prompt research activities in the first place. However, this book is important, we argue, because room must be made for considerations of military research methods in their own right. Thus, this volume intends to speak to practicalities, politics, positions and complexities in an ever-growing and multidisciplinary scholarly landscape characterised by little consensus but much possibility. Our aim has been to do this in a format which provides insights into the range of topics and approaches for those with little experience of military research. For those with greater experience, we hope to provoke fresh ideas, new responses, and alternative approaches to the diverse conceptual, political and personal issues which military research raises.

In the remainder of this introduction we consider these themes in more detail. First, we focus on a key contextual issue, and foreground a discussion of military research methods by considering debates around the terms 'militarism' and 'militarisation'. We also raise questions about the continued relevance (or otherwise) of the identification of military specificity in methodological terms. We suggest that there is indeed a particularity to research in military contexts and on military topics, and explore the reasons why we believe this to be so. Second, we turn to the relationship between the methodological diversity of contemporary research on military-related topics, and more traditional methods and approaches originating in the social sciences in the post–Second World War period. Looking back, we try to explain the dominance of quantitative methodologies in military research, and point to the possibilities opened up, looking forward, by qualitative approaches, including those inspired by and developed in the arts and humanities. The third contextual issue we discuss concerns the position of the researcher and the scale and focus of inquiry. The 'military researcher', we suggest, often inhabits conflicted and contradictory positions vis-à-vis the politics of research. Drawing upon arguments articulated by critical approaches to military studies (e.g. Enloe, 2015; Rech et al., 2015), we build on this assertion and argue that a serious consideration of positionality here is of much broader methodological relevance than hitherto acknowledged. We conclude this introduction by explaining the purpose and structure of the book, and by introducing each section and its chapters.

Militaries, Militarism and Militarisation?

As noted earlier, what concerns the authors in this book are militaries, militarism and militarisation. In this respect, one of our central contentions is that a lack of methodological rigour, variety and reflexivity in military studies corresponds to the lack of clarity with which scholars in the social and political sciences have conceptualised these phenomena. We can begin to explore this contention by offering a more-or-less clear definition of the terms. First, *militaries* might be defined as the organisations authorised by sovereign powers to orchestrate state-sanctioned violence. However this traditional, state-centric definition hides complexities, slippages and overlaps (not least between the state and a variety of nonstate, quasi-military actors). It also obscures a fuller understanding of what militaries are, how they operate, and who and what they are composed of.

Militaries play a complex and adaptive role in the world, and increasingly so. However, in its very essence this expanded role challenges the meaning of 'militaries' implied in a state-centric definition. For example, while militaries are undoubtedly composed of men and women trained to use equipment and techniques which enable them to ensure the security of the state by force of arms, in recent years the rise of private military contractors (PMCs) has radically challenged this notion. Work by Higate (2012a, 2012b, 2013) illustrates that the word 'military' has been appropriated by PMCs to describe a group of trained individuals working to ensure the security of their employers through the threat and use of violence. However, PMCs have no recognised official state-sanctioned mandate, nor are they tethered to the defence of any one nation-state. Rather, the military in PMCs stands for a modus operandi: a set of learned behaviours and skills with weapons and allied equipment that cause civilians to take on the appearance and function of an armed state force, with whom they will often work alongside. PMCs are therefore just one illustration of the slippages that are occurring in the use of military terminologies, but one that also relates to personnel, technology and operations.

Another slippage can be found in much of the work done by traditional, state-run military forces themselves. Humanitarian and emergency relief operations, for instance, are an increasingly common mission for states' military forces, and are carried out in addition to more established roles such as peacekeeping and peace support. British Royal Navy ships now regularly carry humanitarian aid and supplies as standard stores in case of emerging need. Recent deployments by units like these to sites of natural disasters and other emergencies offer a different perspective on what military forces can and might achieve. Yet this diversity of operations causes us to question what a military is and is for in the twenty-first century. What these examples suggest is that our definition of militaries needs to be much more nuanced than the one offered earlier. Take the example of the US military. In 2013, at the same time as one of its aircraft carriers and many of its personnel were deployed to provide emergency relief to the hurricane-ravaged Philippines, other members of its forces were perpetrating drone strikes in the tribal areas along the Afghanistan-Pakistan border – the latter leading Amnesty International and others to accuse the US of international war crimes. Furthermore, sections of the military, at least in the UK, are increasingly involved in skills development and human resource training as part of outreach operations to businesses and universities (Woodward et al., 2015).

Militarism, on the other hand, can be defined in straightforward terms as an ideology which promotes the unproblematic acceptance of militaries and their (often preferential) use in international relations. Related to this, *militarisation* describes the processes and practices which support and enable the (re)production of militarism. Again, these terms are problematic and open to challenge, and a range of scholars have sought to consider and contest their implications (most recently Farish, 2013; Stavrianakis and Selby, 2013). Much of this work has emerged from a burgeoning field of scholarship in critical international relations (IR) and geography around the concept of security and the extent to which this concept (and set of practices) overlaps with and can be used as a alternative to militarism and militarisation. Indeed, Bernazolli and Flint (2009) have suggested that the terms militarism and militarisation should be replaced by the terms 'security' and 'securitisation', which they argue reflect more accurately the increased arming and militarised activities of police forces, as well as the noncombat operations of military organisations, such as the emergency response deployments noted earlier (see also Barkawi, 2011).

The replacement of 'military' with 'security' also illustrates, we suggest, the unease with which some scholars view military terminology. Given the association of military studies with military institutions (which we discuss in more detail later), there is often an eagerness to use the terminologies associated with security and securitisation in order to disassociate contemporary research from traditional military scholarship. While we recognise that issues around, and practices of, security and securitisation are very much in need of critical analysis (see Neocleous, 2008, 2011; Peoples and Vaughan-Williams, 2015), we reject the call to replace the terminology of military with that of security (see Woodward, 2014). Indeed, we contend that the emergence of a security studies and critical security literature requires us to be even more vocal and explicit about research on military forces, their practices and impacts, and to argue that militarism and militarisation remain vitally important terms if our task is to understand and challenge broader questions of power and politics in contemporary society.

The breadth and depth of (principally contemporary) military activities and deployments also means our spectrum of interest here stretches far beyond *just* war. However, while there are myriad subdisciplines that take the social, gendered and cultural constitution of militaries, militarism and militarisation very seriously, there is still little consensus on what, exactly, critical military scholarship should look like. Although recent developments, including the emergence of critical war studies as a conceptual concern within IR (see Barkawi and Brighton, 2011, and Hurst's Critical War Studies book series) and the publication of a new *Critical Military Studies* journal (see Basham et al., 2015), point to exciting new directions, they also imply a further compounding of the 'disciplinarity' of critical military/war studies. Despite this, in the present volume we have sought to provide a range of interventions that are suggestive of a critical military studies and some of its methodological entry points. But this also goes alongside a commitment to cross- and multidisciplinary dialogue, particularly in this case between the social scientists and artists. We have adopted this position because we recognise that warfare is only one (albeit the most newsworthy) facet of what military forces do, the conditions for which are sustained by a much broader set of everyday and often unexceptional practices.

Thus, in compiling this book we have actively sought to align ourselves with an approach which attempts to account for the manifold phenomena surrounding the preparation for war, but not necessarily including it. We do so in part because this allows us to privilege a focus on militaries, militarism and militarisation, terms and activities which, as we've discussed, are much debated. We also do this because critical war studies, as we see it, often fails to account for the breadth of human experiences implicated in and by militaries and militarism (partly because of its preference for theory). Critical military studies, conversely, foregrounds the empirical, focusing on applied and experiential analysis to uncover the range of encounters with the military that pervade our everyday lives.

It is in this conflicted, although vibrant, scholarly landscape that we site *The Ashgate Research Companion to Military Research Methods*. In order to make our argument and take these debates forward in a meaningful way, we argue for the central importance of *method* and a reflexive understanding of how and why research data is sought, gathered, used and presented. In the following section we discuss how methods for undertaking research on and with the military have developed. We offer a brief critique of more traditional approaches to open up space for a discussion of the range of methods articulated in the chapters of this book.

Military Research Methods: From the Traditional to the Critical

Research on the military and military phenomena is not new, but has arguably been neglected relative to other comparable organisations and phenomena of societal importance. Military research as a topic and a discipline needs reinvigorating, especially methodologically, because consideration of the most appropriate ways to account for these phenomena through empirical investigation has, with notable exceptions, been largely absent. The first attempts to account for and understand the attitudes and actions of military personnel were undertaken during and immediately after the Second World War using the relatively new techniques of statistical analysis being developed in sociology (see e.g. Stouffer, 1949; for an overview, see Boëne, 2000). This connection between quantitative methods and military research is also illustrated in the long history of geography's engagements with investigating military phenomena where, traditionally, the development of the tools and techniques of geographical analysis (such as mapping or remote sensing) was undertaken in no small measure for the benefit of military forces (see Woodward, 2004, 2005).

These quantitative methods were innovative in their time and emerged in an academic context where structural functionalism (theoretically) and positivism (methodologically) were in the ascendancy. They were facilitated by the development of practices which enabled the efficient collection of empirical data and the application of statistical techniques for its manipulation through emergent computing technologies. This traditional model of military sociology was, and remains, characterised by a hypothetico-deductive epistemology and a resultant emphasis on positivist methodologies and the development and testing of models of social relations. A number of edited collections give a good introduction to the scope and range of applications of this traditional quantitative sociological approach to the study of the military (see Kümmel and Prüfert, 2000; Caforio, 2003, 2007; Oullet, 2005). That these perspectives have been retained by military sociology over the past four decades, when the social sciences more generally has been marked by a pronounced shift towards methodological pluralism and an increasing scepticism about the claims and limits of quantitative approaches, is notable. For although the legitimacy of qualitative methods as part of the methodological toolbox open to researchers of the military is increasing (see Carreiras and Castro, 2013; Soeters et al., 2014), the fact remains that the study of military phenomena from social science perspectives is dominated by quantitative approaches to a degree unmatched elsewhere. We suggest two possible reasons for this.

First, it is a fact that beyond subdisciplinary areas self-consciously working under the labels of military sociology and military geography, the trend across the social sciences during the post–Second World War period has been for a lack of enthusiasm for studies of military phenomena (Woodward, 2005). This has been matched by an embrace of conceptual approaches informed by structuralist (primarily Marxist), poststructuralist (including some feminist) and interpretivist philosophies of knowledge (and indeed, conversely, antifoundationalist approaches informed by ethnomethodology), and an orientation in anglophone social science towards the idea of research as a tool assistive of strategies promoting greater social justice. Although it would be overstating the case to see traditional military sociology and military geography as complete intellectual backwaters or dead ends, it is notable that the intellectual drivers of social science from the 1960s onwards have been around topics and theorisations far removed from the study of military phenomena, although this is a trend that is changing.

Second, military sociology emerged as, and has continued to be, a subdiscipline highly attuned to the production, development and maintenance of state military organisations, their management, and the enhancement of their operational capacities. As Higate and Cameron (2006) have noted, military sociology has been dominated by an engineering rather than an enlightenment approach to the study of military phenomena, which has the aim of being of contributory benefit to armed forces and associated government military institutions. Governments and managers with an interest in the possibilities offered by social scientific investigation are notoriously keen on quantitative-based, seemingly definitive, results and less certain about the utility of arguments deploying more culturally nuanced or experiential data. This focus on the numerical representation of reality has probably been reinforced by militarily required assessments of troops and their motivations, which tend to be met with the use of quantitative methods of data collection and analysis.

Moreover, the functionalism of much military sociology has additional dynamics which reflect the nature of the military-academic research nexus (Jenkings et al., 2011). It is notable that many social scientists maintain close working relationships with military institutions by either working within or closely alongside them. Note, for example, the relationship between the RAND Corporation, responsible for much defence and security research in the US, and the US military establishment itself (RAND, 2015). In the UK, the close relationship between researchers at Kings College London, a major centre for military sociology, and key Ministry of Defence and armed forces training establishments is clearly mutually beneficial (KCL, 2015). Many military sociologists working within German, Dutch and Belgian contexts also have close military connections through their bases in the Bundeswehr Institute of Social Sciences, the Royal Netherlands Military Academy and the Royal Military Academy of Brussels, respectively. Alongside sociology, anthropologists also have a history of collaboration with the US military in particular (AAA, 2007) – a situation vehemently pursued by critical scholars, particularly in relation to the Counterinsurgency Field Manual (Network of Concerned Anthropologists, 2009), and the Human Terrain System and Bowman Expedition controversies (Bryan, 2010; Wainwright, 2012).

Leaving aside a broader discussion of the political and ethical issues raised by the military-academic nexus (which are significant but beyond our focus here; see Stavrianakis, 2009), military-social scientific collaboration or interaction – by which we mean work carried out together by military and nonmilitary academic actors whose aim is the co-production of knowledge – has a number of outcomes. First, it facilitates access to data, whether primary or secondary. Second, collaboration involves gatekeepers, who by virtue of their role have significant authority and power in shaping research trajectories (this applies, of course, to much social scientific research but is more pronounced with 'total institutions' like the military). Third, collaboration may require accepting military institutional definitions of acceptable methodologies, conceptualisations of the social world that underpin the development of research questions, and understandings of how research fits a broader national interest dictum. This is evident, for example, in processes surrounding UK Ministry of Defence ethical review of research on topics involving serving military personnel and their families. Summarily, the academic-military nexus in social science facilitates research while shaping the types of research being produced.

To be clear, we are not opposed in principle to quantitative methods in military research, and indeed wonder about the possibility for interventions like critical and feminist Geographical Information Systems (GIS; see O'Sullivan, 2006) to make their way into the military methodological lexicon. Rather, we argue that the predominance of quantitative methods in this field reveals a broader politics of knowledge production that can usefully be challenged by adopting critical qualitative research methodologies (see Jenkings et al., 2011).

Such a shift to the qualitative is vital for military research not just in relation to a politics of knowledge, however. It is also vital because there is a distinct lack of work on militaries, militarism and militarisation engaging with subjectivities, experiences and life-worlds. Within social sciences a range of intensive methods, most obviously interviews, ethnography, biography and the like, offer opportunities to investigate the personal and everyday implications of military activities. These approaches enable us to unpack the complexities of our engagements with military forces in more nuanced ways and across a much greater range of scales, sites, encounters, and perspectives than quantitative data allows. As Basham's (2013: 8; see also Enloe, 2000) recent exploration of *War, Identity and the Liberal State* rightly indicates, "there is much critical capacity in looking at the 'humdrum forms' that militarism and militarisation take," just as there is in exploring how warfare and war preparedness insinuate themselves into everyday lives. It is to these themes – intimacies, materialities, gendered identities, positionality, and the like – that many of the following chapters speak.

These qualitative approaches thus enable us to dig beneath the surface of the data generated through quantitative inquiry, to challenge the homogeneity of its results and to uncover hidden stories, revealing more nuanced and richer accounts and a more critical understanding of militarisation. However, while these methods lend themselves well to research seeking to explore the implications of military presence and activities beyond the barbed wire of the military camp, they can be more difficult to execute within the confines of military organisations themselves. As some of our authors discuss, these are challenges that provide opportunities to create exciting research encounters and fascinating data, yet they are not without their difficulties and limitations. This book seeks to illustrate the complexities of engaging with qualitative methods as well as considering the utility of the results that can be gleaned from their use.

Our second point of discussion around critical military research methods concerns cross- and interdisciplinarity. This book is testament to the belief that for a fuller and more critical military studies to be realised (one which adequately accounts for the range of phenomena associated with military life-worlds), we must also consider seriously cross-disciplinary, arts- and humanities-based methods. As Gair Dunlop describes in his chapter in this volume on artistic methods and (soon-to-be) decommissioned military spaces, 'conventional' art is indelibly part of military culture, with renderings of notable victories or regimental colours often adorning the walls of military bases. There is also a growing literature analysing artistic interpretations of military materiel, operations and cultures (see Apel, 2009; Williams, 2014). However, most important for those interested in military research methodologies are the efforts of a number of contemporary artists to engage critically and experimentally with the military establishment (e.g. Paglen, 2009, 2010), with current or past military spaces (see the chapters by Flintham, Dunlop and Wilson in this volume) or otherwise with cultures of militarism and militarisation (Banner, 2004, 2012; Berman, 2004, 2008; Friend, 2013; see also Williams, 2014). The methods these artists employ and the broader possibilities for academic-artist collaborations are, for us, just as important as the art itself. And insofar as arts-based and experimental methodologies are gaining purchase in a range of relevant disciplines (e.g. Driver et al., 2002; Thompson, 2009) we are excited by the possibilities their use by critical military scholars might open up.

In addition to this, we also acknowledge the importance of recognising that military methodologies are not just the preserve of academics. The lived-in worlds of militaries and military cultures are an important and perennial focus for journalists, bloggers, writers of fiction and military memoirists. Archives, reportage, novels and memoirs have long provided the military scholar with a source of empirical material on military campaigns, about specific individuals or both, and will continue to do so. But, of course, academics are not the target audience of these texts, and so the journeys of these materials and their impacts and interpretations within popular culture also offer important opportunities to critically investigate the processes of militarisation that occur in popular cultural worlds. In this volume, we point to the range of methods which might be applied to the outputs of both academic and nonacademic military research activities, along with the possibility for collaborative work with their various producers. In summary, we argue that the study of the military, militarism and militarisation warrants not only multidisciplinary engagements, but also methodological experimentation with and beyond the confines of traditional academic disciplines. In the following section we consider how this vision of critically engaged military methodologies impacts upon the position of the researcher and the scale at which military research is located.

Scale, Positionality and Research Foci

The past ten years or so have seen a marked effort across the social and political sciences to rethink the scale at which scholarly investigation of militaries, militarism and militarisation should take place. The context for this shift can be located in a number of subdisciplines across the social sciences. Pivotal to this enterprise has been the work of Christine Sylvester, whose *Experiencing War* (2011), *War as Experience* (2013a) and *Masquerades of War* (2015; see also the associated book series *War, Politics and Experience*) mark a burgeoning of scholarship which seeks to question dominant frames of analysis, principally those associated with IR but also with sociology (in particular McSorley, 2007, 2010, 2012, 2014; McSorley and Maltby, 2012), geography, anthropology (e.g. Lutz, 2001, 2009) and media and communications studies. As Sylvester (2013b: 671) notes, despite IR's insistence that it knows war, it is nevertheless "historically disinterested in probing the vast expanse of war's ordinary." A genuinely war-focused IR, she notes, should prioritise "looking at the social aspects of war, people and/in/as war" and by "identifying and taking up the marginalized, excluded or hidden social, interactive, moving and changing participants and discourses of war" (Sylvester, 2013b: 671).

The tenor of these engagements is mirrored by work in the field of feminist geopolitics. Here, scholars such as Jennifer Hyndman (2001, 2003, 2007), Jennifer Fluri (2009, 2014), Joanne Sharp, Lorraine Dowler (Dowler and Sharp, 2001; Sharp, 2007) and others have sought to move beyond a "disembodied space of neorealist geopolitics to a field of live human subjects with names, families and hometowns" (Hyndman, 2007: 36). Although not concerned only with matters of militaries, militarism and militarisation, the feminist geopolitical project has as its context a broader attempt by critical geographers to think through the intimate scales of war (Pain, 2015; see also Cowen and Story, 2013), alternative and nonstate securities (Koopman, 2011), the banality of terror (Katz, 2007), the politicisation of the militarised body and the militarisation of childhood and youth (Hörschelmann, 2008; Rech, 2016). These approaches bring into focus an effort to think across and beyond the global and the intimate (e.g. Pain and Staeheli, 2014 and associated special edition), and enable the nuancing and thickening of scholarship that focuses on the strategic and political elements of military research, providing an insight into how processes of militarisation and the production of military power operate at a range of scales from the individual body to the state.

Apart from providing valuable inspiration for this volume, this multidisciplinary rescaling of critical military studies has two significant implications for us. First, it points very clearly to the importance of highlighting the human stories behind militaries, militarism and militarisation. However, it also means thinking seriously about positionality and the role of the researcher. As many of the chapters in this volume demonstrate, we need to think carefully about situating military research amid the life-worlds of researcher and researched and the difficulties inherent to this. Indeed, as some of our authors illustrate, it can be difficult to separate military research from the biography of the researcher (because of personal interest, political inclination, past military service and the like). These are not insurmountable problems, if they are indeed problems at all, although they cannot be overcome by the imposition of rational quantitative methods. Rather, it prompts us to recognise our positions as military researchers, and the value of thinking about and discussing method in order to situate our practices of knowledge production. However, while recognising our own positionality as an important part of critical military research, it does confront us with the question of *our own* militarisation, as some of our authors note. In recognising the human stories intrinsic to militarisation and the operation of military forces, we have to be aware of our own roles in and around these stories and the recursive processes at work that inculcate us, for example, with military knowledge, with the ability to understand military jargon or the capacity to operate effectively within military landscapes. Thus critical military studies must take account of these intersections, and problematise and resolve them every bit as much as we question the problematic objectivity of quantitative data practices.

These positional complexities lead us to the second implication of rescaling critical studies of the military. This concerns the issue of criticality, and specifically, whether being critical of militaries, militarism and militarisation can and should necessitate an antiwar and/or antimilitary stance on the part of the researcher. As we noted earlier, some scholars have actively sought to distance themselves semantically from military research through the adoption of critical war studies or a security studies frame. For those of us who have sought to adopt a critical military studies perspective, the drive to ensure we are not associated with traditional military research can result in or be driven by an a priori antimilitarist stance. In the field of critical geography, a similar move to radically dissociate contemporary scholarship from geography's imperial and military past has led some to offer a distinctly nonviolent vision for critical geopolitics (Megoran, 2008). Although geographers point out the often entangled and conflictual nature of antimilitarism (e.g. Davis, 2011; Woon, 2013) and peace-building (McConnell et al., 2014), it is clear that some in the discipline are committed to "destabilizing, contesting and challenging a killing society . . . [and to building] a broad coalition of academics and activists who are focused on positive peace building practices . . . and on alternatives to war and militarism" (Tyner and Inwood, 2011: 453). What this 'broad coalition' might look like is arguable. However, there are clear opportunities in the field of popular culture where, in the spirit of Ingram's (2011) work, scholars profile the work of dissident artists, hackers and culture jammers (Martin and Steuter, 2010; Stahl, 2010). Engagements with activists and in activism could also offer a productive avenue too, perhaps as part of scholarship around counter-military recruitment (Allison and Solnit, 2007; Harding and Kershner, 2011; Friesen, 2014). Adopting an antimilitary positionality towards topics needs also to be examined reflexively as part of the methodology in the same way that any promilitary inculcation needs to be.

However, while this critical and antimilitarist work offers significant insights, we suggest that it is equally important to recognise the ability to adopt a critical stance that also advocates *working with* militaries (as many of our contributors have done), rather than shunning them, in order to create opportunities to develop the nuanced, rich and intensive methodological engagements discussed throughout this volume. We adopt this stance because we believe there is room to influence change in military institutions – or at least to try and open up possibilities for this – via research and collaboration. This sort of work would necessarily have to be reflexive about positionality, critical about military organisations and practices, yet alive to the complexities and nuances of military forces and processes of militarisation, many of which can only be unpacked and critiqued through close and collaborative engagement (Rech et al., 2015).

The Structure of the Book

In the remainder of this chapter we introduce each of the four sections of the book and the chapters therein, illustrating how this critical approach to military research can be adopted through the use of a variety of methods and from a range of disciplinary positions.

Section 1: Texts

The first set of chapters focuses on how social science and arts and humanities scholars engage with military-related texts, ranging from official archives to personal memoirs. Each of the chapters sets out to discuss a particular methodological engagement and to critique and reflect the author's experience of working with these documentary sources. The first three chapters in Section 1 focus on different experiences of using archival material as a basis for military research. Matthew Farish (chapter 2) begins with a set of reflections on being a civilian social science researcher working with official US and Canadian military archives, often located on active military bases. He discusses a range of challenges, including the practical issues of gaining access to repositories on closed military sites, to intellectual concerns with the partiality and bias of these collections. Emily Gilbert (chapter 3) takes up these ideas through a focus on access to potentially sensitive records. Her work profiles the difficulties faced when attempting to extract information from the US military's financial transaction records relating to death payments to civilians in Iraq. Gilbert succinctly illustrates some of the key issues that research on military texts brings to the surface: freedom of information requests, redaction of material, and the partiality of record-keeping. Completing the focus on the utility of the military archive, Isla Forsyth (chapter 4) discusses the opportunities offered by archival records and the detective work that is required to make best use of these collections. Subsequently, Forsyth reflects on producing biographies for people who, despite not being on the front lines of military operations, nonetheless impacted military policy and practice.

The next three chapters in the Section 1 take us in different directions and focus on published textual material. K. Neil Jenkings and Daniel Bos (chapter 5) describe their experience of using UK newspapers to conduct research on the town of Wootton Bassett (made famous because it was the host for a number of roadside ceremonies which marked the repatriation of the bodies of British service personnel between 2007 and 2010). Jenkings and Bos discuss the complexities of using online search tools, both specialist newspaper databases and specific newspaper search engines, to amass a set of source material that could be analysed for the project. Their chapter provides useful insights into the issues and opportunities that textual methodologies offer. Next, Rachel Woodward and K. Neil Jenkings (chapter 6) discuss the utility of and processes for using published military memoirs as a research tool. Drawing upon extensive experience, they offer reflections on how researchers can engage with these mass-market media, the caveats that must be borne in mind when using this source material to gain insights into military operations, and how these books offer a view into information not available in official military histories. John Beck (chapter 7) discusses how fictional literature has engaged with and represented war and militarism. He offers a unique perspective, suggesting how such works of literature can be analysed to understand how ideas of militarism permeate fictional worlds.

The final chapter by John Schofield and Wayne Cocroft (chapter 8) documents how archaeological work on twentieth-century military sites has not only generated insights into military activities across the century, but also informed archaeological practice more broadly. Military archaeological practice, they argue, offers a broader framework for understanding armed conflict, and suggests how the ground can be read as a text.

Section 2: Interactions

Section 2 brings together chapters under the theme of interactions. The scene is set by Jocelyn Mawdsley (chapter 9), who makes the case for an interpretivist approach to data collection and analysis. Mawdsley illustrates some of the problems that arise when using large-scale data sets to investigate military activities. This is contrasted with a discussion of some of the benefits of using case studies of the relationships between and activities of military research participants. Ross McGarry (chapter 10) recounts a study of the repatriation of the bodies of British service personnel through the small English town of (now Royal) Wootton Bassett. These repatriations emerged as a significant phenomenon in recent British military culture; events which McGarry frames as happening in a 'liminal' space facilitated and sustained by the townsfolk. McGarry makes a differentiation between research at 'long' and 'short' ranges and makes the case for the importance of ethnographic investigations. Amanda Chisholm (chapter 11) then gives a personal and practical account of undertaking an ethnography of private security contractors as a multinational community in Afghanistan. She illustrates the practicalities of gaining and maintaining access in a war zone where issues of race and gender are both embodied in the practices of the researcher and researched, but also constantly negotiated through interactions with gatekeepers and participants and in the requirements of personal security. Chisholm demonstrates that interaction within and beyond straightforward fieldwork activities constitutes reflexive data in itself, and may also be the context in which other data is made meaningful.

Neil Ferguson (chapter 12) turns to the Troubles in Northern Ireland and offers an introduction to the nature of the conflict and the origins of participation in it for paramilitary members on both sides of the sectarian divide. Ferguson's use of an interpretative phenomenological analysis relies on interviews with both open and semistructured phases. This required the utmost discretion with regards to the security of both the interviewers and the participants, highlighting the sensitivity required when dealing with topics of sectarian and political violence. Ferguson notes the stresses placed on all concerned, in terms of access and topics covered, and the benefits of a small team to minimise their impact. The psychological orientation of some research methodologies is also illustrated by Sue Jervis (chapter 13), who broadens the notion of a 'participant' in military institutions in her investigation of military service spouses (a category which includes her) and their relationship to military environments. She explains practical issues such as negotiating access, but also outlines the psychologically informed reflexive research method she adopted involving questionnaires, interviews and participant case studies.

Christopher Elsey, Michael Mair, Paul V. Smith and Patrick G. Watson (chapter 14) look at interaction directly in their ethnomethodological and conversation analysis study of an incident of so-called friendly fire during the second Gulf War. Here, the interactions analysed were those captured on cockpit video and taken from transcripts from a subsequent court martial. Not only do they make the case for a conversation analysis methodology, but they also illustrate the problems of using secondary data without understanding the nature of the interactional practices which go into the production and collection of data. They also suggest that what data is said to be and show might fruitfully be investigated through an understanding of how research participants themselves interpret their own conversational data. Finally, Aaron Belkin (chapter 15), in an impassioned piece, asks us to look at a broader interactional context for military research, and specifically the relationship between citizens and their militarised governments. Using the case of the US, he argues that a normalisation of structured contradictions allows the US government to pursue militarism and militarisation, the barbarity of which, he argues, is portrayed as a noble activity. Belkin urges us to use our research to reveal such structures, and in doing so reminds us that while attending to methodology we should also commit to a dissident and relevant critical military studies.

Section 3: Experiences

Section 3 examines the lived experiences of doing military research. It includes chapters by authors from a range of disciplinary backgrounds who have widely varying personal biographies and equally differing research intentions. The section begins with John Hockey's (chapter 16) reflections on fieldwork conducted with a British infantry platoon, in which the challenges of participant-observation and the ethnographic encounter are explored in reference to both his conceptual framework and personal experiences of military service. This includes a discussion of the presentation of the self in the field, and of the negotiation of this self with research participants. Kenneth MacLeish (chapter 17) uses his experiences as an anthropologist working in Fort Hood, again using ethnographic methods, to tackle quite explicitly what ethnographic writing can bring to wider frames of knowledge about military institutions, and particularly the politics of knowing and understanding. Vron Ware (chapter 18) tells the story of her involvement as a sociologist researching the recruitment of Commonwealth personnel in the contemporary British Army. This includes a discussion of the challenges of writing about military forces and their activities, including a negotiation of the different assumptions and expectations of her diverse readership. Stephen Atherton (chapter 19) considers how, in his research on military masculinities and the places and practices of domesticity, the dynamics of interview encounters generated insights into the role of emotion and ethics in the production of knowledge. David Walker (chapter 20) explores how his engagements with the idea of 'insider-ness' shaped his approach to researching the exit strategies of career soldiers, and uses his experiences of empirical data collection to reflect on the possibilities and limits of insider perspectives.

The final two chapters engage with rather different sites for research. Matthew F. Rech and Alison J. Williams (chapter 21) reflect on attending airshows, and interrogate their personal motivations as critical military researchers *and* willing participants in these military cultural events. Justin Sikora (chapter 22) concludes the section by offering an account of heritage issues on historic battlefields. Specifically, he explores the paradoxes apparent in negotiating sites which, despite their military past, bear scant traces of that military imprint in the present.

Section 4: Senses

In Section 4, the authors consider how various sensory faculties are enrolled by military researchers (and by research participants), and the accompanying politics of sense-making. Beginning with a piece which denotes the breadth of possibility for a multisensory approach to military research, Jane Tynan (chapter 23) investigates the militarised body and uniform design, and implicates the researcher in a complex set of inquiries into representational, visual and material worlds – a theme which recurs throughout the section. In chapter 24, Ian Roderick offers an insight into the multiple (and multiplying) representational spaces in which the military researcher is often implicated, and in doing so provides thoughts on the social semiotics of US Department of Defense military image banks. Next, Daniel Bos (chapter 25) and K. Neil Jenkings, Ann Murphy and Rachel Woodward (chapter 26) demonstrate the porous and unstable boundaries of 'representation'. Bos's chapter explores the on- and off-line worlds of military gaming and the intricacies of player engagement with a simultaneously discursive and material phenomenon. Jenkings, Murphy and Woodward explore the utility of image-elicited interviews, and demonstrate how photography is particularly suited to revealing the British military's visual culture.

Matthew Flintham's chapter (27) on visualising military airspace is the first of four artistic interventions into military research methods offered also by Gair Dunlop (chapter 28), Neal White and Steve Rowell (chapter 29) and Louise K. Wilson (chapter 30). Each chapter explores sight, sound and haptics in different ways, and speaks, notably, to invisible, off-limits or secret military and ex-military spaces. Gair Dunlop explores the methodological strategies he adopted in his work documenting military base closures, which are often protracted periods in which processes of remembering become shared between military and civilian populations. White and Rowell describe a range of research projects under the banner of 'The Office of Experiments' – an artists' collective exploring the relationship between culture and the techno-scientific and military-industrial complexes – offering a photoessay on their overt methodological practices in extraordinary, sometimes off-limits military spaces. Louise K. Wilson, in perhaps the most polemical treatment of the 'sovereign sense' in military research, engages auditory perception and deals with the militarised soundscapes of Cold War Britain. This section, therefore, deals in large part with the senses as the researcher enrols them methodologically. But it also asks how the senses are often themselves militarised, and how regimes of sense-making are co-opted by the military establishment. However, and while we do not engage with this in the present volume, these chapters reflect the fact that senses of pleasure, enjoyment, thrill, desire and their corresponding sensory faculties should be more seriously considered in critical military research. As Joanna Bourke (2000: 1) reminds us, the characteristic act of war is killing – something which is prosecuted by "individuals [who in so doing are potentially] transformed by a range of conflicting emotions – fear as well as empathy, rage as well as exhilaration." Therefore, along with 'fear, anxiety and pain', a focus on senses also implies an equivalent interrogation of 'excitement, joy and satisfaction' (Bourke, 2000), as experienced both by research subjects and military researchers.

References

AAA (2007) *Commission on the Engagement of Anthropology With the US Security and Intelligence Communities.* Arlington, VA: American Anthropological Association.

Allison, A. and Solnit, D. (2007) *Army of None: Strategies to Counter Military Recruitment, End War, and Build a Better World.* London: Seven Stories Press.

Apel, D. (2009) Iraq, Trauma and Dissent in Visual Culture. In Pugh, J. (Ed.), What Is Radical Politics Today? (pp. 92–102), Basingstoke: Palgrave Macmillan.

Banner, F. (2004) *All the World's Fighter Planes.* London: Vanity Press.

Banner, F. (2012) *Harrier and Jaguar.* London: Tate.

Barkawi, T. (2011) From War to Security: Security Studies, the Wider Agenda, and the Fate of the Study of War. *Millennium: Journal of International Studies* 39(3), 701–761.

Barkawi, T. and Brighton, S. (2011) Powers of War: Fighting, Knowledge, and Critique. *International Political Sociology* 5, 126–143.

Basham, V. (2013) *War, Identity and the Liberal State: Everyday Experiences of the Geopolitical in the Armed Forces.* London: Routledge.

Basham, V., Belkin, A. and Gifkins, J. (2015) What Is Critical Military Studies? *Critical Military Studies* 1(1), 1–2.

Berman, N. (2004) *Purple Hearts: Back from Iraq.* London: Trolley.

Berman, N. (2008) *Homeland.* London: Trolley.

Bernazolli, R. and Flint, C. (2009) From Militarization to Securitization: Finding a Concept That Works. *Political Geography* 28, 449–450.

Boëne, B. (2000) Social Science Research, War, and the Military in the United States. In Kümmel, Gerhard and Prüfert, Andreas (Eds.), *Military Sociology: The Richness of a Discipline.* Baden-Baden: Nomos Verlag, pp. 149–251.

Bourke, J. (2000) *An Intimate History of Killing: Face-to-Face Killing in Twentieth-Century Warfare.* London: Granta Books.

Bryan, J. (2010) Force Multipliers: Geography, Militarism, and the Bowman Expeditions. *Political Geography* 29, 414–416.

Caforio, G. (Ed.) (2003) *Handbook on the Sociology of the Military.* New York: Kluwer Academic.

Caforio, G. (Ed.) (2007) *Social Sciences and the Military: An Interdisciplinary Overview.* London: Routledge.

Carreiras, H. and Castro, C. (Eds.) (2013) *Qualitative Methods in Military Studies: Research Experiences and Challenges.* London: Routledge.

Cowen, D. and Story, B. (2013) Intimacy and the Everyday. In Dodds, K., Kuus, M. and Sharp, J. (Eds.), *The Ashgate Companion to Critical Geopolitics.* Farnham: Ashgate, pp. 528–542.

Davis, S. (2011) The US Military Base Network and Contemporary Colonialism: Power Projection, Resistance and the Quest for Operational Unilateralism. *Political Geography* 30, 215–224.

Dowler, L. and Sharp, J. (2001) A Feminist Geopolitics? *Space and Polity* 5(3), 165–176.

Driver, F., Nash, C., Prendergast, K. and Swenson, I. (2002) *Landing Eight Collaborative Projects Between Artists + Geographers.* London: Royal Holloway University of London.

Enloe, C. (2000) *Bananas, Beaches and Bases: Making Feminist Sense of International Politics.* London: University of California Press.

Enloe, C. (2015) The Recruiter and the Sceptic: A Critical Feminist Approach to Military Studies. *Critical Military Studies* 1(1), 3–10.

Farish, M. (2013) Militarisation. In Dodds, K., Kuus, M. and Sharp, J. (Eds.), *The Ashgate Companion to Critical Geopolitics.* Farnham: Ashgate, pp. 247–265.

Fluri, J. (2009) Geopolitics of Gender and Violence from Below. *Political Geography* 28, 259–265.

Fluri, J. (2014) Women as Agents of Geopolitics. In Dodds, K., Kuus, M. and Sharp, J. (Eds.), *The Ashgate Companion to Critical Geopolitics.* Farnham: Ashgate, pp. 509–526.

Friend, M. (2013) *The Home Front.* Stockport: Dewi Lewis.

Friesen, M. (2014) Framing Symbols and Space: Counterrecruitment and Resistance to the U.S. Military in Public Education. *Sociological Forum* 29(1), 75–97.

Harding, S. and Kershner, S. (2011) 'Just Say No': Organizing Against Militarism in Public Schools. *Journal of Sociology & Social Welfare* 38, 79–109.

Higate, P. (2012a) The Private Militarized and Security Contractor as Geocorporeal Actor. *International Political Sociology* 6(4), 355–372.

Higate, P. (2012b) 'Cowboys and Professionals': The Politics of Identity Work in the Private and Military Security Company. *Millennium – Journal of International Studies* 40(2), 321–341.

Higate, P. (2013) 'Switching on' for Cash: The Private Militarised Security Contractor as Geopolitical Actor. In McSorley, K. (Eds.), *War and the Body: Militarisation, Practice and Experience*. London: Routledge, pp. 106–127.

Higate, P. and Cameron, A. (2006) Reflexivity and Researching the Military. *Armed Forces & Society* 32(2), 219–233.

Hörschelmann, K. (2008) Populating the Landscapes of Critical Geopolitics – Young People's Responses to the War in Iraq (2003) *Political Geography* 27, 587–609.

Hyndman, J. (2001) Towards a Feminist Geopolitics. *Canadian Geographer* 45(2), 210–222.

Hyndman, J. (2003) Beyond Either/or: A Feminist Analysis of September 11th. *ACME* 2(1), 1–13.

Hyndman, J. (2007) Feminist Geopolitics Revisited: Body Counts in Iraq. *Professional Geographer* 59(1), 35–46.

Ingram, A. (2011) Bringing War Home: From Baghdad 5th March 2007 to London, 9 September 2010. *Political Geography* 31, 61–63.

Jenkings, K. N., Woodward, R., Williams, A. J., Rechm, M. F., Murphy, A. L. and Bos, D. (2011) Military Occupations: Methodological Approaches to the Military-Academic Research Nexus. *Sociology Compass* 5(1), 37–51.

Katz, C. (2007) Banal Terrorism. In Gregory, D. and Pred, A. (Eds.), *Violent Geographies: Fear, Terror, and Political Violence*. London: Routledge, pp. 349–362.

KCL (2014) War Studies Department website. Available at: http://www.kcl.ac.uk/sspp/departments/warstudies/index.aspx (Accessed 10 April 2015).

Koopman, S. (2011) Alter-Geopolitics: Other Securities are Happening. *Geoforum* 42(3), 274–284.

Kümmel, G. and Prüfert, A. (Eds.) (2000) *Military Sociology: The Richness of a Discipline*. Baden-Baden: Nomos.

Lutz, C. (2001) *Homefront: A Military City and the American 20th Century*. Boston: Beacon Press.

Lutz, C. (2009) The Military Normal: Feeling at Home With Counterinsurgency in the United States. In Network of Concerned Anthropologists (Eds.), *The Counter-Counterinsurgency Manual: Or, Notes on Demilitarizing American Society*. Chicago: Prickly Paradigm Press, pp. 23–37.

Martin, G. and Steuter, E. (2010) *Pop Culture Goes to War: Enlisting and Resisting Militarism in the War on Terror*. Lexington: Plymouth.

McConnell, F., Megoran, N. and Williams, P. (Eds.) (2014) *Geographies of Peace*. London: I. B. Tauris.

McSorley, Kevin (2007) Review Essay: The Mark of Cain: War, Media and Morality. *The War and Media Network*. Available at: http://eprints.port.ac.uk/3859/ (Accessed 28 January 2016).

McSorley, Kevin (2010) The Reality of War? Politics Written on the Body. *New Statesman*. Available at: http://www.newstatesman.com/blogs/cultural-capital/2010/06/war-politics-bodies-body (Accessed 28 January 2016).

McSorley, Kevin (2012) Helmetcams, Militarized Sensation and 'Somatic War'. *Journal of War and Cultural Studies* 5(1), 47–58.

McSorley, Kevin (2014) Towards an Embodied Sociology of War. *Sociological Review* 62(2), 107–128.

McSorley, Kevin and Maltby, S. (2012) War and the Body: Cultural and Military Practices. *Journal of War and Cultural Studies* 5(1), 3–6.

Megoran, N. (2008) Militarism, Realism, Just War, or Non-Violence? Critical Geopolitics and the Problem of Normativity. *Geopolitics* 13, 473–497.

Neocleous, M. (2008) *Critique of Security*. Edinburgh: University Press/McGill-Queens University Press.

Neocleous, M. (2011) *Anti-Security*. Ottawa, ON: Red Quill Press.

Network of Concerned Anthropologists (2009) *The Counter-Counterinsurgency Manual: Or, Notes on Demilitarizing American Society*. Chicago: Prickly Paradigm Press.

O'Sullivan, D. O. (2006) Geographical Information Science: Critical GIS. *Progress in Human Geography* 30(6), 783–791.

Oullet, E. (2005) *New Directions in Military Sociology*. Whitby, ON: de Sitter.

Paglen, T. (2009) *Blank Spots on the Map: The Dark Geography of the Pentagon's Secret*. London: Penguin.

Paglen, T. (2010) *Invisible: Covert Operations and Classified Landscapes Hardcover*. New York: Aperture.

Pain, R. (2015) Intimate War. *Political Geography* 44, 64–73.

Pain, R. and Staeheli, L. (2014) Introduction: Intimacy-Geopolitics and Violence. *Area* 46, 344–347.

Peoples, C. and Vaughan-Williams, N. (2015) *Critical Security Studies* (2nd edition). London: Routledge.

RAND (2015) RAND Corporation Website. http://www.rand.org/ (Accessed 10 April 2015).

Rech, M. F. (2016) Children, young people and the everyday geopolitics of British Military recruitment. In Benwell, M. and Hopkins, P. (Eds.), *Children, Young People and Critical Geopolitics*. Farnham: Ashgate, pp. 45–60.

Rech, M., Bos, D., Jenkings, K. N., Williams, A. and Woodward, A. (2015) Geography, Military Geography and Critical Military Studies. *Critical Military Studies* 1(1), 47–60.

Sharp, J. (2007) Geography and Gender: Finding Feminist Political Geographies. *Progress in Human Geography* 31, 381–387.

Soeters, J., Shields, P. M. and Rietjens, B. (Eds.) (2014) *Routledge Handbook of Research Methods in Military Studies*. London: Routledge.

Stahl, R. (2010) *Militainment, Inc.: War, Media, and Popular Culture*. London: Routledge.

Stavrianakis, A. (2009) In Arms Way: Arms Company and Military Involvement in Education in the U.K. *ACME* 8(3), 505–520.

Stavrianakis, A. and Selby, J. (Eds.) (2013) *Militarism and International Relations: Political Economy, Security, Theory*. London: Routledge.

Stouffer, S. (1949) *The American Soldier: Combat and Its Aftermath*. Volume II. Princeton, NJ: Princeton University Press.

Sylvester, C. (2011) *Experiencing War*. London: Routledge.

Sylvester, C. (2013a) *War as Experience: Contributions from International Relations and Feminist Analysis*. London: Routledge.

Sylvester, C. (2013b) Experiencing War: A Challenge for International Relations. *Cambridge Review of International Affairs* 26(4), 669–674.

Sylvester, C. (Ed.) (2015) *Masquerades of War*. London: Taylor and Francis.

Thompson, N. (Ed.) (2009) *Experimental Geography: Radical Approaches to Landscape, Cartography, and Urbanism*. Brooklyn, NY: Melville House.

Tyner, J. and Inwood, J. (2011) Geography's Pro-Peace Agenda: An Unfinished Project. *ACME* 10(3), 442–457.

Wainwright, J. (2012) *Geopiracy: Oaxaca, Militant Empiricism, and Geographical Thought*. Basingstoke: Palgrave Pivot.

Williams, A. J. (2014) Disrupting Air Power: Performativity and the Unsettling of Geopolitical Frames Through Artworks. *Political Geography* 42, 12–22.

Woodward, R. (2004) *Military Geographies*. Oxford: Blackwell.

Woodward, R. (2005) From Military Geography to Militarism's Geographies: Disciplinary Engagements With the Geographies of Militarism and Military Activities. *Progress in Human Geography* 29(6), 1–23.

Woodward, R. (2014) Military Landscapes: Agendas and Approaches for Future Research. *Progress in Human Geography* 38(1), 40–61.

Woodward, R., Jenkings, K. N. and Williams, A. J. (2015) *The Value of University Armed Service Units*. London: Ubiquity Press.

Woon, C.Y. (2013) Precarious Geopolitics and the Possibilities of Nonviolence. *Progress in Human Geography* 38(5), 654–670.

SECTION 1

Texts

2

REFLECTIONS ON RESEARCH IN MILITARY ARCHIVES

Matthew Farish

In the summer of 2008, I travelled to Alaska for the first time. Since World War II, the territory – and then state – has been the scene of tremendous activity for the US military, and I was seeking sources on the Cold War–era radar construction and environmental research conducted by the Air Force. While I made plans to visit several libraries and archives, including the exceptional northern collection at the University of Alaska Fairbanks, a key purpose of the trip was to work in the 3rd Wing History Office at Elmendorf Air Force Base, not far from downtown Anchorage.[1] Gaining entry to the base was surprisingly straightforward. Before landing in Alaska, I had corresponded with an archaeologist whose office was at Elmendorf. One morning, she picked me up at my university residence in her Subaru station wagon, and vouched for me as we passed through the gates of the base. At the small History Office, my exchanges were almost exclusively with one friendly staff member who had previously worked for the National Park Service. These interactions were small reminders that the US Department of Defense, the subject of most of my research over the last decade, is a vast and complicated institution.

Still, I recall a distinct sensation of estrangement at Elmendorf. Having been raised in a large Canadian city, in an upper-middle-class neighbourhood, with a distant American cousin as my sole connection to the armed forces, I knew remarkably little about military affairs before I entered graduate school. As a child, I wasn't permitted to possess military-themed toys, although I managed to see my share of violent American films, from *Missing in Action* (1984) and *Predator* (1987) to *Navy Seals* (1990) and *Under Siege* (1992). But these mediocre movies did not fully prepare me to encounter Elmendorf's 'mall', complete with the large retail store, off-limits to visitors, known as a Base Exchange, or BX. I saw a kiosk selling bumper stickers with slogans like "I didn't fight my way to the top of the food chain to eat vegetables." At the nearby Burger King, one of the few options for lunch, I ate my veggie-burger while surrounded by tables of uniformed personnel, as Fox News blared on several wall-mounted televisions. (I remember initiating a conversation with my host about the relatively unknown governor of Alaska, Sarah Palin; weeks later, she was chosen by

Senator John McCain to be his running mate in that year's presidential election.) It was my first encounter with what Chalmers Johnson called the "base world": a combination of familiar and dramatically unusual, little-known elements that I encountered again the following year in Alabama, and one – as Johnson stressed – that is present in hundreds of additional US facilities around the globe (Johnson, 2004; Gillem, 2007).

The same uncanny condition clouded my research practices at Elmendorf. My tasks and habits resembled those I have performed in other, nonmilitary repositories. I sat at a table in a reading room, sifting through and occasionally photographing documents pulled from a set of filing cabinets arrayed around the outside of the room, and making steady notes on my laptop. With the exception of detailed internal histories of the Alaska Air Command, the material I saw was of a recognizably scattered sort: folders stuffed with newspaper clippings, perfunctory letters and memoranda, and reports on individual operations or the construction of facilities.

Occasionally, unsure of the History Office's scope or sequence, I would pose a question to my host. As is the case in many small archives, I was the only visitor present that week, and he was able to respond quickly or point me to a new set of documents. But the reading room was adjacent to another room holding classified or more sensitive materials, and I was not permitted to enter that space. Most archives are premised on layers of access, but there was something notably palpable about this arrangement. Working in military facilities induces immediate and unavoidable confrontations with the making, sharing and storing of knowledge, and the hierarchies that make this knowledge available to some and not to others.

Research at Elmendorf, then, was meaningfully if not completely distinct from my time at other, nonmilitary archives, including state and university repositories. While my tenure on the base was inevitably unique, owing to my project and my identity, I suspect that I am not the first scholar to encounter such differences. Despite a growing body of what can broadly be called 'critical' research on militaries, militarism and militarization, and despite a wealth of sophisticated writing on archives, reflections on military archives – on their origins, mandates and the experiences of encountering, reproducing and analyzing the knowledge they store – are rare. In part, this is due to the lack of interest in methodological or even philosophical questions for one strain of military historian, but that no longer seems like much of an excuse. Military archives are of course themselves state archives, and this chapter is preoccupied with the parallels and intersections between military and state records, but also the discontinuities.[2]

My time at Elmendorf and other similar sites suggests that the distinguishing features of official military archives are limited access to material for visitors lacking clearances and narrow collections that are nonetheless made to seem *definitive*. These qualities are not unique to military repositories, but the corresponding fences and walls of bases themselves, and the secondary but substantial arrangements of secrecy and security within those bases, signal their specific profundity. For researchers, the result, I propose, is a feeling of inconsequentiality, but also a sharp sense of one's own self in relation to the subjects and politics of research. In my case, working at military archives has, among other outcomes, led to additional reflection on my own privilege and the frustrations inherent in critical research on military geographies. What follows is therefore less of a primer than an invitation for further assessment – of my own words, and more importantly of military archives themselves.

Productions of Violence

It is now widely acknowledged that archives are just as multifaceted and idiosyncratic as any other context for fieldwork. Even so, there is no escaping the role of archives as both a locus and a reflection of authority, a role so obvious that it should dispense any belief that archives contain "raw" data (Withers, 2002: 304). In an effective, elegant assessment of the archive as a "place," Miles Ogborn notes that treating archives as sites (within networks) where knowledge is "produced, stored and reused" means that their histories are inextricable from histories of modern "state formation." These latter histories concern the exercise of power and control within states, but also beyond their nominal limits. Colonial histories, of course, traverse these boundaries (Ogborn, 2011: 89).

All of this temporal and spatial confusion is another reminder that rather than expelling conflict to an external, international realm, or drawing firm lines between violent and peaceful times and spaces, it is the "normalcy of war," or the "military normal," that should be considered (Cowen and Gilbert, 2008: 6; Lutz, 2009). With respect to military archives, these circumstances require two seemingly contradictory moves on the part of researchers: a recognition that such archives and their missions have, like other facets of militarization, been made to seem ordinary; and a refusal, meanwhile, to treat them as merely strange or even irrelevant sites of study. Only by acknowledging the scope of militarization's normalcy – its presence at the heart of social life in countries like the United States – can this militarization be troubled. As I argue later, however, normalcy does not equal visibility. Militarization thrives on a profoundly spatial combination of presence and absence (MacDonald, 2006; Forsyth, 2014), and archives are both part of this mixture and potential sites for its analysis.

Imperial archives, Ogborn adds, do not simply store the records of empire; they are the *products* of empire, started as states sought to create, manage and hold knowledge concerning colonized spaces and societies (Ogborn, 2011: 89–90; see also Richards, 1993; Hevia, 2012). They were, as Ann Laura Stoler has subtly argued, both registers of colonial anxieties and sites for the literal containment of those anxieties, "arsenals" that could be "reactivated to suit new governing strategies" (Stoler, 2009: 3). While the intentions and actions of militaries might be distinguished from the broader spectrum of colonial government, military archives do affirm two intertwined 'certainties' that conceal similar anxieties: a state's naturalized obligations of defence, and the need, often couched in terms of this defence, to roam violently abroad – and in some cases to stay there at a scale approaching permanence. The military globalism practiced by the United States during and after World War II was accompanied by feverish, haphazard attempts, resonant of earlier colonial efforts, to collect extant information on strategically vital places, and to generate new information, destined for the same collections, on those places (see Barnes, 2005; Farish, 2005). Material remnants of these archival efforts exist in places like the library at Air University on Maxwell Air Force Base in Alabama.[3]

In light of their associations with colonial archives, it is worth noting that the records held in military archives are, to a substantial degree, mind-numbingly detailed and replete with technical language that for visitors verges on the incomprehensible. In many collections, individuals often exist only as signatures on reports or letters, as generic participants in military exercises, or more interestingly, as elusive authors of reports on experiments, enemies, environments and equipment. Military archives do house personal papers, but these tend to belong to prominent officials, and their content can seem equally anodyne. Across the collections I have studied, few hints of daily life in military spaces, of the emotions of soldiers, of the nuances of war's human geographies and of acts of killing, are present. I am generalizing, to be sure. Regimental diaries and logs from ships and planes are obvious counter-examples that have been thoroughly and skilfully used by many scholars (even as the respective limitations of these sources must also be acknowledged).[4] My suspicion, however, is that in military archives, the consolations of historical distance are particularly pronounced, and this distance directly permits a greater diversity and a greater number of available sources. In the United States, this situation is unquestionably related to the vast apparatus of secrecy associated with the Cold War (and specifically the nuclear) security state, a condition – in which some information is infamously "born classified" – that shows no signs of ceasing (see Galison, 2004).

The sources I have encountered in and extracted from military archives are rarely valuable alone. Even when subjected to a seemingly sophisticated reading, they benefit, in terms of their positions in narratives and arguments, from juxtaposition with other texts, including media stories, oral histories, memoirs, forms of popular culture and period publications written by scholars conducting military-sponsored research. At intervals I need to remind myself that these other texts are *proximate* to the endless unit histories or operations reports held in military archives.[5] Moreover, the prosaic qualities of the latter are, one might say, calculated; they are effective devices for the separation of militaries from civilian realms on the one hand, and the reduction of military activity to bureaucracy on the other.

Authority and Visibility

In addition to their functions as "venue[s] for the localization of knowledge," archives also serve as proving grounds for those who wish to reproduce and reposition that knowledge in another validated form. Part of the authority of archives, in other words, lies not just in what they hold, but the potent demand that researchers should or must use these holdings to justify claims about the past (Ogborn, 2011: 88, 92). This is a sobering reminder for those who approach archives, especially military repositories, intending to work against the grain in some manner: it is crucial to deliberate on the compromises made just by entering a facility, reading its records and incorporating them into one's prose.[6] After all, these records amount to histories of violence, however obliquely represented, and to employ this history can be a source of professional credentials. Some of my own publications have been commended by peer reviewers for my use of unusual sources – sources that are rare, at least, within my immediate intellectual community. In part this is due to the intriguing and quite stark separation of military geography, and militaries more generally, from the various spheres of professional geography – since the middle of the twentieth century, at least (Farish, 2009). But if the use of unfamiliar sources ultimately engenders 'expertise', questions linger: what sort of proficiency does this amount to, and at what cost?

Excruciating silences populate military archives. Stories of the dead and wounded are omitted, restrained or romanticized, depending on who those individuals and communities were. But in addition, as Ogborn suggests, discussions about saving, destroying or ordering material invite yet broader queries: "what ideas of permanence, and what stability of material allow the archive to come into being?" If stories need to be "stilled in order for them to be effectively archived," and if these stories are then granted *regularity*, the very premise of a military archive, and the use of that archive, is fraught (Ogborn, 2011: 89; see also Withers, 2002: 304).

For example, the Canadian Directorate of History and Heritage (DHH), a Department of National Defence entity rather ambiguously separate from the National Archives, is "mandated to preserve and communicate military history," but also to "foster pride in a Canadian military heritage." This double obligation is not altogether distinct from the efforts of state archives, although it would be erroneous to conflate nationalism with military 'pride'. But the history recorded and produced at DHH has another purpose: "to deepen professional knowledge of policy evolution, grand and military strategy, operations, and tactics, in the contexts of social, technological and infrastructure change."[7] What is a researcher with different intentions to make of this objective, in relation to the prose that she hopes to produce? If, as the critic Scott McLemee (2013) claims, users of archives learn "to frame questions that the archive knows how to answer," while hopefully "remaining open to . . . secrets and surprises," sites like DHH, by design, permit a slender range of questions, contain few surprises and essentially dare visitors to produce work that lacks military utility. This does not mean that different pathways are unavailable to creative researchers. Ogborn reminds us that it is easy to overemphasize archival systematicity or a simple relationship between archives and "projects of power" (Ogborn, 2011: 91; see also Withers, 2002: 305). But in military archives, where even disorder can come to seem deliberate, the challenge is steep – and this predicament is too often ignored.

The ambition of DHH to shape 'Canadian identity' seems primarily to refer to the activities of its employees. If I ruminated on this sweeping premise at all in 2004, when I first visited the facility, a nondescript building in a rather desolate part of southern Ottawa, I did so in a different vein. Still, it was impossible to ignore the preponderance of traditional military historians working there. I stood out, a student employee informed me, simply because I was a geographer, even if I was requesting similar documents. Military history, like military geography, often seems severed from the rest of its host discipline. This is presumably because of the work that military historians do, or are understood to do. But as with military geography, military history certainly continues to be produced, often in institutions that (like the work itself) are dismissed or ignored by other academics. This dismissal is a mistake on three counts: (1) such scholarship can be more nuanced than is initially presumed; (2) it is invariably connected to other fields, such as environmental history and political geography; and (3) put simply, facilities like DHH should not be left to military historians alone.

The last plea became especially acute after the arrival in 2006 of a federal government in Canada led by the Conservative Stephen Harper (who finally won a majority of seats in 2011). All such governments are keen to shape national identity, but Harper and his associates directly approached this task through the language and symbolism of military heritage. The instances are multiple, from a revised citizenship 'study guide' to celebrations of past wars as arduous but ultimately redemptive instances of nation-building (see Mackay and Swift, 2012). For me, the most intriguing example was the prime minister's annual trip to various locations in the Canadian north, a journey coordinated to coincide with a major 'sovereignty exercise' called Operation Nanook (Dodds, 2012; Farish, 2013b). This was not just a demonstration of military efficacy. Nanook's designation as an expression of sovereignty implied a historical claim to territory, and to the use of that territory as a natural stage for Canadian military activity. These claims, and the name itself, call forth a record of militarization in the region that reaches back to at least World War II, and related colonial 'exercises' that are much older. Much of that military activity, at least, is documented in the collections of DHH.

The spectacle of Operation Nanook is also a reminder that other stories of militarization, including those of delayed and dispersed "slow violence" (Nixon, 2011), are rarely present in popular accounts, but can still be found in military archives. Indeed, the relative obscurity of DHH – and more broadly, the banality and anonymity of many materials in these repositories – indicates that what Edward Said (quoted in Nixon, 2011: 6) called the "normalized quiet of unseen power" may paradoxically be traced to that most visible of state institutions. This is precisely because militaries are also *invisible*: they possess an unrivalled ability to deem a landscape socially or environmentally irredeemable and then enclose it from scrutiny, an act that may additionally mean removing or classifying information related to that space (Paglen, 2010b).

Studying these "blank spots," in Trevor Paglen's (2010a) phrase, is thus exceedingly difficult, perhaps requiring the unorthodox methods of an "experimental geographer" like Paglen himself. And yet, as his mixture of prose and photography shows, the world of military secrecy is inseparable from our own. Seemingly inaccessible histories can be approached creatively, leading to new or repurposed ways of seeing these histories. More fundamentally, in respects that are frightening and unjust, military geographies, like the sources in military archives, lie adjacent to and overlap with the quotidian places that we inhabit (even as that 'we' is assuredly not homogenous). Rather than simply bringing sunlight to bear on classified documents or landscapes, Paglen's more significant accomplishment has been to show that the making of a secret, whether in paper or in soil, is merely the attempt to *manage* this proximity – an attempt, like any hegemonic process, that can never be fully, finally successful. While the emphasis of scholars labouring in military archives should not exclusively be on practices of secrecy, the role of an archive as a crucial contributor to military powers of visibility and territoriality bears consideration, before, during and after research visits.

Faces of the Military Archive

Military archives are nominally public institutions, concerned with the maintenance of knowledge that will, in various forms and contexts, be made (more) public. But these are general and even misleading characterizations. As the case of DHH suggests, military archives wish to promote a very focused form of public knowledge, while concurrently providing other knowledge to departments of defence and inhibiting yet another type of knowledge. Some records may be declassified rapidly, owing to the comparatively diminutive size of collections and the presence of a nearby staff person responsible for such appeals, but the overwhelming impression is that attempting to push beyond unclassified records is futile, at least relative to the span of most academic projects. Requests I have submitted to declassify Cold War–era US Air Force sources never earned a reply; my follow-up correspondence was answered promptly, with a promise to investigate the delay, and then the channel went silent again. As a result, I have concentrated on – and perhaps justified my use of – material that is already unclassified. This is itself a blurry category: visits to multiple sites have yielded reports that were restricted in other locations.[8]

Meanwhile, digitization – for instance, through the US Defense Technical Information Center – is also expanding the number of available documents and opening new avenues for inquiry.[9] For researchers, including those unable to travel to archives, the benefits of digitization, closely tied to declassification, seem clear. I regularly encounter references to an obscure source – a long-forgotten RAND Corporation study, an article in a military periodical or a report from a military laboratory – and then find it within seconds using a search engine. But such discoveries do not reveal much about the selective processes of declassification and digitization. And while I have employed these sources in my writing, I have also collected far too many of them, a familiar problem of glut exacerbated by the military penchant for detail. More crucially still, the online proliferation of military sources may draw attention away from more traditional repositories and the material they house that is not fully digitized (a huge percentage at present). Finally, digitization may limit some of the personal unease associated with research on militaries, but it also invites detachment. I remain devoted to the significance of a physical encounter with a place, where possible, and this holds in the case of a visit to a military archive.

Inside these archives, as in corporate equivalents, researchers also possess various degrees of access, to collections, spaces, technology and assistance. Some of this is formalized, but staff members also make informal judgements along similar lines. At one hushed US Army archive, initial dialogues between researchers and staff invariably began with a recitation of military résumés: in the case of the researcher, his own service (it was a predominantly masculine space), or at least that of one or more ancestors. Regardless of the circumstances – and some of the stories visitors went on to tell were distressing – this was a specific type of interaction, complete with jargon and deference, that did not mirror the combination of politeness, bemusement and mild bafflement I encountered. Historians employed by the US military, it should be noted, occasionally deploy alongside combat troops.

Other restrictions prevent some researchers from even entering military archives: excessive, slow bureaucracy and citizenship requirements are two of the most pervasive. As a Canadian citizen, I have been able to conduct research at a number of military archives in the United States, but often only after completing hours of paperwork. I was once denied entry to a Massachusetts site because my visit did not accord with the public relations mission of the facility – or some such excuse. Even so, I have been fortunate. Scholars carrying different passports, or lacking certain affiliations, would not have received an equivalent welcome. And I know that some researchers have been denied entry to military archives by dint of their scholarship alone. The fact that I have not met much of this resistance reflects the mandate of institutions and the nature and tone of my initial inquiries, but it is perhaps also an indication that my writing is not public *enough*.

Discomforting Conclusions

The image of the academic toiling in obscurity is a familiar caricature. But the juxtaposition of an individual project with a behemoth like the Pentagon, at least if the researcher views the project as in some way out of step with institutional priorities like fostering military pride or improving strategy, can be unnerving and depressing – all the more so if you believe that access to military archives is contingent on your obscurity. Perhaps it is understandable, then, when researchers humoured by the military, taking their cue from embedded journalists, describe their stimulating conversations with officers, or their turns as observers at war games, in excited tones. This is to put it cynically, perhaps. But it is also to name the dilemma precisely, and to indicate that there is no perfect critical sanctuary, either.

Some measure of solace can come from solidarity across the modest community of scholars who have shared experiences of working in military archives and reflected on these experiences. Understandably, I hope that this group will grow and diversify. But it is imperative to ponder why the community is small, given the contemporary and historical significance of an entity like the US Department of Defense. Academic research should not necessarily induce comfort, but I have repeatedly wondered during my stints at military archives: who would encounter these spaces differently? It is presumptuous to attempt an answer, but the relevance of the question remains.

Notes

1   The Office is now part of Joint Base Elmendorf-Richardson, after Elmendorf merged with the US Army's Fort Richardson in 2010.
2   Of course, national archives often feature military holdings: the National Archives of Canada houses a huge record group (RG 24) from the Department of National Defence, while the US National Archives and Records Administration facility in College Park, Maryland, has a separate section of its reference room devoted to military inquiries (a testament to the frequency of these inquiries, but also an arrangement that deserves deliberation). My focus here, though, is on smaller facilities that reflect the unusual and noteworthy determination of militaries to maintain their own archives. My discussion is limited to the United States and Canada, owing to the boundaries of my own experience.
3   See http://www.au.af.mil/au/aul/lane.htm.
4   Thanks to Alison Williams for reminding me of this diversity.
5   For one attempt at this intertwining, see Farish, 2013a; for another, inventive approach, see Krupar, 2013.
6   I thank Oliver Belcher for stressing this point.
7   See the DHH website: http://www.cmp-cpm.forces.gc.ca/dhh-dhp/adh-sdh/index-eng.asp.
8   For a related situation, see Aid, Matthew M. (2006) 'Declassification in Reverse: The U.S. Intelligence Community's Secret Historical Document Reclassification Program', *National Security Archive*, 21 February 2006, http://www2.gwu.edu/~nsarchiv/NSAEBB/NSAEBB179/.
9   The Defense Technical Information Center website is http://www.dtic.mil/dtic; the National Security Archive can be found at http://www2.gwu.edu/~nsarchiv/.

References

Barnes, T. J. (2005) Geographical Intelligence: American Geographers and Research and Analysis in the Office of Strategic Services, 1941–45. *Journal of Historical Geography* 32, 149–168.
Cowen, D. and Gilbert, E. (2008) Introduction: The Politics of War, Citizenship, Territory. In Cowen, D. and Gilbert, E. (Eds.), *War, Citizenship, Territory*. New York: Routledge, pp. 1–32.
Dodds, K. (2012) Graduated and Paternal Sovereignty: Stephen Harper, Operation Nanook 10, and the Canadian Arctic. *Environment and Planning D: Society and Space* 30, 989–1010.

Farish, M. (2005) Archiving Areas: The Ethnogeographic Board and the Second World War. *Annals of the Association of American Geographers* 95, 663–679.

Farish, M. (2009) Military and Geography. In Kitchen, R. and Thrift, N. (Eds.), *International Encyclopedia of Human Geography*, Volume 7. Oxford: Elsevier, pp. 116–121.

Farish, M. (2013a) The Lab and the Land: Overcoming the Arctic in Cold War Alaska. *Isis* 104, 1–29.

Farish, M. (2013b) The Arctic Is No Place for Military Spectacles. *Globe and Mail*, 13 August. Available at: http://www.theglobeandmail.com/globe-debate/the-arctic-is-no-place-for-military-spectacles/article13725994/.

Forsyth, I. (2014) Designs on the Desert: Camouflage, Deception and the Militarization of Space. *Cultural Geographies* 21, 247–265.

Galison, P. (2004) Removing Knowledge. *Critical Inquiry* 31, 229–243.

Gillem, M. (2007) *America Town: Building the Outposts of Empire*. Minneapolis: University of Minnesota Press.

Hevia, J. (2012) *The Imperial Security State: British Colonial Knowledge and Empire-Building in Asia*. Cambridge: Cambridge University Press.

Johnson, C. (2004) *The Sorrows of Empire: Militarism, Secrecy, and the End of the Republic*. New York: Metropolitan Books.

Krupar, S. R. (2013) *Hot Spotter's Report: Military Fables of Toxic Waste*. Minneapolis: University of Minnesota Press.

Lutz, C. (2009) The Military Normal: Feeling at Home With Counterinsurgency in the United States. In Network of Concerned Anthropologists (Eds.) *The Counter-Counterinsurgency Manual: Or, Notes on Demilitarizing American Society*. Chicago: University of Chicago Press, pp. 23–37.

MacDonald, F. (2006) Geopolitics and 'the vision thing': Regarding Britain and America's First Nuclear Missile. *Transactions of the Institute of British Geographers* 31, 53–71.

Mackay, I. and Swift, J. (2012) *Warrior Nation: Rebranding Canada in an Age of Anxiety*. Toronto: Between the Lines.

McLemee, S. (2013) The Unforgotten. *Inside Higher Education*, 25 September. Available at: http://www.insidehighered.com/views/2013/09/25/review-arlette-farge-allure-archives.

Nixon, R. (2011) *Slow Violence and the Environmentalism of the Poor*. Cambridge, MA: Harvard University Press.

Ogborn, M. (2011) Archive. In Agnew, J.A. and Livingstone, D. N. (Eds.), *The Sage Handbook of Geographical Knowledge*. London: Sage, pp. 88–98.

Paglen, T. (2010a) *Blank Spots on the Map: The Dark Geography of the Pentagon's Secret World*. New York: New American Library.

Paglen, T. (2010b) *Invisible: Covert Operations and Classified Landscapes*. New York: Aperture.

Richards, T. (1993) *The Imperial Archive: Knowledge and the Fantasy of Empire*. London: Verso.

Stoler, A. L. (2009) *Along the Archival Grain: Epistemic Anxieties and Colonial Common Sense*. Princeton, NJ: Princeton University Press.

Withers, C. W. J. (2002) Constructing 'The Geographical Archive'. *Area* 34, 303–311.

3

FROM DECLASSIFIED DOCUMENTS TO REDACTED FILES

Tracing Military Compensation

Emily Gilbert

I stumbled into my research on the military. One morning, while living in Australia, I happened upon a short article in the newspaper entitled "Troops Give Cash to Afghan Victims" (Banham, 2009). The gist of the article was that new legislation had been passed that would allow senior officers in Afghanistan to make cash payments of up to AUD 250,000 to 'inadvertent' civilian victims of the Australian Defence Forces. I was immediately intrigued. Surely it was not usual practice for soldiers to carry wads of money on the battlefield? Other questions came quickly to mind: How were distinctions between civilians and insurgents being made? Why were these monies being paid out? And what were the effects of these payments?

These questions dogged me. I had already undertaken a considerable amount of research on money, but it mainly had to do with the nationalization and deterritorialization of currencies. I had never researched the military and I was a bit hesitant to do so. Frankly, the idea was daunting. But I could not stop being interested in these payments. I found myself pushing further ahead with the research, sifting through government documents and media reports. As I learned more about the Australian case, a more complex picture began to emerge. Other militaries were also making payments. Indeed, compensation had become widespread across the international forces in Afghanistan in Iraq, and had even been taken up by NATO and the UN. The US, however, was the most heavily invested, with thousands of payments having being made, totalling millions of dollars. And while monetary payments were proliferating during the 'war on terror', it turned out that they actually had a longer genealogy.

What started as a nagging research question thus soon developed into a full-blown research project that is still ongoing. This transition was largely made possible by coming across thousands of pages of documents on military compensation made available through Freedom of Information Act (FOIA) requests in the US. This raw data provided a rich archive for analysis. In this chapter I will discuss the sources that I have used, the method of analysis, and my reflections on doing research on the military and what makes it unique. My background in thinking differently about money lent itself well to an analysis of the ways that money is being weaponized and the military is becoming monetized. But to get to the point where I was able to make an informed critique, I had to overcome my hesitancy in taking up research on the military and move away from ingrained assumptions that I had about the civilian-military divide.

The Archive

The data used in the early stages of this research consisted of thousands of pages of declassified documents, dated between 2003 and 2006, which were released in April 2007 as the result of an FOIA request made by the American Civil Liberties Union (ACLU). A second request was made by the ACLU in 2010. Altogether, over 35,000 pages of documents have been released, much of which is publicly available on the ACLU website.[1] This includes original records on compensation sought through the Foreign Claims Act (FCA) and through the Commander's Emergency Response Fund, as well as in criminal investigations and courts martial.[2] The ACLU has organized the documents according to the cases to which they refer, and has provided summaries of each case.[3] The stacks of paper and data from CD-ROMS that were provided by the government were organized into distinct records in an online database, much as a traditional archive would be organized into folders and boxes. The sorting of the information was undertaken to make the materials more easily accessible and understandable to the public, with short summaries for each to help explain the context of the documents. Not every case file contains the same information, but among the documents that might be included are incident reports, witness testimonies, photographs of the incident, and army records regarding how much compensation has been approved and paid out. These original documents provide rich insight into the army's documentary practices, but they also contain moving accounts, particularly those by witnesses, of the violence of war.

The ACLU's interest in acquiring this data was, in the words of Nasrina Bargzie, one of their attorneys, so that they could "pull back the veil of secrecy on the issue of civilian casualties and to try to bring some transparency and accountability to this realm."[4] Indeed, the documents provide poignant testimony about civilian deaths and casualties which are otherwise notoriously difficult to acquire in wartime. Obtaining the records was not straightforward, however. While the US Army did make documents available when requested, the ACLU had to file a lawsuit with the Defense Department to ensure compliance. Many of the documents that were eventually provided are redacted (although there is still quite a lot of information that can be gleaned from them, as will be discussed in more detail in the following section). Subsequent attempts to gain access to this information have been even more fraught. Reporters working with the *Nation*, a left-leaning weekly journal, submitted a follow-up request through the FOIA that would cover the period between 2008 and 2011. Although over 1,000 pages were eventually released, this time the documents were so heavily redacted that the information included was "neither inclusive nor comprehensive" (Turse, 2013).

This tightening of control around access is pervasive. In the UK, the *Guardian* newspaper submitted a Freedom of Information request to the Ministry of Defence for records on military compensation by the British Armed Forces up until 2012. Journalists at the newspaper have commented that the information made available was not very revealing. Ben Quinn notes that "in a change of policy, the MoD has decided against giving details of how much was paid out for individual incidents, despite being criticised in the past for a lack of transparency in relation to the payments" (Quinn, 2013). The *Guardian* has summarized the content of the documents and made the spreadsheets available on its website, but not the original records.[5] Nonetheless, looked at alongside the US examples, there are some rich comparisons that can be drawn across the many cases.

These records have been foundational to the research that I have undertaken. I have also sought to acquire data on Canadian military practices, but have been stymied in my attempts. The Canadian Department of Defence has made it clear that it considers the details regarding compensation to be private matters, and maintains that it does not want to jeopardize the security of recipients (Friscolanti, 2011). Only snippets of information have been leaked out by way of a report to the Receiver General in 2010 which revealed that CAD 650,000 was paid out in Afghanistan between 2008 and 2009 (Moore, 2010). In another media article, Michael Friscolanti has reported that the amount of money that was disbursed increased in 2010, but very little other information is provided. In 2010 I submitted an Access to Information request to the Department of Defence, which was rerouted to the Department of Justice. Over 200 pages of photocopied documents were eventually released to me, but the materials are so heavily redacted that there is barely any information legible (see Figure 3.1).

The difficulties I and others have faced in acquiring access to information on military compensation are not atypical. Matters deemed to be important to national security are often hidden from public view or manipulated, especially during times of war. This has certainly been the case during the war on terror. Joseph Masco illustrates that in the US there has been a dramatic increase in the control over information since the 9/11 attacks. The number of records that have been made classified – and not available to the public – has risen dramatically. Indeed, the cost of classifying information in the US has doubled to more than $11 billion a year (Shane, 2012; see also Crampton et al., 2014). Perhaps even more troubling is the rise in the number of documents that have been designated as noncirculating (Masco, 2010). This designation has the same effect as rendering a document classified, in that it is not available to the public, but the decision to remove an item from circulation is made without the same legal controls and oversight as to why confidentiality has been enforced (Masco, 2010: 446–447). Effectively, Masco argues, the official record is increasingly being censored, with that censorship itself taking place in more and more clandestine ways.

Making information secret and the management of "the public/secret divide through the mobilization of threat" are mechanisms through which state power is constituted (Masco, 2010: 450). The state uses the veil of secrecy to act in ways that might not be palatable if the public were to know what kinds of actions are being carried out in their name. For Masco, these mechanisms are indicators that the state "implicitly recognize[s] citizens as a potential barrier to state security policies" (Masco, 2010: 442). And indeed, when news of compensation payments trickles into the media, there is invariably a negative public response. But why? The military is routinely criticized for not attending to civilian casualties, or worse, for objectifying violence with terms such as 'collateral damage'. As Judith Butler notes, civilian lives are not recognized, and their injuries are deemed to be "ungrievable" (Butler, 2009). Compensation would seem, on the surface at least, to be an example of the military and the state taking civilian casualties seriously. It is certainly touted by the military as a mechanism for recognizing the harm inflicted in war, and thus as exemplary of its compassion and humanity. Yet, if military payments are so benevolent, why is there the need to keep it quiet? Why has

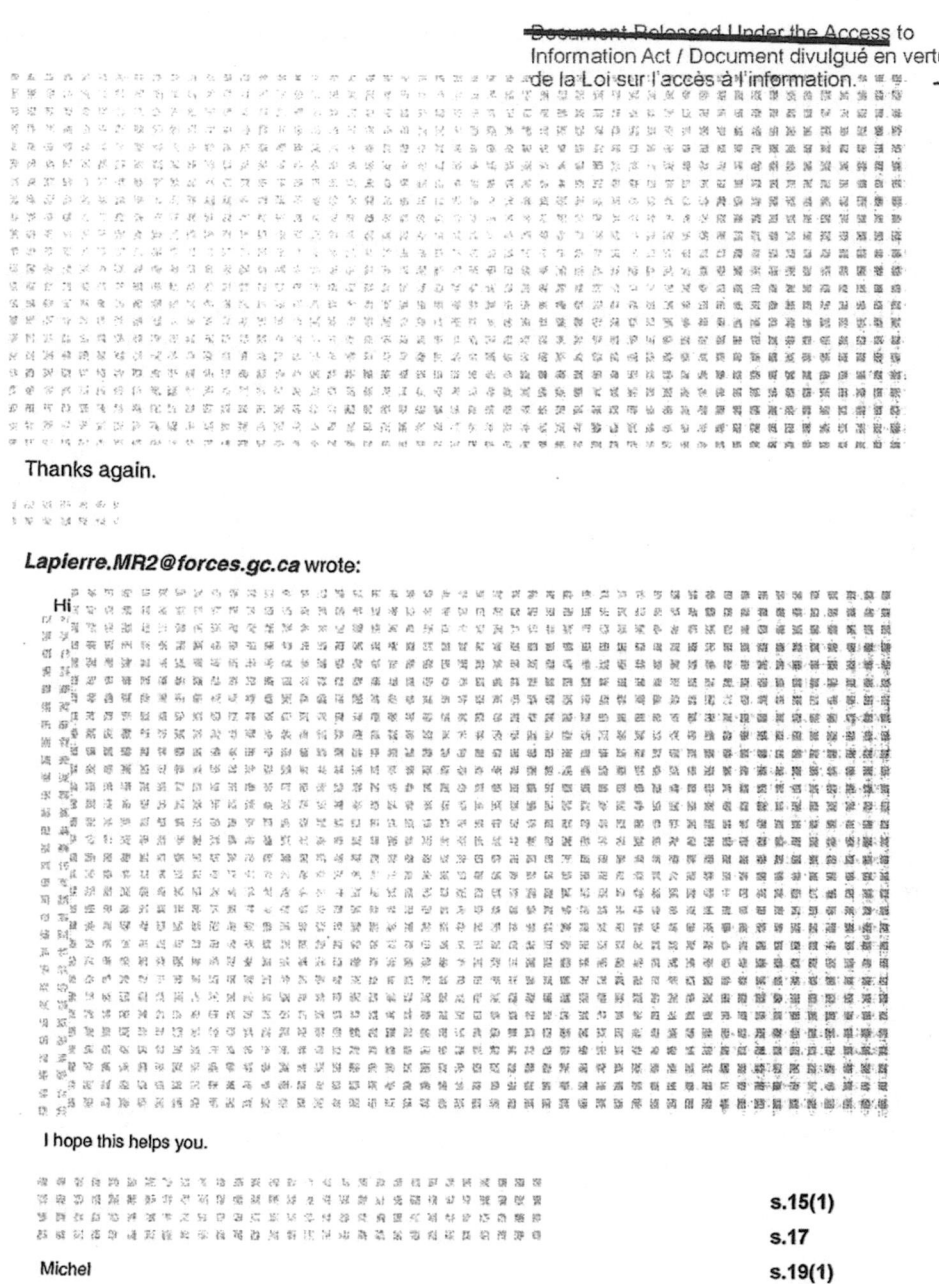

Thanks again.

Lapierre.MR2@forces.gc.ca wrote:

Hi

I hope this helps you.

s.15(1)

s.17

Michel

s.19(1)

s.23

6

Figure 3.1 "Redacted Correspondence on Solatia." Document released by the Department of Justice to the author, under the Canadian Access to Information Act.

there been public aversion to these payments when information about them hits the media? These are questions that I have sought to address in my research.

Policies around national secrecy, however, pose enormous challenges to researchers, perhaps especially those who wish to research the military, as its actions are regularly construed as matters of national security (see also Anaïs, 2013; Howell, 2013; Salter, 2013). This is especially troubling given the growing interoperability between security and military agencies, but also the greater role of private sector and public institutions such as universities. These extended assemblages mean that a wider net of secrecy is being cast. As Crampton et al. (2014: 203) caution, "the mixture of interests and secrecy represented in this nexus threatens the liberal democratic principles of US political life" – principles such as accountability, transparency and conflict of interest. Although some information might be made available through freedom of information policies, this process has its own challenges. As already noted, the records that are released might be redacted. There can also be significant delays in getting the documents. In my own case, I waited nearly a year to receive my data, delays which became common practice in Canada under the Conservative government of Prime Minister Stephen Harper. Indeed, the country recently was ranked 55th in the world (out of 93 states) for upholding freedom of information, falling behind Mongolia and Colombia (Beeby, 2013).

Soliciting information through a freedom of information request can also be tricky because it can be difficult to know what one is looking for, and what documents to request. How does a researcher know what she wants before she sees what documents are available? How can she predetermine what information might be relevant? And how can she ever be sure she has received the documents that are most important to the research? Those with any inclination towards paranoia might also wonder whether information is deliberately kept hidden, even when a request is made. These are issues that all researchers involved in questions to do with national security must struggle with. In my case, being able to sift through the ACLU documents was enormously important, as it gave me an understanding of what kinds of records I might be looking for in the Canadian context. Key terms such as 'condolence' and 'solatia' were also helpful, even though it turned out that the Canadian Armed Forces uses the designation 'ex-gratia'. My original request for information was made to the Department of Defence in Canada, because that is the division responsible in the US; it turned out that the right Canadian institution was the Department of Justice. Still, I found as I continued to push forward with my research that even glimpses of otherwise hidden information can provide a valuable foundation for analysis, especially when read alongside public documents, government reports, media analyses, and other kinds of information, as I discuss in the following section.

Analysis

In *Along the Archival Grain: Epistemic Anxieties and Colonial Common Sense*, Laura Ann Stoler makes an argument for attending to the documentary traces that accompany state rule. Her focus is specifically on the colonial rule of the Netherlands Indies in the nineteenth century, but her arguments for exposition ring true in the contemporary moment. She describes her project as working to "identify the pliable coordinates of what constituted colonial common sense in a changing imperial order" (Stoler, 2010: 3). Archives are repositories that can reveal the truth-claims of a moment in time. The very process of archiving functions as "a force field that animates political energies and expertise, that pulls on some 'social facts' and converts them into qualified knowledge, that attends to some ways of knowing while repelling and refusing others" (Stoler, 2010: 22). Through the gathering of information these truth claims get legitimized and reinforced: the archive itself thus becomes a site of power. As noted earlier by ACLU attorney Nasrina Bargzie, the compilation of the ACLU archive was undertaken in precisely this same spirit: as a mechanism for making visible the number of civilian casualties in war. Thus, as a response to the *lack* of disclosure of information by the state, we might consider the ACLU as a counter-archive, even if the materials that it contains were all made available by the state.

Collected materials – whether in an archive or counter-archive – should not be approached as having a fixed ontological essence. Rather, researchers need to be attentive to the ways that archives seek to fix dominant narratives in ways that might not correspond to acts or events (Stoler, 2010: 4). Mining the archive is thus as much about tracing the said as the unsaid, examining what is taken for granted and ruminating on what is overlooked or omitted (see also Tonkiss, 2004). Furthermore, in addition to reading the archive along the grain, it is also important to read *against the grain*. Edward Said describes this approach as a contrapuntal reading. In an oft-cited quote he suggests that archivists must have "a simultaneous awareness both of the metropolitan history and of those other histories against which (and together with which) the dominating discourse acts" (Said, 1993: 51). This means attending to the voices that are distorted or are rendered silent in any trove of documents. The invocation of a contrapuntal reading is not just an appeal to plurality, but an attempt to draw out articulations of power in the constitution of the archive. Inasmuch, there is a strong echo with Michel Foucault's genealogies of power/knowledge and his excavation of history to render visible the processes that make something sayable or legible (Gordon, 1991). For while he attended to the ways that practices become normalized and normative, Foucault also strove to identify the contingencies through which truths cohere, and thus to denaturalize their very taken-for-grantedness (Foucault, 1980).

Foucault, Said and Stoler all focus their attention on historical archives. Their reflections, however, are equally relevant to contemporary documents, and the techniques that they use to interrogate their materials are equally relevant. These techniques are often characterized as 'critical discourse analysis', which involves an analysis or written or oral texts "and their power to shape 'situations, objects of knowledge, and the social identities of and relationships between people and groups of people'" (Anaïs, 2013: 196). Drawing from theories of deconstruction, this involves a close analysis of texts (and images) for the devices that they employ (e.g. metaphors, rhetoric, syntax), and the conditions around their production and consumption. This might include exposing the ways that a single text is burdened with assumptions, contradictions and/or silences, or interrogating the relationship between texts – their intertextuality – and the ways that discourses rub up against one another. Whether the focus is on one text or many, the point is to dwell on the texts to understand what kinds of social and political effects they produce and reproduce, what they legitimize, and what they make possible. It also means, as Seantel Anaïs astutely observes, addressing the text to understand what kinds of questions it presumes and what kinds of solutions it offers (Anaïs, 2013: 196).

These techniques are more difficult to mobilize when dealing with redacted documents. Details are hidden. Whole sections might be blacked out. Fragments of information are disconnected. Yet although redacted documents can be frustrating for their lack of transparency, they are also, as Anjalie Nath argues, productive of critical reading practices. They "change how you read, what words you dwell on, [or how you] infuse a mundane word with suggestions" (Nath, 2014: 25). Nath suggests that a redacted text necessarily propels a kind of contrapuntal reading that theorizes "beyond words" as "[w]e look for logical continuities and wonder what lies beneath the redacted space" (Nath, 2014: 26). Leaps of logic may be required to guess at what no longer can be seen, but this is not much different from other archival analysis, in that the archive is always incomplete, and the researcher is always engaged in a process of reconstruction. It does, however, make the triangulation of sources and the search for intertextuality all the more necessary so that the researcher is able to decipher whatever clues are legible on a redacted document.

In my research, one type of document in the archive proved to be particularly telling: the "Purchase Order-Invoice-Voucher" form. This is a standard US government form (SF 44), which is multipurpose and pocket-sized, and which is regularly used to record small purchases. In this case, it is used to record the payment of $2,500 made upon the death of someone's father (see Figure 3.2).

The name of the "payee" has been redacted: all that is known is that the location was in Yousifiyah area, Baghdad. This is affirmed in the memorandum issued by senior command which approved the payment, and which has been included alongside the SF 44 form in the online ACLU archive. The payment was made at Al-Mahumudyah, a Sunni city south of the capital, which was the central focus of counterinsurgency campaigns. We can also locate the general area of the incident in that the payor is identified as the "15th FIN BN, North Victory" – the 15th Financial Battalion, at Camp Victory North in Baghdad. We know that the payment was made swiftly after the memorandum was issued: the memorandum is dated March 18, 2005, and payment is recorded on March 23, 2005. The memorandum, however, shows that the incident occurred on January 8 of that year, so over three months passed before a decision was rendered. We thus also get a sense of how the military administration works slowly.

The value accorded to the father is unsettling: is $2,500 all that a life is worth? Yes, as it turns out, that is the standard amount the US pays for a civilian death, as is set out in government accountability reports (see Gilbert, 2015a). But it also the use of the SF 44 form to record this amount that is unsettling. The generic form highlights the stark commensurability that is being made, as death is recorded under "supplies and services." The SF 44 does not contain any further information on the individuals to whom it refers, the incident that led to death, or the reasons why the payment is being made. Some of this information, however, can be gleaned by the accompanying memorandum. We learn that the claimant's father was shot in the return of fire of a US convoy. Payment is justified as an expression of "sympathy for the unfortunate loss. Support will positively influence both the community and local Iraqi leaders." The payment is thus being rationalized as an expression of the compassion of the MNF (multinational forces), so as to help win the 'hearts and minds' of the local population.

On the redacted SF 44 form that has been made public, the names of all US personnel are blocked out. The victims and the perpetrators are made anonymous. The incident is thus

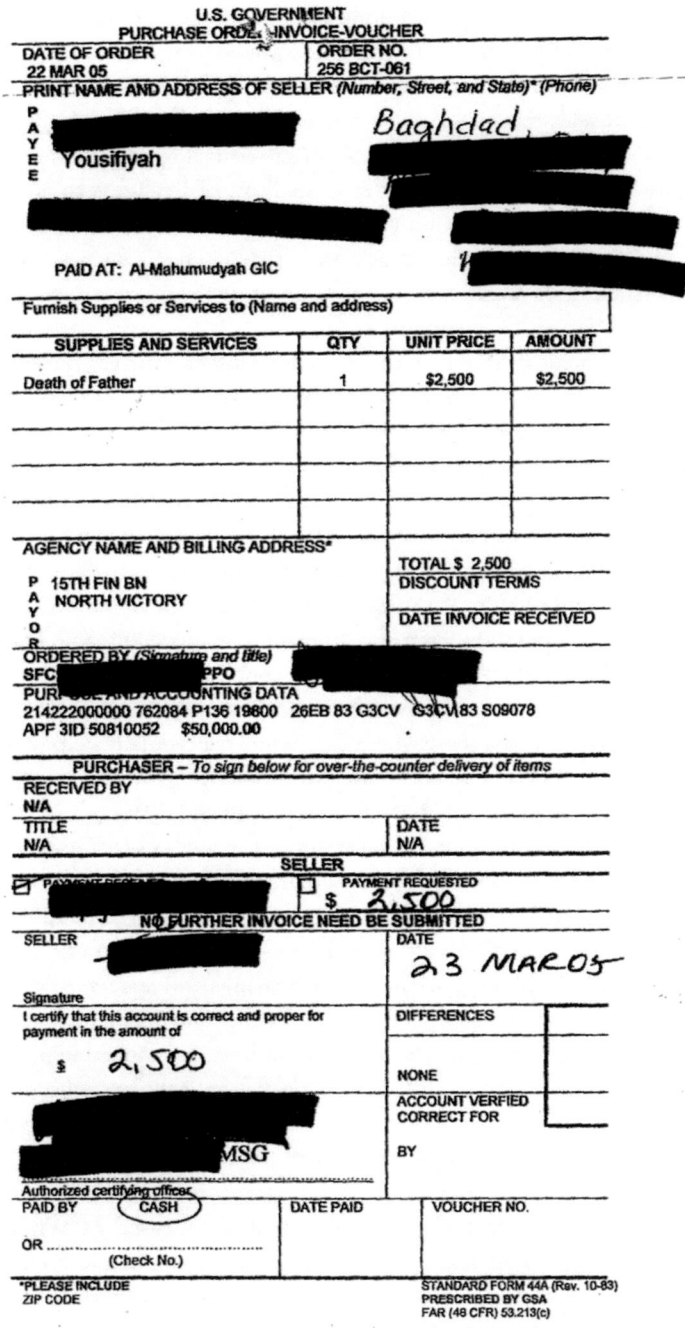

SUPPLIES AND SERVICES	QTY	UNIT PRICE	AMOUNT
Death of Father	1	$2,500	$2,500

001655

Figure 3.2 "Death of Father." US Government Purchase-Order-Invoice, dated March 22, 2005. Made available through a US FOIA request by the ACLU, available here: https://www.aclu.org/human-cost-civilian-casualties-iraq-afghanistan.

depersonalized. We do, however, know that the file passed through the hands of many. The money was ordered by a project purchasing officer (PPO), who oversaw the case, and it was approved by a commanding officer. Each of these individuals signed the forms. The accompanying memorandum also reveals that the file was reviewed by a military judge advocate who would have determined whether a payment was appropriate according to military procedure. The signatures, despite being blacked out, render a small bit of liveliness to an otherwise standardized form, in which most of the information appears to have been typewritten in advance. A few other snippets of other information are written in hand, such as the date, the location of Baghdad, and a double affirmation of the amount, which were probably noted by the PPO.

The overall effect of the generic SF 44, with its predetermined entries and bureaucratic accounting, is to emphasize the routinization of death in war. But what is also apparent from these forms is that compensation for injuries and/or death is also being routinized in war. While some very limited forms of compensation have been in operation since World War I, the use of condolence and solatia payments has increased significantly during the counterinsurgency operations in the war on terror (Gilbert, 2015a). Moreover, it is not only the US, the UK, Canada and Australia which are making claims, but also Germany, Italy, the Netherlands, Norway and Poland – although much is less known about these countries. NATO and the UN are also promoting compensation as a way to make war more humane. This denotes a significant shift in how victims are being dealt with in contemporary military operations. Yet, as I have argued elsewhere, while victim payments can provide much-needed short term assistance to victims and their families, the payment of money may actually help to legitimize the 'collateral damage' of war. While there is bureaucratic accounting for the loss of life, there is no accountability for the damage that has been incurred (Gilbert, 2015a).

Understanding the turn to victim payments has required an analysis that extends beyond the records available in the ACLU archive. As Claudia Aradau has argued, widening the archive enables one to think "both synchronically and diachronically" (Aradau, 2013: 183). In my broader work on this project this has meant a detailed examination of other materials, such as media reports; film; blogs; parliamentary hearings and speeches; government policy; policy papers by partisan and nonpartisan think tanks; government accountability reports; papers and statements by nongovernmental organizations; and some narratives by former military personnel (e.g. Tracy, 2008). The work of legal scholars who have examined the preliminary documents that were released has also been very helpful (Witt, 2008; Sitaraman, 2009). There is also a rich array of military documents that are in the public domain, including doctrinal statements, blogs and journals, annual accounts and financial reports, and even some training materials. Even information released by WikiLeaks has been helpful because it has affirmed many of the fragments found in the archive. These other sources have provided a crucial corroborating and contextual backdrop for interrogating the ACLU files, and for understanding their importance in the war on terror. Indeed, victim payments are so interesting precisely because they are but one poignant example of how the military has been turning to monetary solutions in its counterinsurgency operations (Gilbert, 2015b).

Analyzing records – whether historical or contemporary – is a productive encounter. The researcher does not simply seek to describe the contents of the archive, but to draw upon the documentary traces to conjure up a more fulsome account of their meaning, their relevance and their political implications. As Luis Lobo-Guerrero describes,

> archival research is not a neutral practice. The researcher is not there to depict objectively what he/she reads and observes since an archive becomes a mediated space between the records and the researcher's imaginary. It demands a creative attitude to understand why and how events were recorded and why were they recorded in their specific manner.
>
> *(Lobo-Guerrero, 2013: 121)*

Archival research is thus a creative practice that animates the flattened remnants that have been recorded while also attending to the liveliness, the voices, that have been silenced. Given the active and pivotal role played by the researcher, it is important that she is reflexive as to how her positionality can influence the research results.

Reflexivity

If archival research is not neutral, as Lobo-Guerrero argues, then how are we to understand the partialities out of which it arises? We can begin to do so by better understanding the positionality of the researcher, that is, the social and ethical values that shape the researcher's perspective. This demands a self-conscious reflexivity of the research process that situates the researcher within the field and the project, not as separate from them. This thoughtful awareness of and reflection on research questions, methods and analysis has become relatively commonplace across the humanities and social sciences. Notably, however, as others have remarked, there has been a tendency for social research on the military to be less reflexive, with a few notable exceptions (Higate and Cameron, 2006: 219).

The reasons for this lack of reflexivity in military research are not clear. But I concur with Paul Higate and Ailsa Cameron that reflexivity can bring transparency and accountability to the research project (Higate and Cameron, 2006: 220). In my own case, reflecting on my early disinclination to embark on this research project has been crucial to unpacking some of the preconceptions that I brought to it. To be honest, my disinclination arose out of ardent antiwar politics and, at best a suspicion of, and at worse an antipathy to, the military as an institution. I saw the military as an institution separate and distinct from civilian life, and I wanted to maintain this separation. Naively, I even thought that the less attention directed towards it, the less of an influence it would exert on the rest of society. This may be a particularly Canadian approach of a privileged class, since the military has tended to be downplayed — until recently.

As I was contemplating embarking on this research, things began to change. My naive conceit of the military's relative lack of influence in Canadian society began to fray. Over the last decade the country has become increasingly militarized. The federal Conservative party, in office from 2006 to 2015, successfully galvanized public support, in part through the jubilant militarism that it promoted. Canada has played a significant military role in Afghanistan and Syria, and more surreptitiously has provided support to the invasion of Iraq. At home, the military has explicitly targeted activists and Aboriginal peoples as potential threats to infrastructure. Academics have punctured holes in the benign narratives of peacekeeping that had infused the national imaginary (McKay and Swift, 2012). Sherene Razack's book on the disastrous Canadian military role in Somalia is but one example (Razack, 2004). These events and accounts forced me to acknowledge and confront the interpenetration and interdependence between military and civilian spheres that I had sought to deny.

At the same time, it also became evident to me that academics have played a special role in the war on terror. For example, the US military's Human Terrain System – a project tasked with providing local, cultural knowledge in areas of deployment – enrolled university anthropologists on foreign missions (e.g. González, 2009; Price, 2011). Geographers have also participated in data gathering for the war on terror, particularly with respect to the Bowman Expeditions, which were funded by the US Army's Foreign Military Studies Office, to monitor indigenous peoples in Oaxaca, Mexico (Bryan, 2010; Cruz, 2010; Wainwright, 2013; see also Crampton et al., 2014). These blatant examples of academic conscription have resulted in considerable pushback regarding the dangers of placing civilians in the field, and the ethics of research that will be used to subjugate populations. Yet to suggest that this complicity is new is disingenuous. During World War II and the Cold War, universities and academics were key players in military research and development, as part of the military-industrial-academic complex (Giroux, 2007). With the turn to the war on terror and the identification of new security threats such as climate change, this complex has deepened and extended in new and troubling ways (Gilbert, 2012).

It became increasingly obvious that my assumptions that the military and civilian spheres are distinct is a liberal conceit. Indeed, as Celso Castro observes, "The civilian is a military invention. I am only a civilian in relation to the military and when I am classified by them as such" (Castro, 2013: 11). This not only prompted me to rethink my presumptions, but also provoked me to better understand the centrality of the military to civilian life by taking up research that was explicitly focused on soldiers and war. The decision, however, did not suddenly make the project easy. I continued to find the task daunting. I was deeply anxious about how little I knew about the simple aspects of military life (e.g. military chains of command, weaponry, historic battles). And I was nervous that my lack of knowledge would be immediately apparent in my research and writing. And even as I recognized the central role of the military in contemporary society, I continued to think of the military as a black box, and of soldiers as different from me. I had no idea about whom to contact about setting up an interview, or how to approach this process to better ensure their interest and participation.

These fears and apprehensions have affected my research. Even though the original framing of my research included interviews with military personnel, I have been slow to get going with this. The mystification that I continued to have about the military was paralyzing. Rather than moving forward with the interviews, I sought to read every scrap of relevant materials first so that I would not look foolish. Only now, after several years of arms-length research, do I feel comfortable with the thought of interviewing people about the practices of military compensation. I have been fortunate to have had some unanticipated conversations with military personnel along the way. But whether I am able to find soldiers willing and able to speak with me in formal interviews is as of yet unknown. But I know that not having access to first-hand accounts of how military compensation is made, its rationale, and its practices (e.g. training) is a limitation of the research that I have completed thus far. Starting interviews earlier would also have likely saved countless hours of looking for and chasing after information and reading between the lines. It might also have opened up new avenues of research that would have pushed the research forward in different but productive directions.

Conclusions

The military is a fascinating site for research. As an institution it has its own logics, routines and ideals that infuse those who are within it. But it is not separate from the rest of society. In fact, our militaries are constitutive of how we understand what it means to be a civilian, just as war governs our understanding of peace, even as all of these taxonomies are becoming more blurred and harder to sustain. On a personal level, my research on the ways that lives are valued (or not) by the military apparatus in times of war has helped me to understand the blurring of combatant and civilian, and war and peace. More broadly, this research has shed light on the violence that is inflicted on civilians in war and the mechanisms that are being deployed to alleviate the harm that is incurred, as militaries seek to refashion themselves as humanitarian organizations. As I have suggested, however, compensation has a particular valence when distributed by soldiers, and can actually work to devalue life precisely through its monetization, and through the lack of accountability that is attributed to the cause of the damage (Gilbert, 2015a). I suspect that this is why the military and the state have resisted making information about victim compensation public.

In the early decades of the twenty-first century, militaries around the world are rising in importance in response to new geopolitical fractures. In light of the globalization of militarism, much more attention needs to be directed towards the armed forces and the power that we accord them. To this analysis, researchers and others ought to bring with them all of their critical tools along with a recognition of their own responsibility for what takes place in the military and in the name of the military. For academics, this means paying special attention to the ways the institutions that we are embedded in, such as universities, are playing a significant and troubling role in the global military-industrial-academic complex. Only in so doing will we be able to better understand the reach of military power, but also the more nuanced ways it is sustained within broader society. This does not require an abandonment of antiwar politics. Indeed, I continue to remain suspicious of and apprehensive of military bureaucracy, hierarchy and authority. It does, require, however, coming to terms with one's own complicity in the constitutive role that the military plays in everyday life, on and off the battlefield.

Acknowledgements

Many thanks to the editors for inviting me to contribute to this edited volume. The editors and reviewer provided excellent feedback on my chapter. Alison Williams deserves special thanks for comments and guidance. I am very appreciative of all the work that the American Civil Liberties Union has done to collect and make public the databases on civilian compensation. Funding for this research was made possible by a grant through the Canadian Social Sciences and Humanities Research Council.

Notes

1 The documents are available on the ACLU website, 'The Human Cost: Civilian Casualties in Iraq and Afghanistan': http://www.aclu.org/civiliancasualties.
2 The FCA is available to foreign nationals who have suffered injury, death or property damage as a result of a noncombat scenario. The Commander's Emergency Response Fund is used to make condolence or solatia payments, and is often used when a FCA claim is denied. For more information on these funds and the payments see Gilbert, 2015a.
3 Interview with attorney Nasrina Bargzie and Iraq Body Count: https://www.iraqbodycount.org/analysis/qa/aclu-ibc/.
4 Ibid.
5 The documents acquired by the *Guardian* are available here: https://docs.google.com/spreadsheet/ccc?key=0AonYZs4MzlZbdE9hbTh3LTBVVExickZKMnBvVENIOWc#gid=0.

References

Anaïs, Seantel (2013) Objects of Security/Objects of Research: Analyzing Non-Lethal Weapons. In Salter, Mark and Mutlu, Can E. (Eds.), *Research Methods in Critical Security Studies*. New York: Routledge, pp. 195–198.

Aradau, Claudia (2013) Infrastructure. In Salter, Mark and Mutlu, Can E. (Eds.), *Research Methods in Critical Security Studies*. New York: Routledge, pp. 181–185.

Banham, Cynthia (2009) Troops Give Cash to Afghan Victims. *Sydney Morning Herald*, Thursday, 2 July.

Beeby, Dean (2013) Feds Dispute Canada's Dismal Ranking in Report on Freedom of Information. *Globe and Mail*, February 24. Available at: http://www.theglobeandmail.com/news/national/feds-dispute-canadas-dismal-ranking-in-report-on-freedom-of-information/article9011887/

Bryan, J. (2010) Force Multipliers: Geography, Militarism, and the Bowman Expeditions. *Political Geography* 29, 414–416.

Butler, Judith (2009) *Frames of War: When Is Life Grievable?* London: Verso.

Castro, Celso (2013) Anthropological Methods and the Study of the Military: The Brazilian Experience. In Carreiras, Helena and Castro, Celso (Eds.), *Qualitative Methods in Military Studies: Research Experiences and Challenges*. New York: Routledge, pp. 8–16.

Crampton, Jeremy W., Roberts, Susan M. and Poorthuis, Ate (2014) The New Political Economy of Geographical Intelligence. *Annals of the Association of American Geographers* 104(1), 196–214.

Cruz, M. (2010) A Living Space: The Relationship Between Land and Property in the Community. *Political Geography* 29, 420–421.

Foucault, Michel (1980) *Power/Knowledge: Selected Interviews and Other Writings, 1972–1977*, trans. Colin Gordon. New York: Pantheon Books.

Friscolanti, Michael (2011) What's a Life Worth? *Maclean's*, 10 January. Available at: http://www2.macleans.ca/2011/01/10/whats-a-life-worth/

Gilbert, Emily (2012) The Militarization of Climate Change. *ACME: An International E-Journal for Critical Geographies* 11(1), 1–14.

Gilbert, Emily (2015a) The Gift of War: Cash, Counterinsurgency and 'collateral damage'. *Security Dialogue* 46(5), 403–421.

Gilbert, Emily (2015b) Money as a 'Weapons System' and the Entrepreneurial Way of War. *Critical Military Studies*. 1(3): 202–219.

Giroux, Henry (2007) *The University in Chains: Confronting the Military-Industrial-Academic Complex*. Boulder, CO: Paradigm.

González, Roberto J. (2009) *American Counterinsurgency: Human Science and the Human Terrain*. Chicago: Prickly Paradigm.

Gordon, Colin (Ed.) (1991) *Power/Knowledge: Selected Interviews and Other Writings 1971–1977*. New York: Pantheon.

Higate, Paul and Cameron, Ailsa (2006) Reflexivity and Researching the Military. *Armed Forces and Society* 32(2), 219–233.

Howell, Allison (2013) Medicine and the Psy Disciplines. In Salter, Mark and Mutlu, Can E. (Eds.), *Research Methods in Critical Security Studies*. New York: Routledge, pp. 129–132.

Lobo-Guerrero, Luis (2013) Archives. In Salter, Mark and Mutlu, Can E. (Eds.), *Research Methods in Critical Security Studies*. New York: Routledge, pp. 121–124.

Masco, Joseph (2010) Sensitive But Unclassified Secrecy and the Counterterrorist State. *Public Culture* 22(3), 433–463.

McKay, Ian and Swift, Jamie (2012) *Warrior Nation: Rebranding Canada in an Age of Anxiety*. Toronto: Between the Lines.

Moore, Dene (2010) Cda Shelled Out $650K Over Two Years to Civilians Caught in the Crossfire. *News 1130*, 6 September. Available at: http://www.stalbertgazette.com/article/GB/20100906/CP02/309069934/-1/SAG0806/cda-shelled-out-650k-over-two-years-to-civilians-caught-in-the&template=cpArt

Nath, Anjalie (2014) Beyond the Public Eye: On FOIA Documents and the Visual Politics of Redaction. *Cultural Studies Critical Methodologies* 14(1), 21–28.

Price, David H. (2011) *Weaponizing Anthropology: Social Science in Service of the Militarized State*. Oakland, CA: AK Press Distribution.

Quinn, Ben (2013) MoD Compensation Log Illustrates Human Cost of Afghan War. *Guardian*, 1 January. Available at: http://www.theguardian.com/world/2013/jan/01/mod-compensation-log-afghan-war

Razack, Sherene (2004) *Dark Threats and White Knights: The Somalia Affair, Peacekeeping, and the New Imperialism*. Toronto: University of Toronto Press.

Said, Edward W. (1993) *Culture and Imperialism*. New York: Knopf.

Salter, Mark (2013) Expertise in the Aviation Security Field. In Salter, Mark and Mutlu, Can E. (Eds.), *Research Methods in Critical Security Studies*. New York: Routledge, pp. 105–108.

Shane, S. (2012) Cost to Protect US Secrets Doubles to Over $11 Billion. *New York Times*, 2 July. Available at: http://www.nytimes.com/2012/07/03/us/politics/cost-to-protect-us-secrets-doubles-in-decade-to-11-billion.html?_r=0

Sitaraman, Ganesh (2009) Counterinsurgency, the War on Terror, and the Laws of War. *Virginia Law Review* 95, 1745–1840.

Stoler, Laura Ann (2010) *Along the Archival Grain: Epistemic Anxieties and Colonial Common Sense*. Princeton, NJ: Princeton University Press.

Tonkiss, F. (2004) Analysing Text and Speech: Content and Discourse Analysis. In Seale, Clive (Ed.), *Researching Society and Culture*. London: Sage, pp. 367–382.

Tracy, Jonathan (2008) *Compensating Civilian Casualties: "I Am Sorry for Your Loss, and I Wish You Well in a Free Iraq."* A Research Report Prepared for the Carr Center for Human Rights Policy and Campaign for Innocent Victims in Conflict.

Turse, Nick (2013) Blood Money. *Nation*, 19 September. Available at: http://www.theinvestigativefund.org/investigations/iraqafghanistan/1845/blood_money

Wainwright, Joel (2013) *Geopiracy: Oaxaca, Militant Empiricism, and Geographical Thought*. London: Palgrave Macmillan.

Witt, John Fabian (2008) Form and Substance in the Law of Counterinsurgency Damages. *Loyola of Los Angeles Law Review* 41, 1455–1482.

4

BIOGRAPHY AND THE MILITARY ARCHIVE

Isla Forsyth

If, as Osborne (1999: 56) states, 'archives have beginnings but not origins, they are both controlled by gatekeepers and worked upon, are never innocent,' then military archival sources have in the process of production almost certainly already undergone a particularly fierce round of editing and checking. Official military documents are informative in content, pared in delivery and commonly sealed from prying interpretation, until their period of confidentiality has expired. They may be rich in detail (recording precisely the 'wheres', 'whens' and 'hows' of battle plans, fields and operations), yet for documents that record so much action and bloodletting they are often dry in tone. While soldiers' actions are accounted for, the embodied experience in these sources is left to linger uncommented. Thus, the challenge is how to get under the clipped language, to consider the experiences as well as the strategic impacts. Some researchers have managed successfully to negotiate these issues by looking elsewhere for viable sources. Woodward, Winter and Jenkings (2010) have used soldiers' photographs taken during tours of duty in recent conflicts to tell stories about the role of the soldier in active military spaces and territories. They have also explored the image of the soldier in contemporary print media, and Woodward and Jenkings (2011a, 2011b) have considered the role of the military memoir to the social production of the contemporary British military. These studies employ diverse military sources to tell an embodied account of the military experience, and they suggest the possible means to draw on creative and lively military archives.

In order to explore and narrate a cultural history of WWII British military camouflage, I extended the reach of the military archive beyond the National Archives, the Imperial War Museum and the Royal Engineers Library and Museum, and beyond official reports and documentation. I also included art galleries, zoological museums and military training grounds and materials such as art, personal correspondence, photography, dioramas, film, taxidermy, scientific publications and expedition reports (see Forsyth, 2014a, 2014b). The reason for adopting this approach was because I wanted to narrate a biography of camouflage that explored and accounted for the situated and also embodied nature of technological innovation in military conflict. Therefore I decided to focus predominately on the life of one of WWII's British camoufleurs, eminent zoologist and skilful artist Dr Hugh Cott. To achieve a rich and replete biography of camouflage I studied Cott's scientific biography entwined with the fragmentary biographies of other camouflage practitioners, including artists, animals and even a magician, in order to trace and account for the sites and spacings of camouflage's life-path from the late nineteenth century into the Desert War. Thus, researching the military through the archive became an exercise in channelling 'creative energies' (McGeachan et al., 2012: 169), piecing together diverse, distributed and discrete remnants in order to illuminate the complexities of military knowledge, practice and technology.

In order to explore the means by which historical military research can narrate engaging and meaningful biographies, this chapter alights upon three archival items – a zoological photograph, personal correspondence between Dr Cott and his mentor Professor Graham Kerr and the military authorities, and a museum diorama – each uncovering something of the history and biography of WWII British military camouflage. Therefore, the chapter will first explain what an expanded approach to the military archive entails. Second, it will explore how a biographical framing of military research allows for enriching, more-than-human narratives to be told, before narrating three vignettes from the archive. Overall, this chapter reveals how combining official military documentation with personal biography can contextualise and account for technological innovations in warfare, which can enrich and make tangible military histories and biographies.

Archives

Steedman (1998: 77) describes the archive as a place empty of the past; indeed, it contains the past's material fragments, but the past does not 'in fact live in the record office, but is rather, *gone*'. Thus archives are often depicted as being inhabited by disjointed presences and haunted by absence. However, although archival practice can be an exercise in piecing together the fragmented, it can equally be a task of sifting through an overabundance of material presences from the past. The archive, whether characterised by fragments, excess, dispersal, concentration, thoroughness or scrappiness, deserves attention as a place for making history.

That history is created within the archive, not found, is a sentiment recognised by Osborne (1999: 55), who states:

> the archive is there to serve memory, to be useful, but its *ultimate* ends are necessarily indeterminate. It is deposited for many purposes; but one of the potentialities is that it awaits a constituency or public whose limits are of necessity unknown.

The archive becomes less a site for retaining the past and more a space for creative potential in the telling of the past. But furthermore, the archive and all those who keep, construct, contribute to, preserve and utilise it are in a continual process of making history. The archive enacts a temporal collectivism (and of course also at times a collective eradication) upon each object held in its possession. And so, where once the archive was viewed by scholars as little more than a 'storage space' (Lorimer, 2009), as Mayhew (2007: 24) states there is now recognition that 'evidence itself is constructed at a certain time and place.' Accordingly, researchers have begun to focus on the *doing* of research and the dynamism of the archive. In the specific context of the museum, Alberti (2009: 143) argues that 'the biography of an object did not halt when it reached a museum, and nor was its meaning frozen – rather, it was subject to considerable work while in the collection.' This dictum could be applied equally to objects within the archive, which, after deposition in some archives, are continually worked upon, by ordering and preservation in active afterlives (Ogborn, 2004). Even if lying dormant, their potential to shape and add to stories spanning time and lives are only one pencil-filled request form away.

However, as well as being a place for potential stories and storytelling, the archive is also a place that condenses the power inherent in inclusion and exclusion. Withers (2002: 303) explains that the archive can be a site 'of authority and meaning', compiled of 'selected and consciously chosen documentation from the past' (Steedman, 1998: 75), as it is populated with the carefully selected, edited and censored, and void of the intentionally omitted. But, Withers also warns not to oversimplify the archive as being a straightforward expression of power because, along with the consciously preserved and carefully catalogued, there also resides haphazard remnants, whose presence and purpose remain an indefinite mystery. The archive can be ordered and systematic, yet it can also be ambiguous: its contents revealing, enigmatic, even at times cryptic, requiring the researcher's close scrutiny, disclosing the objects and their own desired intentions for those objects. Just as the materials are worked upon, so too is the researcher, their biographies for a time entwined in a process of becoming in the archive.

This close attention to the archive and its materiality, Lorimer (2009) explains, is linked to poststructuralism and postmodernism, which has generated a more theoretically sensitive appraisal of archival research and also invigorated debate about what once constituted archival practices and sources. Alongside the thorough analysis and thoughtful interpretation of textual documents such as letters, diaries, court records and reports, there has emerged a creative and dynamic, more-than-textual expansion of what counts as an archive and what objects it includes, so that images, scents, sounds, textures and feelings are now also quarry for the historical researcher. This extension of historical sources and new sites that constitute the archive offers a means effectively to navigate the absences in historical research, giving alternative means and new ways actively to enrol gaps and disjunctures. This piecemeal gathering or 'make-do methodology' has led to research which experiments with innovative storytelling, aiming to capture the diverse and myriad lived experiences in history more fully and viscerally (Lorimer, 2006: 515).

Such experimentations with make-do methodology have produced engaging research. The absence of conventional documentary sources is supplemented by creative research with material encounters in order to attempt to recover 'the embodied practice and places of their making' (Patchett, 2008: 18). The common thread connecting these diverse studies is the resourceful use of more-than-textual objects and an extension of the lives that can be explored in historical geographical research. This is appealing for a study focusing on the entwined biographies of a camoufleur (Dr Hugh Cott) and a military technology (camouflage), which is narrated through a dispersed and varyingly abundant and scrappy archive of texts, images, sounds and senses, thus making military archives more lively and biographies richer.

Biography

Biography has the potential for resurrecting in new forms past lives and narratives; it is a means to mediate between memory and history, and to allow different perspectives on these narratives to be explored (Bensaude-Vincent, 2007). Rose (2000) explains how biography entwines three lives: the person at the centre of the work; the researcher who works from traces that are left, to reshape and recreate the first life for the third; and the reader who interprets this final work. It seemed for this study that biography could be a way of producing a cultural-historical geography of camouflage that allowed for such openness and fallibility. Dr Hugh Cott (1900–1987) provides a means of accessing the diverse facets that made up the character of British military camouflage. Most importantly, he was an expert in biological camouflage, and author of what is still regarded as a standard text on the subject. His scientific expertise could address an absence in much of the camouflage literature on this area and allow an exploration of the more-than-human nature of scientific practice in the study of camouflage. Cott was also a gifted artist, his black-and-white pen drawings demonstrating an artistic appreciation of the nature of camouflage and the need to understand its visual techniques of eluding exposure. Finally, Cott was a WWII camoufleur who participated in official camouflage committees and military camouflage training and served in the army as a camouflage officer throughout the war. Thus the micropolitics and lived experience of camouflage in war and battle can be investigated and an innovative approach to military biography, both human and nonhuman, can be narrated.

Classic biography, as Rhiel and Suchoff (1996) explain, can be described as the writing of a life, often of 'great men', traced and retold chronologically through a linear structure. This traditional approach is characterised by the view of an individual as the creator of meaning, which the biographer can access through 'an empirical, material basis', thus, writing stories that 'reflect a lived experience' (Roberts, 2002: 6). The focus in a classic biography is to know and understand the life of the person at the centre of the study. However, since the cultural turn biography has been influenced by postmodernism and the focus has increasingly been placed on plurality, previously unheard voices and the biography's 'performance' or 'collaboration' with and between the individual at the centre of the biography and the biographer (Rhiel and Suchoff, 1996). Contributing to this shift in biography, geographers have identified the potential of framing biographies not through the temporal, by simple chronology, but through the various spatialities of a life. Lorimer (2003: 283) has suggested that we need not narrate subjects' entire life histories as 'a fixed arc that begins, happens and ends', but instead construct 'more mobile biographies told through different episodes or moments happening within the longer context of a life'. Therefore the history of camouflage drew attention to moments, places, sites, networks and spacings in Cott's life, not in an effort to close in around the individual and know him fully, but rather as a means to extend the breadth of military camouflage and to cast greater clarity on its history. Thus, it was a geographical biography of camouflage through Cott.

Appadurai (1986: 3) has argued that 'commodities, like persons, have social lives,' so a biography of objects can reveal that, like people, they too are subject to transformation. Camouflage, as a technology and aesthetic, has a biography, but it is a very difficult life to pin down and explain. Its beginnings are embedded in nature, its multiple selves are intricate and conflicting, residing in nature (on skin, fur and scales) and in war (on wings, fabrics and metals) – both protective and predatory. Therefore, to cover such a complex series of lives is a daunting prospect. Pile and Thrift (1995: 1) explain that 'there is the difficulty of mapping something that cannot be counted as singular but only as a mass of different and sometimes conflicting subject positions.' They suggest that life can be accounted for in a more fluid conception of wayfinding, navigating between mapping and tracing, explaining how

> as bodies move they trace out a path from one location to another. These paths constantly intersect with those of others in a complex web of biographies. These others are not just human bodies but also all other objects that can be described as trajectories in time-space; animals, machines, trees, dwellings and so on.
>
> *(Pile and Thrift, 1995: 26)*

Camouflage thus can be viewed as a military technology with a life history which traces multiple and diverse trajectories, a technology shaped through intricate relationships between, and set in networks of, people and nonhumans, politics, science and art. Hoskins states that 'objects are both real and historical'; they may be inanimate, but their relations are animated within and by human lives (Hoskins, 2006: 81–82). To drive this idea further, a biography of camouflage could reveal objects *animated* by human lives that in turn *animate* human lives, influencing the embodied and visceral experiences of battle.

A question then arises of how to design a biography of camouflage, scrutinised through its practitioners. Throughout the study, the sites of convergence between Cott's life and camouflage's biography were explored, and this spatially mobile structure was used in order to gain insight into the nature of the development of military camouflage. These human-centred biographies, as research practice, narrative technique and structural design, were employed to draw a series of trails that entwine and diverge, which collectively *fleshed out* the biography of camouflage.

Vignette 1: A Photograph/Biological Beginnings

In the University of Glasgow Archives, Cott's correspondence with his mentor the zoologist Professor Graham Kerr is carefully preserved. At times these letters between friends and colleagues are illustrated with field notes and photographs. One image sent to Kerr by Cott, of a partridge in a nest, recalled to mind a story that a former student of Cott's had recounted on a visit to Cambridge Zoology Museum. After finding the photograph of the partridge I went back to the transcript of my interview with Cott's former student. I then attempted to conjure the event that led to the photograph being taken and what it could reveal to me about Cott's science and his thought on military camouflage. Therefore, I studied the image, the interview transcript and Cott's expedition reports which described how he observed animals in nature as a means to make sense of and give shape to his camouflage beginnings. This photograph and the story associated with its production seemed an effective way in which to narrate Cott's perspective on the development of military camouflage.

One Monday afternoon in the tea room at the Cambridge University's Zoology Department and Museum, Cott told a colleague that he had spent a large proportion of his weekend in a hide, studying and watching a partridge nest. Cott was used to spending hours waiting, watching and observing nature: that was his science. To observe, note and collect, his patience for his quarry was almost boundless. From this method he observed in his book on adaptive coloration in animals: 'So accustomed are we to reject what the eye sees in nature, so dull and dead have we become as a result' (Cott, 1940: 2). As zoologist, naturalist and Darwinian, countless field hours disclosed to Hugh Cott the complexity of form, pattern and adaptive coloration in nature. He was a student of nature's concealment, nature's surfaces. The smooth cylindrical body of the snake flattened to the earth by its colouring, the ringed plover scattered to the ground by use of its patterned feathers alone. But today his waiting was in vain. The partridge nest among the grass and leaves tucked in tight under the hedgerow was empty. Had she abandoned it, had she left, had she sensed his presence, was she observing him until his departure. With a final few clicks of his camera, Hugh Cott stopped waiting, stopped watching and rose to leave his hide (Figure 4.1).

But later she did reveal herself to him, the partridge he had been waiting for all those observant hours. In the dark room she began slowly to reveal herself, feather by feather. 'Come and

Figure 4.1   Partridge.

Source: H. B. Cott, Glasgow University Archives DC7.

enjoy Nature with me. By means of what is only a picture, I may be able to remind you of beauties you have not fully perceived' (Johnson cited in Cott, 1959: 20). As he developed the photographs, the partridge took form, and there she sat in the nest. The photograph revealed the partridge and her camouflage to be a shadowland of interweaving, interpenetrating, patterns, colour and forms and entwining of ecological relations. As Cott was the student, the partridge had provided him with a master class in concealment, in surfaces. This understanding of nature's near perfection in the skill of deception Cott knew had importance beyond the scientific. His hours in the field were spent during the 1920s and '30s when brooding clouds of war were gathering over Europe. Scientific and military knowledges Cott realised needed to be interwoven to result in efficient military camouflage, like bird and background.

Cott was not the first to realise the importance of camouflage to warfare. Since WWI camouflage had begun to be adopted and developed in a systematic way. This had resulted in the military looking beyond its boundaries to incorporate usually distinct and discrete knowledges out with. In Britain this had seen the employment of artists such as Solomon J. Solomon and Norman Wilkinson, and correspondence and advice from the zoologist Graham Kerr and from the US artist and naturalist Abbott Thayer (see Behrens, 2002; Robertson, 2002). Their contributions had seen some successful inventions in concealment, nets to cover artillery, to observation posts, although most probably the most noted of inventions was the 'dazzle' or 'Easter egg' design of ships to confuse submarines as to their direction of travels.

What Cott had correctly perceived in his research on biological camouflage was that by the cusp of WWII the need for effective camouflage was urgent. There was a growing threat in the clouds as more advanced and efficient aeroplane and observation technology was being developed, and battle was being envisioned vertically as well as horizontally. Therefore, camouflage would have a central role in designing defence from aeroplane bombardment and the outwitting of the reconnaissance camera.

Cott's partridge story proved a useful means to communicate the ways in which natural history knowledge had military utility. It demonstrates that knowledge incorporated into the military and utilised to develop strategies and technologies of warfare has its origin beyond the battlefield. The roots of technological innovations and their routes into military strategy are worthy of tracing as they reveal that the military has scope beyond its immediate sphere; it is influenced and influences other knowledges and practices. In order to shape this narrative I had to bring different archives and texts – photographs, memories and scientific studies – into dialogue. This demonstrates that to write rich military biographies, the military archive should be composed of multiple and diverse sites and materials.

Vignette 2: Letters/The Personal is Political

In order to explore the role of social relations, tensions and at times competing knowledges and practices in the development of military technology, the faltering fate of the Camouflage Advisory Panel in WWII is illuminating. Its fractured history can be traced through Cott's correspondence with his mentor Kerr alongside official military and home office documents held at the National Archive.

In the 1930s, as tensions between the European nations grew more and more fraught, the British military began to prepare for the likely event of another war. One of the greatest fears for Britain was the likelihood of enemy attack and invasion from the skies. It was realised that camouflage had the potential to protect against this. In October 1936 the Committee of Imperial Defence (CID) made a recommendation that the home office should take charge of matters concerning the security of vital British factories, key points and landmarks from aerial attack in the event of war (Goodden, 2007).

Home Secretary John Anderson decided in response to establish the Camouflage Advisory Panel (CAP) in order to advise him

> from its wider knowledge and outside contacts, so that policy may not be entirely restricted to the experience and information gained by my officers and may be given every opportunity to embrace the problem which will arise in war-time and to solve them with all the ingenuity that can be collected.
>
> *(Letter from John Anderson to Sir Ernest Swinton,*
> *27 September 1939 – NA HO186/668/4)*

And to

> absorb all ideas from outside and to contribute its own conceptions of what should be done, so as to form a recognisable policy which can be given in the form of advice to this Department.
>
> *(Letter from Mr Galpin from the Air Ministry to Ernest Swinton,*
> *15 September 1939 – NA HO 186/668/3)*

Although Anderson's central aim was to have a body of experts in the field of visual literacy who could guide him on military camouflage matters, he also had another aim in setting up the CAP. Anderson hoped to change the prevailing and potentially destructive attitudes between different established camouflage experts. This attitude had emerged from WWI camouflage initiatives, through an acerbic dispute between the scientist Professor Graham Kerr and the artist Norman Wilkinson over the invention and origins of dazzle patterning for the admiralty. The home secretary hoped that the CAP would soothe grievances between differing factions and, further, create a space for collaboration. In a letter the home secretary set out his aspirations for the panel (Letter from John Anderson to Mr Swinton, 6 January 1940 – NA HO 186/174/2).

> There are conflicting theories: there is, for example, the biologist school and the artistic school. What we must secure is that the whole matter is tackled vigorously on a sufficiently high plane and that all the petty jealousies and rivalries, which are largely the result of the manner in which the whole thing has grown up, are eliminated.
>
> *(Letter from John Anderson to Mr Swinton,*
> *12 January 1940 – NA HO 186/174/3)*

Anderson recognised that in a time of national threat, 'imposed cooperation with other disciplines' was required (Barnes and Farish, 2006: 814). Therefore, the military was required to forge networks and connections with sources of outside knowledge and those who had skills, such as art and biological sciences, outside its usual remit. He turned to the Research and Experiments branch to compile a list of proposed members for the CAP who would be useful in providing the diverse and specialist skills required. Among the list submitted were artists such as Norman Wilkinson and Paul Nash, paint manufacturers, military academics and other professions whose talents had been successfully drawn upon during WWI (Letter from deputy undersecretary of state from Research and Experiments branch, 13 June 1939 – NA HA 188/668/1).

Scientists were also included on the list since Anderson had acknowledged that developments in military camouflage had, so far, possessed too limited a scientific input. This included Cott, who had established himself as an authority in the field through the publication of his 1940 book *Adaptive Coloration in Animals*, by conducting a camouflage experiment on an aircraft hangar, and by delivering a lecture on camouflage to the Royal Engineers. Cott was eager to be used by the military in order to adapt and develop their military camouflage by adhering to the principles of biological camouflage. This move to the boardroom saw Cott rely greatly on his mentor the zoologist Graham Kerr, who had in WWI attempted to encourage the military to adhere to biological camouflage for their modern war requirements. On the cusp of another war Kerr was determined that Cott should achieve where he had failed: the military adoption of biological principles of camouflage.

Glasgow University Archives reveal that from the moment of Cott's acceptance letter to join the CAP there was a rapid and frantic correspondence between the two zoologists. Kerr meticulously filed Cott's letters and made copies of his own responses. Every action the panel took was relayed from Cott to Kerr, and attitudes of the military service departments and the other members of the CAP towards camouflage were scrutinised. For Cott, Kerr was a close ally and mentor from whom he sought advice and guidance at every turn. For Kerr, Cott's appointment to the panel was an opportunity to position Cott and thus science at the helm of military camouflage initiatives. This close relationship between the two scientists and Cott's reliance on Kerr's opinion and advice saw the threading of WWI anxiety and tensions into the CAP. Old grievances resurfaced.

The personal correspondence was read alongside the CAP's meeting minutes, which are held at the National Archives. Between the official and the personal I examined whether Cott and Kerr's exasperation at the CAP and their perceived grievances were founded. Although the tone of the official documents was cooler and calmer, the delay in establishing the first meetings of the panel and the lack of coherence or action taken by it did demonstrate why Cott was becoming exasperated, if not quite giving credence to his perception that science was being overlooked. On the panel Cott did not take any decision or action without Kerr's considered opinion or appraisal, and so Cott, only with Kerr's approval, resigned (letter from Cott to Kerr, 30 March 1940 – Glasgow University Archives DC6/737). Using a resignation letter, penned by Kerr 'almost word for word on the lines of the draught which you so kindly suggested yesterday' (Letter from Cott to Kerr, 27 March 1940 – GUA DC6/732).

Cott wrote to the home secretary setting out his perceived failings of the CAP, which in the main he attributed to lack of coordination between and within the service departments; lack of attention to the scientific principles of camouflage; and the panel not being fully briefed on camouflage initiatives that were being carried out (Letter from Cott to John Anderson, 27 March 1940 – GUA DC6/733).

Cott's resignation from the CAP was accompanied by a very public criticism of the military's attitudes towards developing camouflage, in the form of an article in *Nature* by Cott and Kerr, who by now was a MP raising the issue in the House of Commons.

The personal correspondence between the two zoologists in regards to Cott's experience on the CAP reveal that military technological development is situated within social and political networks which influence their biographies. Reading official documentation alongside Cott's and Kerr's correspondence demonstrates that personal relations and tensions influence military initiatives. The question to ask of this material is not whether Cott was correct to be frustrated by the CAP or whether the CAP was effective. It is to ask in what ways the networks and assemblages of people, knowledge, technology and politics influence military history. By shifting focus and reordering narratives away from solely the implication and outcomes of military technologies, military histories become firmly placed in their cultural, social and political context. As a result, the small scale of the personal and large scale of military strategies come into dialogue in order to expose the complex character of military assemblages and for the consequences (however long-lasting or transitory) of state-sanctioned violence to begin to be traced.

Vignette 3: Diorama/Lasting Effects of War

Cott's camouflage archive consisting of letters, photographs and official military documentation had given me much detail about his experiences of developing military camouflage. I had a sense of Cott's motivations and temperament, and yet he still seemed distant. I sought material that would tell me how war had influenced and impacted him. When I arrived at the Cambridge Zoology museum I enquired about whether there were any remaining members of staff in the museum or the department who knew Cott. The archivist helpfully put me in touch with a former colleague of Cott's and one of his students who now worked in the department. I contacted both and was able to interview them. After conducting these interviews I tried to frame my own impressions of Cott gleaned from the archive of letters, artwork and scientific writings with the memories that had been recounted to me. My drive to gain a firmer grip on Cott and his experience of military camouflage led me to search for others who had known him. I also enquired at Glasgow University's Zoology department (Cott had worked there briefly in the 1930s), where an emeritus professor revealed he had been taught by Cott as an undergraduate at Cambridge, and he managed to put me in touch with other zoologists who had been tutored by Cott. I thus interviewed numerous students who were willing to share their memories of Cott. It was by combining oral history with objects from Cott's past that I got the clearest glimpse at the lasting effects of being enrolled in warfare.

A former colleague of Cott's (Ken Joycey) offered to take me on a tour of the Cambridge Zoology Museum where the museum's objects and displays provided the catalyst to memories. The most fruitful object on this tour was tucked away at the rear of the museum. There stretched a long, low diorama depicting a shoreline of sea birds, flanked on either side by corresponding yet rather tired looking taxidermy specimens. This was one of the few remaining displays attributable to Cott, a remnant of a 1950s zoology and museum display. Ken revealed that it had been constructed, under the careful direction of Cott, by several artists with whom he had served during WWII. It seems that on returning to university Cott had been instrumental in keeping together a 'gang of camouflage mates' by commissioning artwork and dioramas for the museum. Ken recalled that the dioramas were not only teaching aids during the postwar period of field trip austerity but also material objects and outcomes of friendship between old comrades in war. 'For Cott', he mused, 'they were the fruits of the war' (Figure 4.2).

Natural history dioramas can be insightful objects of study, revealing something of the natural world, and ourselves too. Wonders (2003) has described how the placing of species can convey unspoken messages that animals contribute to and are enrolled in depictions of national identity. Cott's coastal diorama can be read in this vein, but it also offers a more personal and melancholic construction of British identity and memory. On closer inspection, alongside the oyster catchers there is evidence of military material presences on the beach: a tangled, rusted piece of barbed wire, an empty artillery box, a lone black boot washed onto the strand (Figure 4.3). This model beach strewn with the discards of war may be absent of human beings, but their capacity for brutality and violence is clearly evident. The diorama not only communicates a vision of Britishness through birds; it also reveals the intrusion and threading of militarism through landscape, art and science in postwar Britain, articulated by people who had witnessed, participated in and shaped the war. Although Cott never wrote or spoke of his military experiences, his shoreline offers a very personal insight into the lingering of war on memory. For a moment unease crept close: looking upon the diorama, Cott's experiences of war are both less and more tangible; he took firmer form, but alongside lay the realisation that hours of archival sifting had yielded volumes of information on Cott's science and his involvement in the development of military camouflage. Of Cott's own embodied experiences and lasting memories, however, little can be known. The interviews conducted with people who had known Cott, such as Ken Joycey, helped me to gain a better sense of Cott, but it was also through close reading of objects such as the coastal diorama that I could glimpse the lasting and personal impacts of war. Therefore, it is by including objects and memories and considering the absences within an archive that some of the more personal and unsettling aspects of military histories can begin to be traced and accounted for.

Figure 4.2   Coastal diorama.

Source: Cambridge Zoology Museum, author's own image.

Figure 4.3   Discarded artillery box.

Source: Author's own image.

Conclusion

What can be glimpsed from entwining these three objects from Cott and camouflage's extended military archive is that military histories are replete with rich personal narratives that were influenced by broader geopolitical imperatives and actions, but which in turn shaped in small and subtle ways the processes, conduct and outcome of warfare – in this instance through the development and innovation of camouflage technology. Yet biography can be a troublesome method. Historical studies which take lives at their centre continually grapple with a subject who forever evades firm grasp. Terrall (2006: 306) explains that biography 'in some manner brings back to life someone from the past, known to the present only through material traces left behind'. The tangible remnants of a life provide a connection with the past, but this is a connection only allowing partial access, and inevitably there will always be 'something missing' (Terrall, 2006: 307). Therefore, one problem is that the person presented in the written biography is not in fact revived but created. Perhaps biography is the practice of a life reborn to be a monstrous being; tacked, sewn, and glued: a composite body created by a few of the many multiple selves that once constituted a lively individual. Another difficulty is that the visceral and somatic elements of the biographical life become muted through text.

Lee (2005: 8) suggests that all too often, 'biographers try to make a coherent narrative out of missing documents as well as existing ones; a whole figure out of body parts.' This process to smooth over cracks can result in a flat narrative. Rose (1995: 416) suggests that in historical work, researchers should 'articulate boundaries, distinctions, and disjunctures instead of erasing them'. Applying this approach, the disjointed contents of an archive can result in a disjointed narrative, a stitched, folded, occasionally embroidered or sometimes threadbare biography. Through this process, a life's material traces become 'somewhat like a cubist picture' (Richards, 2006: 302), in that absences and presences jostle for attention, and form may be viewed where there is discontinuity, and discontinuity may mask smooth surfaces. But this cubist vision can be embraced: Cott will have always proved unknowable in life, so focusing on the spatialities of a life, where interest alights more on abutting 'surfaces' than on chronological continuities, may disrupt and complicate the narrative, emphasising its ultimate unknowability – but this can only add insight to the account of camouflage's development.

The role of the biographer can hence be to allow room for these multiple perspectives and voices to be included in both Cott's biography and camouflage's narrative. As such, the moments or scenes of camouflage and Cott in the desert presented in this biography are a composite of remaining shards, revealing the jarring and conflicting selves of human and technological life. Memory, history, geography and storytelling are marshalled to make Cott's biography, and are less an attempt to rescue *him* from the archives or history and more a will to retrace and recount parts of his life and something of his selves. This biography works to reveal that

> to tell a story, then, is to *relate*, in narrative, the occurrences of the past, retracing a path through the world that others, recursively picking up the threads of past lives, can follow in the process of spinning out their own.
>
> *(Ingold, 2007: 90)*

The subsequent narrative will be openly incomplete and forever mutable and transforming. Cott's biography was a tool to interpret a more complex life, a means to explore some of the spaces and spacings of the development of military camouflage technology. In summary, an extended military archive and attention focused on the biographical (human or nonhuman) is used alongside and complementary to official military documentation. The weaving of the two can make military histories richer, disjointed, tangible, visceral, less knowable and less certain. Extending the military archive to include the voices of those who witnessed, experienced and were affected by war not only contextualises military histories in their cultural, political, social and temporal settings, but it also forces attention to the myriad scalar lasting consequences of war and global histories conceived through human ingenuity and violence.

References

Alberti, S. (2009) *Nature and Culture: Objects, Disciplines and the Manchester Museum*. Manchester: Manchester University Press.

Appadurai, A. (1986) *The Social Life of Things: Commodities in Cultural Perspective*. Cambridge: Cambridge University Press.

Barnes, T. and Farish, M. (2006) Between Regions: Science, Militarism, and American Geography from World War to Cold War. *Annals of the Association of American Geographers* 94(4), 807–826.

Behrens, R. (2002) *False Colors: Art, Design and Modern Camouflage*. Dysart, IA: Bobolink Books.

Bensaude-Vincent, B. (2007) Biographies as Mediators Between Memory and History in Science. In Soderqvist, T. (Ed.), *The History and Poetics of Scientific Biography; Science, Technology and Culture, 1700–1945*. Aldershot: Ashgate, pp. 173–184.

Cott, H. (1940) *Adaptive Coloration in Animals*. London: Methuen.

Cott, H. (1959) *Uganda in Black and White*. London: Macmillan.

Forsyth, I. (2014a) The Practice and Poetics of Fieldwork: Hugh Cott and the Study of Camouflage. *Journal of Historical Geography* 43, 169–185.

Forsyth, I. (2014b) Designs on the Desert: Camouflage, Deception and the Militarization of Space. *cultural Geographies* 21, 247–265.

Goodden, H. (2007) *Camouflage and Art; Design for Deception in World War 2*. London: Unicorn Press.

Hoskins, J. (2006) Agency, Biography and Object. In Tilley, C., Keane, W., Kuchler, S., Rowlands, M. and Spyer, P. (Eds.), *Handbook of Material Culture*. London: Sage, pp. 81–82.

Ingold, T. (2007) *Lines: A Brief History*. London: Routledge.

Lee, H. (2005) *Body Parts: Essays in Life-Writing*. London: Chatto & Windus.

Lorimer, H. (2003) The Geographical Field Course as Active Archive. *Cultural Geographies* 10(3), 278–308.

Lorimer, H. (2006) Herding Memories of Humans and Animals. *Environment and Planning D: Society and Space* 24(4), 497–518.

Lorimer, H. (2009) Caught in the Nick of Time: Archives and Fieldwork. In DeLyser, D. Herbert, S. Aitken, S., Crang, M. and McDowell, L. (Eds.), *The Sage Handbook of Qualitative Geography*. London: Sage, pp. 248–273.

Mayhew, R. (2007) Denaturalising Print, Historicising Text: Historical Geography and the History of the Book. In Gagen, E., Lorimer, H. and Vasudevan, A. (Eds.), *Practising the Archive: Reflections on Method and Practice in Historical Geography*. London: Royal Geographical Society, Research Group 40, pp. 23–36.

McGeachan, C., Forsyth, I. and Hasty, W. (2012) Certain Subjects? Working With Biography and Life-Writing in Historical Geography. *Historical Geography* 40, 169–185.

Ogborn, M. (2004) Archives. In Pile, S. and Thrift, N. (Eds.), *Patterned Ground*. London: Reaktion, pp. 240–242.

Osborne, T. (1999) The Ordinariness of the Archive. *History of the Human Sciences* 12(2), 51–64.

Patchett, M. (2008) Tracking Tigers: Recovering the Embodied Practices of Taxidermy. *Historical Geography* 36, 17–39.

Pile, S. and Thrift, N. (1995) *Mapping the Subject: Geographies of Cultural Transformation*. London: Routledge.

Rhiel, M. and Suchoff, D. (1996) *The Seductions of Biography*. London: Routledge.

Richards, J. (2006) Introduction: Fragmented Lives. *Isis* 97, 302–305.

Roberts, B. (2002) *Biographical Research*. Buckingham: Open University Press.

Robertson, F. (2002) Dazzle Painting: The Art of Deceit in War. *Journal of the Scottish Society for Art History* 7, 7–12.

Rose, G. (1995) Tradition & Paternity: Same Difference, *Transactions of the Institute of British Geographers* 20, 414–416.

Rose, G. (2000), Practising Photography: An Archive, a Study, Some Photographs and a Researcher. *Journal of Historical Geography* 26, 555–571.

Steedman, C. (1998) The Space of Memory: In an Archive. *History of the Human Sciences* 11(4), 77.

Terrall, M. (2006) Biography as Cultural History of Science. *Isis* 97, 306–313.

Withers, C. (2002) Constructing 'the geographical archive'. *Area* 34(3), 294–302.

Wonders, K. (2003) Habitat Dioramas and the Issue of Nativeness. *Landscape Research* 28(1), 89–100.

Woodward, R. and Jenkings, N. (2011a) Reconstructing the Colonial Present in British Soldiers' Accounts of the Afghanistan Conflict. In Kirsch, S. and Flint, C. (Eds.), Reconstructing Conflict: Integrating War and Post-War Geographies. Farnham: Ashgate, pp. 115–131.

Woodward, R. and Jenkings, N. (2011b) Military Identities in the Situated Accounts of British Military Personnel. *Sociology* 45(2), 252–268.

Woodward, R., Winter, T. and Jenkings, N. (2010) 'I Used to Keep a Camera in My Top Left-Hand Pocket': The Photographic Practices of British Soldiers. In MacDonald, F. Hughes, R. and Dodds, K. (Eds.), *Observant States: Geopolitics and Visual Culture*, pp. 143–165.

5

ANALYSING NEWSPAPERS

Considering the Use of Print Media Sources in Military Research

K. Neil Jenkings and Daniel Bos

It would seem almost impossible to overstate the importance of the media to today's armed forces. This applies especially but not uniquely to Western-style democracies. For example, the media representations of the Vietnam War at the breakfast tables of the American public led to their disenchantment with the conflict and a steady erosion of their support. This was a major factor, although not the only one, in the eventual US withdrawal from Southeast Asia (Huebner, 2008). Hence the relationship between media and the military receives continuing interest from academics (Caruthers, 2011; Maltby, 2012). Research has sought to unpack the significance that various news media play in legitimising and framing the military and their activities to civilian populations (Thussu and Freeman, 2003). This relationship is constantly changing due to technological transformations, shifting media-military relations and evolving media production and distribution practices, all of which have significant effects on our understanding of and relationship with the military. Television, and the arrival of 24/7 news, has implications on the way conflict is presented (Hoskins, 2004). Also the expanding use of social media has begun to blur the line between producers and audiences of news. Outlets, such as blogs, discussion forums, video hosting websites and other social media platforms, have provided alternative channels for disseminating news and discussions about the military.

While aware of this wider context and media environment with its issues of intertextuality, we want to explore the use of print media and specifically the role of print newspapers as a data resource. Despite the enhanced affordances of digital online media and the transformation of production, distribution and consumption behaviours of news media, we maintain that the analysis of print media remains a key medium for scholars interested in publicly disseminated representations of the military and its activities in the mass media. Print newspapers have been affected by the digital revolution in ways that have implications for related research methodologies, which we shall highlight.

In this chapter we examine the issues of research into military phenomena and their representation using print newspapers. Previous research has been reticent in divulging the practical steps undertaken in locating, collecting and storing print media materials (exceptions include Roy et al., 2007; McGarry and Ferguson, 2012). To help tackle this oversight, we focus on the practical measures undertaken on a research project conducted by the authors and others, which analysed UK newspaper representations of military repatriations that took place in the town of Wootton Bassett between 2007 and 2011. An outline of this project will be given, providing an examination and explanation of the research strategy, the process of data collection and its subsequent storage implemented by the project team. Given the limits of what we can cover in this chapter, we will focus on these elements rather than our specific analysis and findings (see Jenkings et al., 2012). Following this example, we will provide a discussion which will expand on the practicalities and important considerations in using print media sources for military research.

News Media and the Military: An Overview

Prior to the 1960s and the Vietnam War, the main form of news media was arguably the printed news found in newspapers and magazines. From the Vietnam War onwards, the television became increasingly more influential. Following the perceived negative impact of the coverage of the Vietnam conflict, the media's presence and the reporting of conflicts has since tended to be more managed by the military. The aim of the military is to keep the reporters and public 'onside' and any restrictions are justified in the name of 'security'. News management of events is difficult at the best of times, as illustrated during the Falklands War, where the media presence was extremely limited and the transmission of news from the frontline to the news office highly controlled. Despite this, the BBC still managed to inform the world (including Argentina) in advance of the British intention to attack Goose Green (Freedman, 2005). Despite newspaper circulation declining in the face of television, and more recently the Internet, the print media are still significant players in the representation of the military and their activities. Indeed, major newspapers now have digital as well as print editions, and the difference between the two, while important, can appear blurred. Nevertheless, significant numbers of printed newspapers are still purchased daily or weekly, and the significance of free newspapers cannot be excluded, especially in the UK.

When, as we have noted, the dominant media representations of the military are now audiovisual – and this applies to news representations as well as entertainment – it may well be legitimate to ask: why study written newspaper representations? The answer is that newspapers are still relevant, and significant, in the contemporary news media landscape, and not just digitally but in their traditional paper incarnations as well. In the UK, the *Sun*, the *Daily Mail* and the *Mirror* daily newspapers alone still sell almost five million copies between them, with the other major newspapers adding well over another two million daily newspapers sold (ABCs, 2014). It is important to note these figures do not include free newspapers (*Metro* has a daily circulation of 800,000 and the *Evening Standard* over 700,000) or regional newspapers. Nor do they acknowledge that a single newspaper is frequently read by more than one person. Thus, despite an overall decline in the sales of newspapers since the 1950s, the figures demonstrate that newspapers still reach a wide audience within the UK. Arguably their relevance has not diminished. While their circulation in paper format has decreased, news media organisations now produce digital versions of their newspapers, which in many instances are freely available and are capable of reaching a global audience.

Newspapers also tend to have much more detailed accounts, especially in length, than audiovisual news which tends to rely on visual images; although of course newspaper stories frequently include photographs (Woodward et al., 2009). Also, newspaper reporting is not separate from audiovisual media but supplements and may even set the agenda for other media representations. With regards to the phenomenon of Wootton Bassett, while we investigated it in terms of printed newspapers, it nevertheless needs to be understood alongside and intertwined with other representations. Indeed, an analysis of one media format is by definition a partial exploration of media coverage. Yet news print research is becoming a less favoured phenomenon in relation to audiovisual news, and indeed as we shall illustrate, becoming more problematic to collect as news archiving focuses on the audiovisual and less so on the written word. In addition, it is also the case that the visual in the form of the still image rather than the moving image is becoming more difficult to research due to the changes in archiving and archiving technologies.[1]

Researching the Military Using Print Media

The newspaper coverage of wars and conflicts, such as Vietnam (Hallin, 1992; Hammond, 1998), the Falklands (Harris, 1983; Adams, 1986) and more recent campaigns in Afghanistan and Iraq (Robinson et al., 2010; Taylor, 2014), have resulted in investigations into the relationship between the media and the military and the implications this has on the representation of armed conflict within society. There are numerous examples of work that demonstrate the continuing importance of the print media in communicating ideas, values and logics of the military and its penetration into social, economic and political life. The journal *Media, War & Conflict* testifies to this enduring interest. It publishes a range of material examining and analysing a range of newspapers in their depiction of war and military violence. Articles have focused on the framing of war and conflict, providing international examples and undertaking qualitative and quantitative analysis to examine the ways armed conflict is framed, rationalised and legitimised through the print media (see e.g. Goddard et al., 2008; Golčevski et al., 2013; Hjarvard and Kristensen, 2014).

The proliferation of online newspaper archives has offered a number of opportunities. For instance, comparative research between differing international newspapers has become more achievable. This comparative research delivers potential insights into the different geographical and cultural perspectives and viewpoints adopted by various newspaper organisations in discussing military activities (Dardis, 2006). Furthermore, it can be illustrative of how media can be inhibited by national political circumstances and the differing practices of production (Dimitrova and Strömbäck, 2008). While offering contemporary printed media sources, online archives are increasingly also providing extensive historical collections of newspaper publications.

It is important to recognise that newspapers not only present textual accounts but also provide visual materials such as photographs, statistics, graphs and maps, all of which are of interest to military scholars. Moving beyond textual analysis, scholars have emphasised the importance of the visual in print media, which frames how war is often understood (Parry, 2010). Certain newspapers offer satirical critiques of military conflict, such as the work of cartoonist Steve Bell in the *Guardian* (analysed by Dodds, 1996, 2007). Visuals used in newspapers help to provide distinct but also complementary materials that present particular narratives about armed conflict.

Our aim in this chapter is to discuss the methodological processes and practicalities in researching news print media texts. The methodologies disclosed by other authors, despite revealing their intentions and rationales, overlook the practical aspects and challenges faced when working with newspapers. Digital database services, such as LexisNexis, own official archives and websites where print media content are stored, as do many newspapers. These databases are a repository for a wealth of contemporary and historical newspaper articles. However, the availability of newspapers and the period of time in which they are retrievable varies between these different databases. Nevertheless, the convenience and the ability to navigate relevant material at the click of a button offers an attractive proposition for scholars.

The digitalisation of the print media has opened up new research opportunities, but also new challenges with regards to how relevant data is located, collected and analysed. We faced a number of challenges and issues when using digital archives of newspapers not reported by other researchers when discussing their methodological practices. As Nicholson (2013: 62) indicates:

> The methodological implications of digital archives – their ability to extend the boundaries of research and answer questions that were previously unanswerable – have remained frustratingly absent from the debate.

The advent of digital archives, although offering a number of opportunities for researchers, also entails a number of challenges that have so far gone relatively overlooked. However, this chapter cannot produce a definitive account of the limits and problems encountered in undertaking print media research. Our research, for instance, looks at the main tabloid and broadsheet newspapers as well as regional newspapers within the UK. We are not talking about nonnewspaper print media, such as *Newsweek* and other similar publications. Nevertheless, we use this chapter as an opportunity to begin to bring to the forefront the methodological implications of current print media research and what these mean for military studies. To do this, we discuss our experiences of using newspaper sources in a research project which examined the media representations of the repatriations of British soldiers through the English town of Wootton Bassett, UK.

Newspaper Coverage of Military Repatriations at Wootton Bassett

Wootton Bassett[2] is a small market town located in Wiltshire, England. The town emerged as a social and media phenomenon at the end of the first decade of the twenty-first century, when it became a temporary site of national mourning for the repatriated dead of the British Armed Forces. Between 2007 and 2011 the repatriated bodies of British military men and women killed in the Afghanistan and Iraq conflicts passed through the town. On their way from RAF Lyneham[3] to the John Radcliffe Hospital in Oxford, the cortèges were met by the local population of Wootton Bassett, who lined the streets. This began in April 2007 when the mayor and a group of local members of the Royal British Legion first paid their respects to the passing hearses. Over the following months, this evolved into a regular event in which crowds of locals, as well as those from further afield, lined the streets alongside grieving families and relatives. These regular events occurred until August 2011, when incoming flights moved back to RAF Brize Norton (see McGarry in this volume).

As the repatriations continued through Wootton Bassett, the events began to receive extensive media coverage. They were regularly broadcast on British television and occupied the pages of regional, national and international newspapers. The coverage revealed various events in which Wootton Bassett became politicised and a place in which the meaning of war and the role of the military were negotiated. Noting this continuing media presence, our inquiry set out to investigate the emergence of this social and media phenomenon and note the significance that these repatriations had on understandings of civilian and military relations. As such, our research strategy set out first to produce a metanarrative of the phenomena by tracing its origins and the key events. Second, the project sought to analyse the portrayal of these repatriations within newspapers, and to assess the wider politics of what was going on and what this said about contemporary militarism.

The use of print media sources, specifically newspapers, allowed us to create a chronological narrative of the Wootton Bassett phenomena. We were able to locate its origins, the key events and the various commentaries that emerged concerning the wider political debates that it encouraged. Our choice of print media sources was also dictated by a number of practical considerations. The content of television coverage, although a prominent media outlet covering the repatriations, was more complex and time-consuming to systematically collect, store and analyse. Newspapers, on the other hand, were relatively easily accessible to the research team and, considering funding constraints and the proposed duration of the project, were deemed more appropriate. We used the metadatabase LexisNexis as our primary source, as it offered the widest and most reliable coverage compared to other databases available at the time. It also provided an increasing number of individual newspaper online archives as secondary resources where gaps existed in the primary source.

Online Newspaper Databases: Using LexisNexis

Online archive services, such as LexisNexis, have become an increasingly popular resource for academics wishing to view, store and analyse a range of newspaper sources. This particular database is a popular resource for academics within the social sciences and offers researchers national, regional and local newspapers from the UK, as well as international coverage. The database allows for individual and collective searches of newspapers to locate articles, which are displayed in text-only format. Here we focused on the main British tabloid and broadsheet newspapers[4] and their Sunday equivalents, and aimed to locate all relevant articles concerning the repatriations at Wootton Bassett between 2007 and 2011. Despite being utilised by a range of researchers, there remains little critical methodological reflection on the use of LexisNexis (Weaver and Bimber, 2008). To redress this shortcoming, we will briefly discuss how we collected and stored our data for the project using LexisNexis.

Collecting Print Media Sources

There are a number of ways to collect newspaper articles. Researchers can buy their own newspapers to build their own personal archive. This would be suitable for small-scale research projects, focusing on specific newspapers and specific issues. This provides the researcher with a hard copy of the article, and situates it within the broader context of the page and the newspaper itself. However, this method can be costly and time-consuming; additionally, certain newspapers may be less easily accessible to the researcher. Other possible sources are newspaper websites where the e-newspaper content is presented and stored. The majority of these websites offer simple search functions that can locate articles. Recent developments have seen newspapers implement paywalls, offering only limited accessibility to previously freely available articles. Thus payment may be requested in order to gain full access to the website and to view articles in their entirety. This presents fiscal as well as 'format variation' challenges for the researcher, which we will discuss later.

LexisNexis, on the other hand, provides an extensive collection of UK national and regional newspapers which we could search for relevant articles. In the first instance, careful consideration was placed on the search terms and the possible impacts this would have on the returning data. We set the time parameters between 1 January 2006 and 31 March 2011. This ensured we captured the beginning of the repatriations coverage and determined a definitive end point for the research. We did not filter our keyword search with 'repatriation' so we would not lose relevant articles that did not use the term. The term 'Wootton Bassett', although encouraging a sizeable search list of 1,153 results, allowed an inclusive search of related materials. As we discovered from the research process, the name Wootton Bassett itself became a metonym for wider legitimisation and justification of British involvement in the Iraq and Afghanistan wars. Subsequently, it was important to examine each article to determine the ways Wootton Bassett was used and the context in which it was presented and referenced. In order to filter out irrelevant articles these results were checked by the authors, who noted their relevance to the project.[5] This selection process reduced the number of articles to 655.

Another consideration was the different forms of written accounts that newspapers offer, such as news reporting, editorials and supplementary magazine articles. In our project, irrelevant articles including stories unrelated to the repatriations, but also image captions and letters from readers, were removed from the database. This was because our main focus was on the views from the journalists themselves. This reduction process was guided by the research aims and objectives and allowed a more manageable dataset.

At the time of the research we were not aware of when the ongoing coverage of the repatriations began. We therefore undertook an examination of local, regional and national newspaper archives to map out the development of coverage. While LexisNexis has an extensive coverage of national broadsheet and tabloid newspapers, local and regional newspapers for the area of Wootton Bassett were restricted in terms of their availability. For instance, the *Swindon Advertiser*, one of the main regional newspapers, only went back to May 2007. Hence it was appropriate to conduct a separate search on the websites of local newspaper publishers to ensure full coverage. Local and regional papers were necessary to the project as secondary sources of data to fill gaps found in the LexisNexis archive. Thus we were able to locate the beginnings of the 'Wootton Bassett phenomenon'. Through the local newspapers, we discovered the first mention of repatriation taking place in Wootton Bassett in spring 2007, whereas coverage in national newspapers did not appear until the following year.

Data Management and Storage

Alongside the collection of print media sources, an important consideration is the management and storage of these materials. Online databases can employ restrictions on the use of materials and this must be built into any methodology. At the time of the research, LexisNexis allowed users to store only 100 articles through their online interface. Due to the amount of data, a temporary archive was created in Microsoft Excel which allowed us to store more articles and additional metadata relating to the newspaper articles we had found. Headings and categories included the title of the newspaper, article title, name of author, date of publication, number of words, and edition number. Hyperlinks to the article were included for each entry; these went directly to the newspaper article on the LexisNexis website. An additional search was performed which located the article on the corresponding online newspaper website. This gave us the ability to note any discrepancies between the versions presented by LexisNexis and the online version presented by the newspaper itself.

The database served as a chronological timeline outlining the development of coverage of the repatriations and its subsequent evolution. A number of key events emerged which were covered by the newspapers. The data was first coded into these key events. This enabled us to group articles and allowed us to analyse the different ways these events had been narrated and understood in the newspapers. The quantification of results revealed the frequency of articles in terms of date and publication. From these results the uneven nature of the media coverage was exposed, with noticeable peaks reflecting significant media attention for key events which could then be interrogated via qualitative content analysis (see Bertrand and Hughes, 2005).

We have provided an example of a specific research project which looked at the reporting of the repatriation of soldiers to the UK from the wars in Iraq and Afghanistan, as described earlier, to illustrate what such a project can look like. Such research relies on coherent methodologies and on clear methods and their application. However, we currently lack a substantial literature on the practicalities of collecting and analysing newspapers, especially in the context of newspapers currently being available in both paper and digital formats. We suggest it is necessary to be aware of the relationship between them, and how this relationship is evolving, when developing a research methodology and data collection strategy. There appears to be a continual shift in what news print media actually is. It is changing all the time; as we experienced, it can change even during the timescale of the project. Consequently the aim of this discussion is to highlight the issues that we feel are useful to bear in mind when undertaking a research project using newspapers (both paper and digital). Finally, we will look at some of the lessons learned, and discuss some of the problems we encountered.

Discussion

We suggest that when beginning a research project using newspapers, there are three main factors that will impact upon the data collection strategy. First, is the research going to look at newspapers that are being published during the research project? Second, is the research looking at previously published newspapers? Third, is it going to include both previously published and emerging newspaper stories? Each of these factors has different methodological and practical issues regarding data collection, storage and analysis. Creating your project's own archive on a daily basis – which is largely straightforward for printed media, if there are sufficient funds to buy all the newspapers and if the rationale for which they are required has been worked out in advance – is a good strategy when researching an emerging phenomenon. Retrospective analysis requiring data collection from archives is more complicated, but this is a common strategy and is necessary when researching phenomenon that have already occurred. Combining the two methods can also be problematic. This is not because there is a problem with analysing past and current phenomena at the same time per se, but because even if current newspapers are collected digitally, the formats of more recent digital archives may not correspond to those of older ones. Digital archives produce databases with varying digital coverage and availability of newspaper titles, varying content collection and format saving strategies, and varying methods of data search in the archives that impact on a systematic search strategy. One may think this can be solved via the use of a meta-archive, but these too are not without their problems.

Limitations of Digital Archives

While we found digital archives – in this case LexisNexis – useful in gaining widespread media coverage of the repatriations at Wootton Bassett, they do present certain considerations and challenges to the research process. Most notable is the fact that the articles retrieved were presented in a text-only format. Within the database, news articles are often stripped from the newspaper layout and reconfigured into a simple text format. Therefore, visuals such as maps, figures or photographs which may have accompanied the original news article are omitted. Furthermore, the placement and context of the article within the newspaper publication itself is lost. Bingham (2010: 230) raises concerns about the process of digitalisation of newspapers:

> There is a danger in this process of forgetting that newspapers were material objects that were bought, read and passed around, and that the location and presentation of individual articles is of central importance in understanding how those articles were received by readers and how much significance was ascribed to them.

With the importance of the visual communicating war, conflict and the military, it is not only the digitalisation of print newspaper archives that offers potential challenges for the study of news images (Woodward et al., 2009). The specific context and form in which articles are presented are also significant towards the overall meaning and reception of the article. What is also lost is the font size. For example, a headline emblazoned across the front page, typical of tabloids the world over – such as 'Gotcha' from the *Sun* on the Falklands War – becomes a line of 12-point font.[6] Furthermore, the photograph is lost from the story, thus what remains is an impoverished version of the original article. Judging the importance allocated to a story is lost, as is its impact, so one loses much of its meaning. Topical cartoons alongside articles are also lost. Accordingly, we need to be aware of what is lost in certain archival formats, and consider whether this restricts the types of analysis possible.

The loss of photographs from newspapers illustrates the decontextualisation problem more widely. The reason for photograph omission is not just data storage (i.e. file size); another reason is that copyright is not always held by the newspapers but by individual photographers or photo agencies. This is important since although photo-journalism may see the words accompanying photographs as 'numbingly irrelevant' (Evans, 1978: 225), the use of photographs with news articles is different, and many news stories gain 'in actuality if the words could be accompanied by a picture' (Giles and Hodgson, 1990: 75). Indeed, John Taylor (1991: 10) states: 'The objectivity of the photograph is notional: the photographs are intricately sewn into the web of rhetoric. They are never outside it, and always lend it the authority of witness.' Thus the database of newspapers held digitally is much impoverished compared to a collection of newspapers in the paper format.

Another issue discovered while using LexisNexis is the consistency and availability of the data. For example, data cleaning of the material we worked with revealed two or more instances of the same article. A majority of articles had various copies referring to different editions of the same newspaper. This led to multiple versions of one single article which had near identical content. The number of editions varied between individual articles. At times during the data cleaning process, up to five editions of the same article were discovered.[7] These included different regional but also national editions. This illustrates the regional processes of production and editing which affects the content and presentation of national newspapers in different geographical contexts. It was unclear why this duplication process occurred, but changes to content, no matter how small, alter the nature of the data and thus need to be carefully considered. Our approach, where possible, was to locate and use the first national edition of each paper, but different analytic requirements may necessitate other approaches.

Developing on this point of the consistency and availability of the LexisNexis database, Deacon (2007) identified levels of omissions of published articles that do not appear in the LexisNexis archive. Deacon (2007) also revealed inconsistencies in terms of the unitization of material in the archive. Sometimes articles with a main body and related subheading were either presented as one unified article or divided into two distinct articles; these issues were resolved in data cleaning.

Conclusion

To conclude, we need to reflect on working with newspapers as source material for researching military phenomena. We begin by looking at the relationship between the military and the media – the military-media nexus – as it illustrates how there is a background context and process to what we eventually read in the newspapers.

The Ministry of Defence takes its media representation seriously and has a large media relations department, as do the three individual services constituting the British Armed Forces: the Royal Navy, Army and Royal Air Force.[8] The same is true of virtually all modern militaries. The simple reason for this is that the modern military requires the support of its home population, or at least a significant part of it, both for the well-being of its members and for sufficient funding. This is important in wartime but arguably even more so in peacetime, when the military's utility is less obvious. Indeed it could be argued that the military itself is almost as keen to be reported on for its nonwar activities, from peacekeeping to disaster relief to royal weddings to sports and expeditions, as for its combat operations. While war reporting is often the most dramatic of the representations of the military in the media, it is certainly not the only type. Military media organisations are not merely reactive but proactive, and the news organisations are fed military stories by the military's media professionals who themselves are reacting reflexively to both events and their media coverage.

Most governments have the ability to censor the press. In the UK this is usually done via official 'D Notices', but there are various informal practices too. As can be expected, the wartime news reporter, their reports and their final publication are much more likely to be under the eyes of various censors than in peacetime.[9] 'Soft' censorship is preferred over formal censorship, and during a conflict the military tries to control access to the theatre of war, and indeed may not just deny access for some but selectively facilitate it for others (i.e. via the embedded journalist). However, due to digital technology reducing the reliance on military communication systems to send copy, even the copy of embedded journalists is hard to censor at the source. Additionally some news organisations will use reports from embedded and nonembedded journalists. It is important to bear these censorship variations, from formal to self-censorship, in mind. Indeed, even though as here our concern is with those articles that have actually managed to make it into print, they do not necessarily reflect all the reports that were written. Furthermore, news stories, while attributed to specific journalists, are produced in collaboration with their editors and other commentators. These other collaborators include organisations that feed journalists and editors stories, often almost fully written and requiring only editing into the house style of the newspaper. Such stories then do not appear first in newspapers but on the websites of organisations such as the UK's Ministry of Defence, a factor which may be worth including in developing a research methodology.

It is important to investigate the news media's reporting of military related activities, whether it is their normalised ceremonial presence in society or their actions in the prosecution of armed conflict. Wootton Bassett repatriations were a hybrid, where the fatalities of armed conflict were being reported as they returned to civil society and the formal ceremonial process then appropriated by its members. It is important to understand these phenomena as the military has the power to kill (and be killed) in the name of the state, and ultimately in the name of its citizens, in conflicts at home and abroad. How those conflicts are reported, and the reactions of civil society to those reports, are important. Through the analysis of newspaper reports we can see how events emerge and are responded to by the population in whose name they are being undertaken. Of course this is not always through the lens of the newspapers themselves, but they are the accounts that the majority of the population will have available to them. Analysing these reports gives us access, although limited, to events that are occurring and the responses to them. In the case of Wootton Bassett, the analysis of the news media allows us to see how and when the phenomenon of the repatriation ceremonies emerged locally and was picked up by the media, publicised nationally and thus fed into the further development of the events. Thus, the media cannot be extracted from the phenomenon it is reporting on. Other media also play a part, but that does not diminish the significance of the paper news media and its digital representations. Analysis of newspapers allows us to see what was reported as happening, how it was being interpreted and who was being reported as saying what about it. Such analysis is key to understanding more fully the nature of the military and its host society, both in and of itself, and in relation to other conflicts. It also makes that analysis available to other disciplinary understandings – not just those of the social sciences, but also of politics and history. A sound methodological grounding and evidence base is necessary to enable any persuasive conclusions regarding the new forms of militarism and militarisation to be drawn.

There are a lot of issues that need to be understood when using newspaper archives. We undertook our Wootton Bassett study on the cusp of the transition to the digitalisation of newspapers, and have reported here what we did, how and why. However, newspapers as a resource are not a static medium, and nor are the ways they are archived and available as researchable databases. They are constantly changing, and this is an important issue in the design of any research methodology.

Notes

1 Early newspapers did not have pictures, and when visual representations were introduced, initially it was prephotograph sketches, engravings and artistic representations from 'war artists'. Photographic representation and the ability to print them in newspapers has developed from basic daguerreotypes, to black and white, and then to colour form. While our project did not focus on photographs as such, photographs are key to understanding many written newspaper reports, whether printed or digital, and accessing these or accounting for their absence is also key to any newspaper-based research methodology.

2 Wootton Bassett was granted royal patronage in March 2011 and officially confirmed in a ceremony on 16 October 2011. Its name is now officially Royal Wootton Bassett.

3 Previously the flights arrived from RAF Brize Norton, however from April 2007 the runway at RAF Brize Norton underwent resurfacing work. Flights were diverted to RAF Lyneham, approximately thirty miles away. After the completion of resurfacing of the runway, flights continued to land at RAF Lyneham to allow the continuation of repatriations through Wootton Bassett. On 18 August 2011 the last repatriation went from RAF Lyneham through Wootton Bassett as incoming flights resumed at RAF Brize Norton.

4 Tabloid and broadsheet used to refer to the physical size of the paper and to correspond to 'entertainment' and 'serious' newspapers in terms of content, respectively. Some so-called serious newspapers now use a tabloid format, and arguably the distinction in content has also become more blurred.

5 Wootton Bassett is an otherwise quiet and unnewsworthy small English town, so it was feasible to search for all articles referencing the town and then manually filter the results. This would have not been appropriate if the repatriations had taken place at a site which generated many and varied news reports.

6 The iconic status of this particular headline is evidenced by its numerous reproductions, including on the cover of Harris's (1983) *'Gotcha!':The Media, the Government and the Falklands Crisis.*

7 To help tackle this issue, a recent addition to the LexisNexis search system is the 'duplicate' option. This groups articles with similar content, and allows the user to distinguish and remove near matches.

8   For example, the British Navy has its own newspaper: http://www.navynews.co.uk.

9   Although wartime and peacetime are simple binary concepts and usually understood as such, they should be used with caution when applied analytically.

References

ABCs (2014) ABCs: National Daily Newspaper Circulation February 2014. *Guardian Media Homepage*. Available at: http://www.theguardian.com/media/table/2014/mar/07/abcs-national-newspapers

Adams, V. (1986) *The Media and the Falklands Campaign*. London: Palgrave.

Bertrand, I. and Hughes, P. (2005) *Media Research Methods: Audiences, Institutions, Texts*. Basingstoke: Palgrave Macmillan.

Bingham, A. (2010) The Digitization of Newspaper Archives: Opportunities and Challenges for Historians. *20th Century British History* 21(2), 225–231.

Caruthers, S. L. (2011) *The Media at War* (2nd edition). London: Palgrave.

Dardis, F. E. (2006) Military Accord, Media Discord: A Cross-National Comparison of UK vs US Press Coverage of Iraq War Protest. *International Communication Gazette* 68(5–6), 409–426.

Deacon, D. (2007) Yesterday's Papers and Today's Technology: Digital Newspaper Archives and 'Push Button' Content. *Analysis European Journal of Communication* 22(5), 5–25.

Dimitrova, D. V. and Strömbäck, J. (2008) Foreign Policy and the Framing of the 2003 Iraq War in Elite Swedish and US Newspapers. Media, War & Conflict 1(2), 203–220.

Dodds, K. (1996) The 1982 Falklands War and the Critical Geopolitical Eye: Steve Bell and the If . . . Cartoons. *Political Geography* 15(6/7), 571–592.

Dodds, K. (2007) Steve Bell's Eye: Cartoons, Geopolitics and the Visualization of the 'War on Terror'. Security Dialogue 38(2), 157–177.

Evans, H. (1978) *Photo-Journalism, Graphics and Picture Editing*. Oxford: Heinemann.

Freedman, L. (2005) *The Official History of the Falklands Campaign: War and Diplomacy*, Volume 2. London: Routledge.

Giles, V. and Hodgson, F. W. (1990) *Creative Newspaper Design*. Oxford: Heinemann.

Goddard, P., Robinson, P. and Parry, K. (2008) Patriotism Meets Plurality: Reporting the 2003 Iraq War in the British Press. *Media, War & Conflict* 1(1), 9–30.

Golčevski, N., von Engelhardt, J. and Boomgaarden, H. G. (2013) Facing the Past: Media Framing of War Crimes in Post-Conflict Serbia. *Media, War & Conflict* 6(2), 117–133.

Hallin, D. C. (1992) *The Uncensored War: The Media and Vietnam*. Oakland: University of California Press.

Hammond, W. M. (1998) *Reporting Vietnam: Media and Military at War*. Lawrence: University Press of Kansas.

Harris, R. (1983) *Gotcha! The Media, the Government and the Falklands Crisis*. London: Faber and Faber.

Hjarvard, S. and Kristensen, N. N. (2014) When Media of a Small Nation Argue for War. *Media, War & Conflict* 7(1), 51–69.

Hoskins, A. (2004) *Televising War: From Vietnam to Iraq*. London: Continuum.

Huebner, A. J. (2008) *The Warrior Image*. Chapel Hill: University of North Carolina Press.

Jenkings, K. N., Megoran, N., Woodward, R. and Bos, D. (2012) Wootton Bassett and the Political Spaces of Remembrance and Mourning. Area 44(3), 356–363.

Maltby, S. (2012) *Military Media Management: Negotiating the 'Front' Line in Mediatized War*. London: Routledge.

McGarry, R. and Ferguson, N. (2012) Exploring Representations of the Soldier as Victim: From Northern Ireland to Iraq. In Gibson, S., and Mollan, S. (Eds.), *Representations of Peace and Conflict*. London: Palgrave, pp. 120–142.

Nicholson, B. (2013) The Digital Turn. *Media History* 19(1), 59–73.

Parry, K. J. (2010) Media Visualisation of Conflict: Studying News Imagery in 21st Century Wars. *Sociology Compass* 4(7), 417–429.

Robinson, P., Goddard, P., Parry, K. and Murray, C. (2010) *Pockets of Resistance: British News Media, War and Theory in the 2003 Invasion of Iraq*. Manchester: Manchester University Press.

Roy, S. C., Faulkner, G. and Finlay, S. (2007) Hard or Soft Searching? Electronic Database Versus Hand Searching in Media Research. *Forum of Qualitative Social Research* 8(3), Art. 20. Available at: http://www.qualitative-research.net/index.php/fqs/article/view/289/635 (Accessed 24 March 2015).

Taylor, I. (2014) Local Press Reporting of Opposition to the 2003 Iraq War in the UK and the Case for Reconceptualising Notions of Legitimacy and Deviance. *Journal of War & Culture Studies* 7(1), 36–53.

Taylor, J. (1991) *War Photograph: Realism in the British Press*. London: Routledge.

Thussu, D. K. and Freeman, D. (2003) *War and the Media*. London: Sage.

Weaver, D. A. and Bimber, B. (2008) Finding News Stories: A Comparison of Searches Using LexisNexis and Google News. *Journalism & Mass Communication Quarterly* 85(3), 515–530.

Woodward, R., Winter, T. and Jenkings, K. N. (2009) Heroic Anxieties: The Figure of the British Soldier in Contemporary Print Media. Journal of War and Culture Studies 2(2), 211–223.

6

THE USES OF MILITARY MEMOIRS IN MILITARY RESEARCH

Rachel Woodward and K. Neil Jenkings

In this chapter we explore the utility of the military memoir as a resource in military research. Military memoirs are the autobiographical narratives written in the first person about a life of military participation, or about a specific operation or conflict within a military career. Military memoirs are more commonly understood as sources of reading entertainment rather than a form of research data; yet, as a resource for understanding phenomena where primary research experience and data collection can be difficult to obtain, they have a specific utility that their initial appearance belies. In this chapter we explore the genre and its evolution, examine how memoirs have been used in writing about war, and look at some of the issues raised by their use as a form of research data.

The Military Memoir as a Genre

The concept of genre implies a unity of intent or shared communicative purpose. Military memoirs have a generic intent to communicate the truth of the lived experience of military participation. This defining feature sounds simple enough, and military memoirs stand alongside other forms of life-writing in claiming their authority as truthful accounts based on the specificity of an individual's experiences. However, this simple idea belies a greater complexity to the genre, and this complexity merits attention because of the ways it may shape approaches to the memoir as a research resource and subsequent analysis.

As a genre, the military memoir has a long tradition in Western models of both autobiographical writing and writing about war. The earliest accounts that can be understood as military memoirs date back to the fifteenth century, in the writings of those engaged in armed conflict and military service, committing their accounts to paper as a document of record or form of curriculum vitae (Harari, 2004). The accounts of Renaissance knights and contemporary army captains, although stylistically different, may have some features in common in terms of communicative intent. However, the feature which marks the big distinction between premodern and modern memoirs is the idea of the memoir as an account of the revelatory nature of war. Harari (2008) identifies this as emerging in the mid-nineteenth century as a result of changing cultural sensibilities about the self. The military memoir from the mid-nineteenth century onwards emerges as a form recognisable in the present as an account of the revelations war brings, about war and about the self, to an individual writing with the authority of having been a 'flesh witness' (Harari, 2008). The military memoir, then, is not only about the lived experience of war, but also about the revelations war brings and the impact of this on the individual and their relations with others. They are not just about what war is, but also about what it does.

We see the emergence of the genre in soldiers' accounts of the Napoleonic Wars (Ramsey, 2011) and developing in accounts of later nineteenth-century conflicts fought to establish a global British empire. But it is really with the end of the First World War and the attempts of its participants to come to terms with their experiences in the 1920s that we see a flourishing of published memoirs. As Fussell (2000) and others have noted, our contemporary understandings and imaginings of the First World War are shaped quite fundamentally by the autobiographical writings of those caught up in the experience. The books of Siegfried Sassoon, Edmund Blunden and Robert Graves are among the best known in the contemporary period, possibly also due to the authors being published poets, but they are also significant for the ways that their interpretations of the First World War in their respective memoirs shaped Fussell's own highly influential analysis. Notwithstanding the critiques of Fussell's analysis, in which trench warfare dominates to the exclusion of other fronts and other experiences, the point remains that what Fussell identifies is the significance of the military memoir in shaping public narratives of war (see also Roper, 2000). In the aftermath of the Second World War, books published in the later 1940s and 1950s were, similarly, highly influential in shaping postwar narratives about national defence and military victory as contingent on the bravery, daring and skill of individual combatants and not just the tactics of generals. Military memoirs, then, are significant for the way in which public narratives of war are established.

In considering the military memoir as a research data source, then, we need to be mindful not only of the text itself and its inherent communicative purpose as a revelatory account based on the authority of lived experience, but also of the wider social functions of such texts in shaping public narratives. There is a further point here, about the need also to attend to the changes in *types* of memoir which occur as distance increases from the events these books describe – compare, for example, P. R. Reid's *The Colditz Story* (first published in 1952) and Eric Lomax's *The Railway Man*, published in 1995. Both are prisoner-of-war narratives from the Second World War, but are completely contrasting in their respective emphases and authorial understandings of the public sensibilities framing the reading of their books. So, while the wars of the twentieth and twenty-first centuries have produced a significant volume of revelatory memoir accounts of the flesh-witness, which in turn are influential on public narratives, there is in fact great diversity within the genre.

A Diverse Genre

It is in this diversity of the genre of the military memoir that we can start to point to some of the issues framing the use of these books as a research data source. This diversity is evident in the range of approaches to writing about the experience of war – the categories of different types of military memoir – and reflects a broader range of approaches within forms of life-writing more generally. We can see this most literally when looking at book covers and the indications they give the potential reader about where they sit within the genre of the military memoir and within nonfiction life-writing (Woodward and Jenkings, 2012a). We can identify the action-adventure stories, written and published as thrilling, absorbing accounts of an individual or small group on a specific mission, often dealing with the complexities of the fog of war and compromise. Classic examples include *Bravo Two Zero*, Andy McNab's account of a mission in Iraq in 1991; Ed Macy's *Apache*, detailing a daring attack aviation rescue in Afghanistan in 2008; and Cameron Spence's story of a Special Forces mission in Bosnia in 1994, *All Necessary Measures*. Each is about a lived experience of war, revelatory for what that experience prompts in individuals and small groups when things go badly wrong.

We can identify trauma and survival narratives, written in the aftermath of war as a means of communicating the experience of the soldier and his (usually his) experiences and efforts to overcome their effects. Examples include Ken Lukowiak's reflections on his 1982 Falklands War experiences in *A Soldier's Song*; Kevin Ivison's traumatic experience as a bomb disposal operative in occupied Iraq in *Red One*; and *Man Down*, Mark Ormrod's story of horrific injury in Afghanistan in 2007 and subsequent rehabilitation. We can identify vindication narratives, written to explain an individual's actions in the face of criticism and censure (sometimes legal) and to make a case for exoneration. Examples include Milos Stankovic's Bosnia memoir *Trusted Mole*, Tam Henderson's Iraq memoir *Warrior* and Tony McNally's 1982 Falklands memoir *Cloudpuncher*.

The distinctions we have drawn here are not absolute, because many memoirs combine elements from the across the genre. For example, Doug Beattie's *An Ordinary Soldier* is primarily an action-driven story of a two-week siege in Garmsir in Afghanistan in 2006 but includes also exploration and explanation of the effects of this experience on him as a soldier and as a man. Both Patrick Hennessey's *The Junior Officers' Reading Club* and Patrick Bury's *Callsign Hades* focus primarily on the revelations of war to a young platoon commander deployed on a six-month tour of Afghanistan, yet are interspersed with action sequences, and a sense of war as both thrilling and dramatic, and mundane and tedious. We can also identify books sitting on the margins of the genre in terms of encounters with violent conflict. Adam Ballinger's *The Quiet Soldier* and Héloïse Goodley's *An Officer and a Gentlewoman* are training narratives. Eddy Nugent's *Picking Up Brass* and *Map of Africa* are humorous antimemoirs which poke fun at elements of the wider genre. Kleinreesink's (2014a) work in categorising the plots of Afghanistan narratives published in Dutch, German, British, US and Canadian contexts shows both the dominance of particular plot types and cross-national differences in their use. It is also notable that although book content (i.e. the text itself) will remain fairly stable across reprints and new editions, book covers vary greatly over time and across different national book markets, in response to marketing trends and new conflicts. So even in the packaging of identical books, there is diversity.

In accounting for the diversity of the genre when using military memoirs for research purposes, it is also useful to consider variations in publishers' practices around the publication of these books. Published military memoirs are, by definition, textual sources which bear the imprints of intervention by publishing houses. Indeed, one reason for studying military memoirs in published book form – rather than also including diaries, letters, field notes and personal manuscripts from archive sources, useful though they are – is to consider both the text itself and the ways in which that text as a commercial product has been shaped, which in turn has a bearing on the memoir as contributory to the public narratives of war. Despite the emergence in greater numbers of self-published books because of the greater availability of the technologies which enable this (particularly in the US; see Kleinreesink, 2014a), the vast majority of English-language military memoirs are published by commercial publishing houses. They thus appear on bookshelves as the end product of a number of interventions.

Publishing houses tend to only publish books which they think will achieve a certain volume of sales, with the volume anticipated or aspired to in turn reflecting the commercial judgements of the publisher, its list and niche and so on (Thompson, 2010). The process of turning a rough manuscript (or a set of ideas and memories) into a published book involves the interventions of commissioning editors, copy-editors, cowriters and ghostwriters, designers, and sales and marketing teams. For the larger publishing houses, investment in such interventions is done solely for the purposes of sales – often very successfully. Ed Macy's *Apache* and subsequent *Hellfire*, both Afghanistan attack helicopter narratives, bear the hallmarks of considerable publisher investment in terms of careful attention to narrative structure, flow and copy-editing, cover and illustration design, and multimedia marketing campaigns. But many memoirs are published in limited print runs by much smaller publishing houses, with far less intervention in the production of the text and reach in terms of sales. For example, Jake Scott's *Blood Clot*, an account of a Parachute Regiment patrol platoon in Afghanistan in 2006 has a fundamentally different look and feel to similar infantry experience stories from Afghanistan published around the same time but by bigger publishing houses (see Woodward and Jenkings, 2012b). This is not to say that the memoirs published by smaller publishing houses are less readable, interesting or useful as a data source. On the contrary, the role of smaller publishers is essential in making certain military stories visible: of the five military memoirs (of nearly 200) written by women of their experiences with the British Armed Forces from 1980 to the present, only one is published by a major publishing house (Sarah Ford's *One Up*, published by HarperCollins).[1] Our point here is that the interventions of the publisher are also part of the picture when considering military memoirs as a research resource.

We can also consider the platform (Thompson, 2010) and status of the author when thinking about the utility of the text. This goes beyond the often remarked variations across time in terms of which voices may or may not be heard, from those of the general and brigadier to those of the common soldier, and includes author statements about the novelty or uniqueness of their position. Both Vince Bramley's *Excursion to Hell* and Nick Vaux's *March to the South Atlantic* (both Falklands memoirs) make distinctive claims on the basis of their seniority or otherwise as soldiers, as do Sandy Woodward's *One Hundred Days* and Barrie Fieldgate's *The Captain's Steward*, respectively accounts of an admiral and the personal steward of a ship's captain.

In summary, there is diversity in the genre in terms of type of narrative, intervention of publisher, status and ambition of author and type of military participation written about. Is this still a genre? We think yes, because underpinning all of them is that they are lived-experience books, attempts by personnel to put into textual form and make available for readers something of what it was like participate in military activities.

Diverse Uses of the Genre

It follows that a diverse genre will have diverse uses: the range of ways of writing about military experience mean that the military memoir provides a great range of possibilities as a resource (see also Kleinreesink, 2014b). We have already suggested the significance of these books in contributing towards dominant public discourses and narratives of war. We must also consider how they serve different academic discourses and analyses. The military memoir as a research resource is not the preserve of any one discipline, and it is useful to consider this diverse genre not just in terms of the range of books within that genre, but also in terms of the range of academic uses of the genre by historians, political scientists, geographers, sociologists, critical international relations and geopolitics scholars, and cultural theorists.

Most obviously, given that military memoirs are accounts of experiences of military activities, usually armed conflict, they have a basic utility as sources of record for facts about such activities and conflicts. They provide the immediacy of the eyewitness account, the detail not necessarily present in the writings of journalists, the personal reflection and nuance of emotional, affective responses absent from the strictly official reports and records, and the possibility for the inclusion into the public record of issues not necessarily welcomed into the public domain by official institutional records. An example here would be Jackie George's *She Who Dared*, an account of attachment to an intelligence unit in Northern Ireland in the 1980s, published by a specialist military publisher and subsequently recalled (somewhat ineffectually) following Ministry of Defence concern about certain aspects of the author's commentary. More usually, certain books become established versions of record: see for example the utility of Sandy Woodward's *One Hundred Days* or Hugh McManners's *Falklands Commando* in providing commander-level and troop-level accounts of the Falklands War, respectively. Military memoirs have utility, then, as sources of factual record, although the ways in which they may be used as such is enormously variable. As sources of record, their utility extends beyond the provision of facts to more abstract ideas as sources of information about the wider geopolitical and strategic imaginaries around war; see for example how General Sir Michael Rose's *Fighting for Peace* or Lieutenant Colonel Bob Stewart's *Broken Lives* informed accounts of the representation of Bosnia and the Bosnian war in the British press (Robison, 2004).

Military memoirs have also proven to be an invaluable source for those seeking to explore the life-worlds of military personnel. Uses include explorations of military masculinities in the context of military identities and their embodiment (Higate, 2000, 2003), changing military missions from active combat to peacekeeping (Duncanson, 2009, 2013) and the relationships between gendered military identities and military landscapes (Woodward, 1998, 2000). They have informed discussions of military personnel's negotiations of ideas of citizenship (Woodward, 2008), of communicative action and task cohesion (King, 2006), of participation in the postwar conflicts of the waning British empire (Newsinger, 1994), the reinterpretation of past wars in the light of the present (Kieran, 2012), and of the strategic development of particular organisational transformations within armed forces (King, 2009).

In their own right, they constitute a phenomenon for investigation, so they are useful not just for what they tell us, but also for what they are. See for example how narratives of trauma and recovery written by participants in the Falklands conflict have informed ideas of trauma, recomposition and memory (Robinson, 2011). Our own work has used them extensively, to talk about ideas around war and the body (Woodward and Jenkings, 2012c), soldier mobilities (Woodward and Jenkings, 2014), participation and memory from the Northern Ireland conflict (Jenkings and Woodward, in press), blue-on-blue or friendly fire (Jenkings and Woodward, 2014b), the censorships of war (Jenkings and Woodward, 2014a), the geopolitical imaginaries of the Afghanistan war (Woodward and Jenkings, 2012b) and the role (particularly as visual material objects) in the construction of public narratives of war (Woodward and Jenkings, 2012a, and forthcoming).

There is a complexity, then, to the genre of the military memoir, and a variety evident even in this short overview in terms of their use (see also Woodward and Jenkings, forthcoming). It follows that there are issues pertaining to the use of memoirs in strictly methodological terms, and it is to this that we now turn.

Methodological Issues in the Use of Memoirs

We focus here on four issues in the use of military memoirs as a research resource. First we discuss those concerning the text itself; second, we consider the paratext; third, we discuss questions around the collaborative activities through which memoirs are produced; and finally, we consider questions around censorship.

Text

At first sight, it appears obvious that the text stands as the source of data when using military memoirs as a research resource, and furthermore that the text provides a variety of types of data, ranging from the factual (dates, locations, objectives, personnel, equipment, workplace practices and protocols) to the experiential (emotions, physical responses, records of conversations and group responses, morale and thoughts about anticipated and actual activities). Beyond this, using quite standard strategies of textual analysis enables the discourses, politics and structures of sensibility within the text to be located and identified. Useful also are the metanarratives provided by the text as a whole and the way social and political ideas are mobilised through this. There are also nuances of language and writing style at play, which may involve technical specificity and terminology, and (quite usually) the use of military-specific slang. None of this is particularly surprising, of course.

Those who use military memoirs, however, do have to be careful in their reading of text, for quite distinct methodological reasons. Military memoirs constitute a form of secondary (rather than primary) data, and as a form of textual data the analyst has to be alert to the circumstances under which this data is constructed. For the military memoir, this is first and foremost an issue about understanding a particular text within the context of the genre and the conventions of the memoir, and this includes considering factors such as the appropriate style and tone of the prose, and the origins of a book in terms of authorial motivations (e.g. catharsis, entertainment, vindication, character development). The career autobiography of a senior soldier is written with a different readership and objective in mind by its author, and published for a different market niche by a publishing house, than a worm's-eye view of life from an enlisted soldier. Compare General Sir Mike Jackson's *Soldier* and his account of leadership in Northern Ireland with A.F.N. Clarke's *Contact*, ostensibly both about the same conflict but written with very different purposes, something reflected in the structuring sensibilities of the text and its format. This is not to dismiss the utility of the military memoir, but rather to point to the need for a methodological awareness when using these texts that is alert to their nuance and specificity as military memoirs.

There is also the structure of the text to consider. This is partly a matter of publishing convention: more radical or experimental styles of life-writing do not tend to be used in military memoirs. But structure is also a matter of changes in memoir-writing conventions over time, and reflects the context and time period in which a memoir was written and not just the date of the events it recounts. The structures of published First and Second World War narratives differ from the narratives of the twenty-first-century Afghanistan war; indeed this is even the case between Falklands War (1982) memoirs written near the same time as Iraq War (2003–2007) memoirs. In addition, Falklands War memoirs written during or after the Iraq War may have increasing similarities with the latter. A current convention is to start the book by putting the reader at the centre of a particularly kinetic event: an explosion, a patrol that comes under fire, an observation during a particular mission. This structure is now almost standard: the reader is given a dramatic opening chapter to get him/her interested, and then in the second chapter the narrative takes the reader back to the beginning of the unfolding events, the soldier's enlistment, possibly a short account of a career trajectory, their position and role, and possibly an introduction to the rest of the unit or troop. Then the sequence of events that leads to the dramatic climax of the book is recounted, followed by reflection on what happened afterwards. This is deliberate: Ed Macy's *Apache* is a textbook example. The author[2] described to us the deliberate care taken by his editors to structure his story, so that the central dramatic event – the rescue of a fallen Royal Marine Commando (thought to be alive at the time of rescue) through unorthodox means using the Apache helicopter's capabilities – is understandable to the reader when it is recounted in the narrative. The reader is systematically introduced, chapter by chapter, to the weapons and remote sensing capabilities of the aircraft, the skills, aptitude, trained abilities and physical and mental requirements of the pilot, the mechanisms for communication between ground troops, operational command and close air support, and the practices of deployment and doctrine which determine exactly how pairs of Apache helicopters are flown. So in terms of the military memoir as a data source, it pays to be alert to the narrative structure of the book, as well as the text itself and its metanarrative, in order to produce a more nuanced and informed reading and richer analysis.

Paratext

The concept of paratext has been given greatest exposition by Genette (1991) to account for the range of framings (some textual, some material) which surround any given text. Genette articulates the idea of the paratext as a threshold to the text, or a zone of transaction between the author and reader, and it may include verbal[3] or other features. The key point is how the paratext frames the text and encourages a particular reading. The paratext, then, includes the textual and the visual, the verbal and nonverbal, the conceptual and the material. Analysis of military memoirs, we argue, necessarily involves a reading of the paratext as well as the text as a means of unpacking the intentions of the text as presented by the author and publisher, and reflexivity on the part of the researcher/analyst about their own engagement with the book.

Paratextual features, as defined by Genette, include anything beyond the text itself. These are multiple and various, including cover design, imagery and font; back-cover and inside-flap dust jacket blurbs and endorsements; acknowledgements and credits; the preface or fore-word, including those written by third parties; maps, photographs and other illustrations; and glossaries of technical terminologies and abbreviations. Take jacket design, including endpa-pers and how they point to the particular market niche. Pen Farthing's *One Dog at a Time*, for example, indicates its niche as a dog book for animal lovers, with its endpapers containing snapshot photographs of the dogs Farthing and his company of Royal Marines rescued during their tour of Now Zad in Helmand province, Afghanistan. Patrick Bury's *Callsign Hades*, about his experiences of platoon command as a junior officer in Afghanistan, is bracketed quite lit-erally by endpapers containing, inside the front cover, a photograph of the whole platoon in fighting gear, and inside the back cover, an image of a soldier walking away from camera, eyed nervously by two small Afghan children. The endpaper photographs neatly frame the text as a story of a young group of soldiers in an uncertain war.

The paratext is not necessarily confined to the physical book itself, especially with the pop-ularity of digital books and multimedia publishing and marketing strategies. For example, Ed Macy (author of *Apache*) set up a website for readers to engage both with himself and to access additional text and imagery. Indeed, versions of his book were produced with hyperlinks to a website for additional information to be accessed during the reading of the original text. Paratext also could include author interviews and book reviews.

Collaboration

The third methodological issue at play in the use of the military memoir as a research resource is how their status as collaborative endeavours, yet ascribed most usually to a single named (soldier) author, can be accounted for in the analysis (see Jenkings and Woodward, 2014c). With any published book (including, of course, this one), there is an element of col-laborative intervention from copy-editors, commissioning editors, manuscript reviewers and so on. What is interesting about the military memoir is the range of collaborative practices which surround their production. This is a genre sometimes mocked for being dominated by accounts put together by ghostwriters in order to serve the propaganda wing of defence ministries (an accusation sometimes made to us by our fellow academics), or dismissed as fab-rications of 'truth' constructed by lone authors to make as much money as possible, as quickly as possible (an accusation communicated to us by military personnel, presumably envious of supposed author sales revenues). What is important here about collaboration is that in order to use these books as a source of data, there has to be some understanding of how these books come to be written in the first place.

Collaboration takes many forms, in ways which may or may not be evident in the final pub-lished text. It starts with the initial impetus to write. Some memoirists start spontaneously by themselves, but many do not. Some are approached by agents. Others might have discussed the possibility with colleagues while the events they wrote about were taking place (Barrie Fieldgate and Ed Macy both did this). Some are prompted by family members with an interest in what their relative, the military operative, had actually done while with the armed forces or as a record for their children by the author. Others are prompted by family witnessing the emotional costs of an operational tour. Doug Beattie's *An Ordinary Soldier* describes the reaction of his wife Margaret to notes that he had made on his return from Afghanistan as a means of coming to terms with violent and traumatic experiences, and the encouragement she then gave to him to write and publish a fuller account.

Collaboration continues through the process of writing, researching and remembering, as the process of writing moves from being an abstract idea to a concrete action centred on a material object and objective. There may be collaboration with colleagues to check the accuracy of recollections of facts, events, conversations, operational details and strategies. This form of collaboration is very significant for authors because of its utility in establishing an idea of the veracity and accuracy of their account. Some report not only the inevitable prompts to memory which recollections with colleagues invoke, but also the additional perspectives on events which this can provide. Family, friends and current and former work colleagues may be incorporated as collaborators through their reading of drafts of text and their input in terms of ideas about content and structure, language and punctuation. With the involvement of a publisher (usually but not exclusively following the production of a draft manuscript), collaborations with professional copy-editors develop with inputs on everything from story arcs to suggestions as to the writing of dialogue. As one of our author interviewees noted, his book was 'a group production, by the end of it'.

In using military memoirs as a research resource, then, we have to be alert to the fact that these are texts produced through collaborative endeavour and move away from the idea that they constitute the output of a single person's labour. These collaborations are usually identified very clearly in the text (including coauthors explicitly named, at least in acknowledgements if not as coauthors). This is not to devalue the texts as usable sources when using memoirs; rather, it adds to an understanding of the text as the product not of a single author but rather of a varied group. At the end of the day, the published memoir is the author's own story – but much like academic writing, delivering that story to a reading public takes input from others, and the use of memoirs in research must take account of this and the role it has in the final text.

Censorship

The fourth methodological issue pertaining to military memoirs as a research resource concerns censorship. There is often a great deal of interest in memoirs as to whether they have been officially endorsed or not, and what might have been (or might not have been) omitted by order of the defence establishment (in the UK, the Ministry of Defence [MoD]). It is an obvious point that in using memoirs as a resource, there are always questions about what might have been excluded – in the context of the reader usually being unable, by definition, to establish what any exclusions might constitute. In the UK, the Defence Press and Broadcasting Advisory Committee (DPBAC; established in 1912 and chaired by a senior civil servant) presents itself as quite distant from acts of what might be termed censorship (Jenkings and Woodward, 2014a). In fact, the DPBAC is not the site of interventions, but it is the public-facing forum where such outcomes are reported. Rather, memoir texts are reviewed by single-service public relations teams with input from additional defence establishment officials with the requisite specialist expertise to judge potential sensitivities of content. There is a specified requirement that authors serving with the armed forces or MoD submit their manuscript for official approval. There have been criticisms from authors about this process, less because they are dismissive of the need for official approval (as individuals with full knowledge of the needs of operational security, authors have an understanding of this), and more for the way in which such oversight might be executed, particularly when those charged with reviewing a manuscript might be deemed by the author to be inadequately qualified for the task. The majority of still-serving authors comply, although we know of at least one case where an individual resigned from his post within the MoD rather than change the finished text of his book. Difficulties for authors can still emerge even if a book has received clearance. Vince Bramley's *Excursion to Hell* was subject to close reading a significant time after publication when events recounted in the book were deemed to constitute offences according to military regulations concerning conduct in war, and an official police inquiry was undertaken.

Our research on authorial practices in the production of military memoirs suggests, however, that the most significant form of censorship around the production of the text is the self-censorship of authors themselves. This is in part a matter of care and respect, because when fatalities have occurred, authors recognise the potential trauma to surviving family members and friends that can be caused by descriptions permanently inscribed in print. Self-censorship is also a practical issue: authors recognise that some issues simply cannot be written about for reasons of operational security or government policy, or because to make statements in print about specific events might in turn lead to censure for other colleagues, particularly those who may still be serving with the armed forces. Thus a research awareness of the censorship practices of authors and publishers, as well as censorship bodies, is key to understanding the production process, the text's final content and what may have been omitted.

Conclusions

In this chapter, we have provided an overview of some of the key issues that the study of military memoirs raises. Having considered the diversity inherent in the genre and the variety of uses to which such texts can be used, we have gone on to explore specific methodological questions raised by issues of text, paratext, collaboration and censorship. We want to conclude by making more general points about military memoirs and research in the social sciences on military forces.

The first of these concerns the possibilities and limitations of cross-national and cross-temporal comparisons between memoirs. We have chosen in this chapter to focus on the context which we know best – contemporary British military memoirs – while raising points which may pertain to other national contexts and other time periods. That said, just as it is a truism that national armed forces as state-centric expressions of power produce, articulate and feed into national narratives about armed conflict in general, specific wars and distinct discourses about military forces, so it is also the case that military memoirs speak specifically to those same national cultures. These national cultures of war are temporally and spatially specific, and we raise as a question the limits of possibilities for international comparisons between different memoir traditions in different contexts. As work by Harari (2008), Kleinreesink (2014a) and Kieran (2012) illustrates, the cultural contexts in which military memoirs are both written and published, and read, are highly specific, reflecting established national war cultures. This leads us to assert the utility and necessity of considering military memoirs as a data source by any research methodology exploring nation-state and national-cultural constructions of armed forces and military power.

Our second conclusion concerns the readership of military memoirs, something which we have not considered in our own research and have not considered here. As a research problematic, it is straightforward enough to define a range of research questions which might explore this issue. These include, for example, questions about purchasing, borrowing and reading practices; questions about textual interpretations and any patterns (or otherwise) of this among different demographic, gendered, cultural or socioeconomic groups; and questions about reader understandings of the validity and veracity of memoir accounts. However, it may be more complex than appears at first sight to actually research such questions with any degree of methodological rigour, something which in turn raises some interesting questions about the limits to research on memoirs.

Our third conclusion is about the significance of memoirs – along with a wide range of other cultural forms and texts – as a legitimate data source and object of study within military research. As should be evident from this chapter and this volume, we work within a model of military research which sees its objectives as questioning not just the constructed, contingent and contested nature of military institutions, forces, cultures and so on, but recognises also the almost limitless data sources and approaches available across the social sciences for exploring such issues. This necessitates being alert to the analytic possibilities which military memoirs facilitate for researchers and commentators. In using memoirs, however, we also need to be alert to the necessary partiality of memoirs as a resource. Military memoirs are nothing more or less than individual accounts of experiences of armed conflict and military participation, and they need to be treated accordingly.

Notes

1 The others are Jackie George's *She Who Dared* (Leo Cooper, 1999), Charlotte Madison's *Dressed to Kill* (Headline, 2010), Chantelle Taylor's *Bad Company* (DRA, 2011) and Héloïse Goodley's *An Officer and a Gentlewoman* (Constable, 2012).
2 Author interview as part of Woodward, Rachel and Jenkings, K. Neil (2009–2011) *The Social Production of the Contemporary British Military Memoir* project, ESRC reference RES-062–23–1493.
3 Genette uses the term 'verbal' in contexts where more commonly the word 'textual' would be used, to refer to the words within the text.

References

Duncanson, C. (2009) Forces for Good? Narratives of Military Masculinity in Peacekeeping Operations. *International Feminist Journal of Politics* 11(1), 63–80.

Duncanson, C. (2013) *Forces for Good? Military Masculinities and Peacebuilding in Afghanistan and Iraq.* Basingstoke: Palgrave.

Fussell, P. (2000) *The Great War and Modern Memory* (2nd edition). Oxford: Oxford University Press.

Genette, G. (1991) Introduction to the Paratext. *New Literary History*, 22, 261–272.

Harari, N.Y. (2004) *Renaissance Military Memoirs: War, History and Identity, 1450–1600*. Woodbridge, Suffolk: Boydell Press.

Harari, N.Y. (2008) *The Ultimate Experience: Battlefield Revelations and the Making of Modern War Culture, 1450–2000*. London: Palgrave Macmillan.

Higate, P. (2000) Tough Bodies and Rough Sleeping: Embodying Homelessness Amongst Ex-Servicemen. *Housing, Theory and Society* 17, 97–108.

Higate, P. (2003) 'Soft clerks' and 'hard civvies': Pluralizing Military Masculinities. In Higate, Paul (Ed.), *Military Masculinities: Identity and the State*. Westport, CT: Praeger, pp. 27–42.

Jenkings, K.N. and Woodward, R. (2014a) Communicating War Through the Contemporary British Military Memoir: The Censorships of Genre, State and Self. *Journal of War and Culture Studies* 7(1), 5–17.

Jenkings, K.N. and Woodward, R. (2014b) Blue-on-Blue in Military Memory and Memoir Accounts. CuWaDis-European Group Workshop to Advance the Study of War, Discourse and Culture. *Accounting for Combat-Related Killings*. Goethe University, Frankfurt, 21–23 July 2014.

Jenkings, K.N. and Woodward, R. (2014c) Practices of Authorial Collaboration: The Collaborative Production of the Contemporary Military Memoir. *Cultural Studies ↔ Critical Methodologies* 14(4), 1–13.

Jenkings, K.N. and Woodward, R. (in press) Serving in Troubled Times: British Military Personnel's Memories and Accounts of Service in Northern Ireland. In Dawson, G., Dover, J. and Hopkins, S. (Eds.), *The Northern Ireland Troubles in Britain: Impacts, Engagements, Legacies and Memories*. Manchester: Manchester University Press.

Kieran, D. (2012) 'It's a Different Time. It's a Different Era. It's a Different Place': The Legacy of Vietnam and Contemporary Memoirs of the Wars in Iraq and Afghanistan. *War and Society* 31, 64–83.

King, A. (2006) The Word of Command: Communication and Cohesion in the Military. *Armed Forces & Society* 32(4), 493–512.

King, A. (2009) The Special Air Service and the Concentration of Military Power. *Armed Forces & Society* 35(4), 646–666.

Kleinreesink, E. (2014a) *On Military Memoirs: Soldier-authors, Publishers, Plots and Motives*. PhD thesis, Erasmus University Rotterdam.

Kleinreesink, E. (2014b) Researching 'The Most Dangerous of All Sources': Egodocuments. In J. Soeters, J., Shields, P.M. and Rietjens, S. (Eds.), *The Routledge Handbook of Research Methods in Military Studies*. London: Routledge, pp. 153–164.

Lomax, E. (1995) *The Railway Man*. London: Jonathan Cape.

Newsinger, J. (1994) The Military Memoir in British Imperial Culture: The Case of Malaya. *Race and Class* 35(3), 47–62.

Ramsey, N. (2011) *The Military Memoir and Romantic Literary Culture, 1780–1835*. Farnham, Surrey: Ashgate.

Robison, B. (2004) Putting Bosnia in Its Place: Critical Geopolitics and the Representation of Bosnia in the British Print Media. *Geopolitics* 9, 378–401.

Robinson, L. (2011) Soldiers' Stories of the Falklands War: Recomposing Trauma in Memoir. *Contemporary British History* 25(4), 569–589.

Roper, M. (2000) Re-Remembering the Soldier Hero: The Psychic and Social Construction of Memory in Personal Narratives of the Great War. *History Workshop Journal* 50, 181–204.

Thompson, J. B. (2010) *Merchants of Culture: The Publishing Business in the Twenty-First Century*. Cambridge: Polity Press.

Woodward, R. (1998) 'It's a Man's Life!': Soldiers, Masculinity and the Countryside. *Gender, Place and Culture* 5(3), 277–300.

Woodward, R. (2000) Warrior Heroes and Little Green Men: Military Training and the Construction of Rural Masculinities. *Rural Sociology* 65(4), 640–657.

Woodward, R. (2008) 'Not for Queen and Country or any of that shit. . .': Reflections on Citizenship and Military Participation in Contemporary British Soldier Narratives. In Gilbert, E. and Cowen, D. (Eds.), *War, Citizenship, Territory*. London: Routledge, pp. 363–384.

Woodward, R. and Jenkings, K. N. (2012a) Military Memoirs, Their Covers, and the Reproduction of Public Narratives of War. *Journal of War and Culture Studies* 5(3), 349–369.

Woodward, R. and Jenkings, K. N. (2012b) 'This place isn't worth the left boot of one of our boys': Geopolitics, Militarism and Memoirs of the War in Afghanistan. *Political Geography* 31, 495–508.

Woodward, R. and Jenkings, K. N. (2012c) Soldiers' Bodies and the Contemporary British Military Memoir. In McSorley, M., Maltby, S. and Schaffer, G. (Eds.), *War and the Body*. London: Routledge, pp. 152–164.

Woodward, R. and Jenkings, K. N. (2014) Soldiers. In Adey, P., Bissell, D., Hannam, K., Merriman, P. and Sheller, M. (Eds.), *The Routledge Handbook of Mobilities*. London: Routledge, pp. 358–366.

Woodward, R. and Jenkings, K. N. (forthcoming) *Bringing War to Book: The Social Production of the Contemporary British Military Memoir*.

7

A MILITARY DEFINITION OF REALITY

Researching Literature and Militarization

John Beck

Wars, soldiers, battles: literature is full of these. Armed conflict is among the earliest and most powerful subjects in literature and continues to preoccupy writers across the spectrum of genres and forms.[1] During the twentieth century, however, the once clear divisions between war and peace, military and civilian life, have been eroded as weapons and military equipment have become embedded in national and international economies; as entire populations have become potential targets; and as weapons of mass destruction have irreversibly placed an existential burden not just upon the warrior class but upon humanity itself. Literature concerned with these matters is no longer straightforwardly a literature of war but rather a literature of militarization; further, given the collapsed distinction between military and civilians spheres, it might be more accurate to say that literature, along with everything else, has been militarized. Can we now say that there is an object of analysis called militarized literature – literature that has itself been organized for war? Such a literature would need to be distinguished from a militaristic literature, since militarization does not necessarily presuppose a celebration or endorsement of military values. How would a militarized literature be identified as such? What might be the conditions under which it is produced? Would such a literature have distinctive features or be specific to particular times or places?

These questions are important, I think, because they begin to probe into an increasingly grey area in contemporary culture produced by the folding of military into civilian affairs. The twenty-first century has seen the proliferation and normalization of many military devices, strategies, and values in civilian life, including the use of digital surveillance technologies and drones, domestic police forces using war zone equipment and tactics, and the familiarity in public discourse with, for example, what might be meant by 'enhanced interrogation techniques'.[2] These examples are highly visible and controversial, although increasingly the infrastructure of daily life, not least in terms of communications technologies, is bound up with, and available to, military and security services. There is no clear line separating the apparatus of civilian society and the military; technology, media and language are coextensive and codependent. To refer to a militarized literature, then, is not to speak of something outside and other, nor to think of militarization as something imposed or forced upon literature; instead, it is to recognize militarization as an integral part of the contemporary world and the cultures produced within it. From what position can it be possible for literature to show the militarized condition of everyday life? In other words, if everything is militarized, how can militarization be seen and represented as such?

It is this grey area that I am interested in addressing here, since it refuses the convenient compartmentalization that would bracket military issues as distinct from ordinary life and its notionally nonmilitary concerns. The notion of a militarized literature, however, does open up a range of problems for research: How might such a field of studies be located and defined when militarization itself is grounded in the elimination of boundaries? What are we looking for in so-called militarized texts? What interpretive strategies are appropriate for understanding such texts? These questions may not be easily answered, since I think it is the difficulty of defining a field of militarized literature that is itself most revealing of how a normalized militarization works, in literature and beyond.

War and the Death of Literature

For scholars working within existing disciplinary specialisms, conventional compartmentalization by genre, period, language and/or geography might suffice to frame consideration of militarization as a subcategory or thematic focus. Further, the familiar identity-based categories of class, race and gender have become a convenient, if not unproblematic, means of framing inquiry. Such practical constraints, however, may narrow the focus but get no closer to locating where militarization is and how it might operate upon and give shape to the literary text itself. Is the objective to locate the phenomena of militarization at the level of content, through an examination, for example, of British xenophobia and the production of a 'martial spirit' in British popular culture in the decades before the Great War (Eby, 1988)? Or, less straightforwardly, perhaps, is the aim to somehow locate militarization at the level of form – to understand the process of militarization as more than (or less than) a set of empirically verifiable historical actions and practices? Is militarization, at least as it might be apprehended in literature, not so much about accounts of social organization and the transformation of institutions, habits and relations, and more about the production and reception of a climate, an atmosphere – about the shapes of experience, the ways language is flooded with attitudes and values through the endlessly nuanced mutations of sentences, clauses, words and the ordering of words, and their relation to the things they describe? To apprehend the shapes and forms militarization might take in any given literary work, then, would require not only a careful understanding of the historical circumstances surrounding the production and reception of the text but also a sense of the text as somehow a manifestation of militarization, made out of the syntax and vocabulary of the militarized condition.

This, at least, I think, is the dilemma faced since the end of World War II, when the twin cataclysms of the mid-twentieth century – the Holocaust and the dropping of atomic bombs on Hiroshima and Nagasaki – each produced a profound and much cited crisis of representation. After Auschwitz, Theodor Adorno famously concluded that writing poetry is barbaric, and 'even the most extreme consciousness of doom threatens to degenerate into idle chatter' (1983: 34). For Jacques Derrida, nuclear war is all talk, since it is an event that can only be described, never experienced; idle chatter (noninstrumentalized, nontargeted discourse) is here, paradoxically, what saves us. Nuclear war is, according to Derrida, 'fabulously textual' (1984: 23) – a fable – because it has not happened; continuing to talk about nuclear war serves to indefinitely postpone the moment when talk might become action. In each case – the failure of language, the necessity of language – violence has become unspeakable, unwritable, a condition to be approached, if at all, through indirection. Adorno's and Derrida's comments have themselves become so commonplace in discussions of twentieth-century violence that the force of their assessment has also become normalized and conventional: the idle chatter that gestures towards language's failure. Since total annihilation is modern militarization's unspeakable referent, the end to which all preparedness is obliquely directed, to think about literature and militarization is to address the conditions of possibility for representation within a situation where language itself has become irredeemably degraded (barbaric) and yet utterly indispensable.

No consideration of the literary and cultural consequences of militarization in its post-Hiroshima forms can ignore the ways in which the conception of language itself is radically recalibrated by modern war and the world it has created. The language of militarization is, as a result, a destabilized language of disenchantment, of gaps, ruptures, dissemblance, insecurity, loss and incompleteness; of paranoia, anxiety and dread. It is a language drained of confidence in the power of words to represent the world or to participate in its affairs. It is a language always already contaminated, complicit with the structures of domination it endlessly reproduces, a language suspicious of itself and its loyalties.

The signifying function of language may be irreparably broken after World War II, but it is with the violence of the Great War that the dissolution of the distinction between war and peace is most decisively identified. If the insecurity of language becomes normative after World War II, the initial damage is done during the earlier conflict. Remember the old saying, advised Sigmund Freud: '*Si vis pacem, para bellum*' ('If you wish peace, prepare for war'). But then, remembering the times, he offers a paraphrase: '*Si vis vitam, para mortem*' ('If you wish life, prepare for death') (Freud, 1918: 72). The times are times of war – these are Freud's parting words in the second of a pair of essays published in 1915 under the title *Zeitgemäßes über Krieg und Tod* and in English translation in 1918 as *Reflections on War and Death*.[3] It is in literature, Freud claims, that 'compensation for the loss of life' has been traditionally sought (46); in literature the unthinkable can be rehearsed, threats approached, and identification with the dead and dying allowed, without exposure to mortal danger. 'We die,' he writes, 'in identification with a certain hero and yet we outlive him and, quite unharmed, are prepared to die again with the next hero' (46–47). The war, however, 'must brush aside' the security of this conventional response to literary suffering: 'Death is no longer to be denied; we are compelled to believe in it' (47). Here, the dilemma of literature as doubly insupportable and essential might be positioned along the same contradictory axis as Freud's peace-and-war, life-and-death formulation: if you wish for literature, prepare for death.

War Standards

Speaking in 1917, President Woodrow Wilson, who promised but failed to keep the US out of World War I, wondered what it means to go to war. It means, for Wilson, in an assessment that recalls Freud's sense of a war fought with 'no future and no peace after it' (1918: 12), an attempt 'to reconstruct a peacetime civilization with war standards, and at the end of the war there will be no bystanders with sufficient peace standards left to work with. There will be only war standards' (Wilson in Morison and Commager, 1956: 466). Wilson is astute enough to recognize that modern war mobilizes the entire civic sphere, but he also appears to believe that once mobilized, there can be no return to peace. On the threshold of entry into modern war, the US, for Wilson, is about to become a warfare society, living by war standards that, once established, cannot be revoked. In this respect, the literature of the so-called lost generation (Hemingway, Fitzgerald, Eliot et al.) produced in the aftermath of World War I can only deliver a damaged and incomplete representation of what used to be called peacetime. The famously disenchanted, broken forms of writing that dominate the literary history of the 1920s are rarely directly at war itself but about the irreversible circumstances encountered after being ripped out of history as it was previously understood: this is not really postwar literature at all, and might be more accurately be called postpeace, written as it is by the only standards left to work from: war standards.

There can be no return from the barbarism of war standards, as Adorno knew, although some continued to post warnings. For example, the sociologist C. Wright Mills, in his influential 1956 examination of the entwined political, economic and military interests running Cold War America, *The Power Elite*, had cause to cite Wilson's 1917 observation, anticipating as it does Mills's own assessment of the post–World War II order. 'American militarism', Mills elaborates, 'in fully developed form, would mean the triumph in all areas of life of the military metaphysic, and hence the subordination to it of all other ways of life' (2000: 365). Like Wilson, Mills sees that while militarization might be driven by wars real and anticipated, the most profound threat to civil society is not war itself but the far wider social, even metaphysical – indeed, transcendental – dimension of the all-encompassing warfare state. For Mills, militarization produces no less than 'a military definition of reality' (2000: 186). After World War II and the invention of nuclear weapons, what Freud and Wilson already intuited during World War I had come to pass (although Mills is careful to choose the provisional 'would' over the more fatalistic 'is'): the emergence of a new reality underwritten by the naturalized standards of war and embodied in the permanent, irreversible threat of a device capable of global destruction. It is here that militarization in its nuclear form, as William Chaloupka argues, 'organizes public life and thought so thoroughly that, in another era of political theory, we would analyse it as an ideology' (1993: 1).[4] Nuclearism here is like an ideology but more so, since no political change, however radical, can uninvent the bomb: physics has fused with politics to produce a society governed by permanent threat that is at once existential and managed day to day by a military-industrial state (or in Eisenhower's famous articulation, 'complex', a word that carries extra weight by suggesting conspiracy and psychological dysfunction as well as the tightly knitted vested interests signalled by Mills).

Militarization and Invisibility

With the collapse of the Soviet Union, the agonistic binary thinking of the Cold War dissolved even as the US military was busy rebooting organizationally and technologically via its so-called revolution in military affairs (RMA). This was first seen in action with the use of satellites delivering real-time information during the US invasion of Panama in 1989, accelerated during the Gulf War and in Kosovo, and aggressively normalized during the post-9/11 wars in Afghanistan and Iraq (Orr, 2004: 453).[5] George W. Bush's notorious announcement in October 2001 that 'every American is a soldier' might have been timely, pace Freud's *Zeitgemäßes* (literally, 'that which is in keeping with the time'), but he was really only stating baldly (and locally – he might just as easily have said 'every person') the irreversible truth of the post-Hiroshima condition.

A militarized or militarizing society, as we have seen, tends to dissolve distinctions – between civilian and soldier, between geospatial distinctions (when the planet is the target, there is no front to defend), and between all modes of social and economic activity (business, education, politics, law and so on) and military concerns. As Paul Virilio argues, 'the *battlefield*, the *domain*, even the *front line* have all been superseded. [. . .] War is everywhere, but the front is nowhere' (2005: 89; emphasis in original). If Virilio is right, then representations of war are merely the manifest content of – and a function of – a militarized reality so complete that it is for all intents and purposes invisible – which is to say that there can be no literature of militarization, only literature.

The dream of critique – that the concealed truth can be lifted into view through rigorous analysis – is itself put under question by the demolition of the distinction between war and peace, since there is no longer an outside from which to view the wreckage, no language not always already contaminated by the jargon of militarized thinking. Nonetheless, just as there is a vast literature on war, there are many distinguished novels and nonfiction narratives that deal with militarization and militarized environments. Environmental writing has become a key component, in recent years, in thinking through the challenges of accounting for the legacies of war, especially in the context of advanced, especially nuclear, weapons testing. Terry Tempest Williams (1991), Rebecca Solnit (1994) and Ellen Meloy (1999), for example, have addressed the often untraceable damage upon land and people by decades of nuclear testing in the western US and the ways in which places have been appropriated, sealed off, bombed and irradiated. Critical interrogation of this kind of internal colonization has contributed to a broader sense of militarization as an extension and continuation of longer colonial projects. Recent writing by Asian Americans and Pacific Islanders, for example, such as Chris Perez Howard's *Mariquita: A Tragedy of Guam* (1986) and Nora Okja Keller's *Fox Girl* (2002), addresses the impact of US military presence in what Cynthia Enloe describes as the 'militarized interconnectedness' of the Pacific Rim (Enloe, 2000: 85; Kim, 2014: 155).

These works remain committed to a discourse of exposure inasmuch as they continue to seek a route, however compromised, out of the totalized discourse of militarized thinking. In a different register, and with a much more formally self-reflexive awareness of the collapsed categories of the postpeace world, Thomas Pynchon's fiction, especially *The Crying of Lot 49* (1966) and *Gravity's Rainbow* (1973), directly addresses post–World War II American culture as a thoroughgoing militarized totality. Likewise, Don DeLillo's many novels, taking their cue from Pynchon, often seek to make visible the militarized underpinning of semiotic systems intended to normalize conflict, such as the deliberately unsubtle yoking of football and Cold War game theory in *End Zone* (1972) or, at much greater length, in what DeLillo calls the 'counterhistory' of the Cold War outlined in *Underworld* (1997).

The latter novel contains, for example, an extended set-piece describing an afternoon in the life of a suburban 1950s family, the Demings, whose material environment is apparently entirely constructed out of the Cold War military-industrial enterprise. The push-buttons on household gadgets are from the same companies that make guided missiles; teenage Eric, 'jerking off into a condom' (DeLillo, 1997: 514) is reminded by the prophylactic of the 'Honest John' surface-to-surface missile, and so on. DeLillo is particular heavy-handed in this section of the novel, the Demings and their world delivered as a parody of how, by the 1990s, the fifties had become as commodified and caricatured as the household items that now serve as nostalgic shorthand for a patently phony sense of the era as one of innocence about to be lost.

Molly Wallace explains, in an insightful discussion of the Demings episode, that 'if the commodity is the "chief ideological prop" of national self-definition in Cold War ideology, an ideological weapon emblematic of a capitalist economic system, the commodity is also intimately connected to the material weaponry of the Cold War' (2001: 374). Pynchon made the same point more succinctly in *Gravity's Rainbow* – the 'true war is the celebration of markets' (1973: 105) – but, to extend Wallace's point, the problem DeLillo has set himself here is not just to show how 1950s consumption is inseparable from the arms race, but to expose the commodification of history itself. By bracketing the fifties as the time without irony, DeLillo reveals the way, post–Cold War, a simplified representation of the overtly militarized recent past can serve, prophylactically, to deny the continuation of perhaps less obvious forms of naturalized militarization in the present. When war is everywhere, Virilio writes, 'we lose our direct view of things' (2005: 90). It is through indirection – the parodic exposure of the lost world of the 1950s as blindly militarized – that DeLillo shows the strategies (in this case, irony) through which the present claims to know better than to repeat the mistakes of the past. So just as a novel explicitly about conflict – a war story – might be said to confirm a clear distinction between war and peace (the desire for war to reside in the story and be contained by it, as Freud suggests), a narrative about militarization offers the position of an outside from which militarization can be located there and then (in the fifties world without irony) instead of here and now (where we know better). The satire in the Demings section of *Underworld*, then, is not directed at the 1950s at all; the target is the disavowed legacy of militarization that continues to shape the present but can only be obliquely approached.

If Freud is right, World War I marks the end of literature as a mode of redemption, a space where it possible to die and die again through an empathic identification with imaginary forms of human tribulation. DeLillo knows better than to try and reinstate literature as a medium of truth directly told, but he does indicate ways in which literature can expose peace as perhaps the foundational fiction of a militarized society. What DeLillo takes from Pynchon, and from other Cold War–era writers like Philip K. Dick, is the sense of everyday life as itself a fictional construct, made of discourse and shaped in order to dissemble. The act of reading, accordingly, moves beyond the reading of literary texts and outward to include ordinary life in all its forms: media, events, behaviour and relationships become significant as they are taken to be signs of a metalanguage that manufactures the fiction of a normal, civilian, nonmilitarized world. Life becomes a system of encrypted messages that, like literature, must be read against the grain in order to yield an understanding of its deep militarized encoding. The common identification of Pynchon, Dick and DeLillo as writers concerned with the paranoid imagination is, on these terms, inaccurate and unhelpful inasmuch as they insist upon the hiddenness of militarized society as a truth that can only be obliquely told through fiction; the pathologizing of suspicion is itself a function of militarization-as-normative.

Showing as Hiding

Nowhere is the strategy of indirection employed by DeLillo more vividly in evidence than in a curious book published in 1984 by the celebrated American nonfiction writer John McPhee, *La Place de la Concorde Suisse*. So well camouflaged by its title is the subject of this book that the British publisher, Faber, decided to rename it *The Swiss Army*, presumably in the hope that the work might gain readers by rendering visible its contents. Yet the opacity of the title is entirely to the point, since McPhee's book, as the opening line makes clear, is about radical indirection: 'The Swiss have not fought a war for nearly five hundred years, and are determined to know how so as not to' (1984: 3). *La Place de la Concorde Suisse* is McPhee's account of time spent with the Section de Renseignements, the information patrol in a unit of the Tenth Mountain Division of the Swiss Army. Famously neutral, Switzerland nonetheless maintains robust and battle-ready armed forces, drawing around 95% of its military personnel through compulsory service. It was the Swiss concept of a fighting nation – what McPhee cites at one point as the Swiss self-description as a country with 'an aptitude for war' (54) – that provided the model for the Israel Defense Forces (Greenberg, 2013).

The Swiss Army, as McPhee observes, is a 'civilian army,' and most of the time its soldiers are to be found 'walking around in street clothes or in blue from the collar down' (3). Writing for a US audience – and more specifically, for readers of the *New Yorker*, where McPhee's book began as a series of articles published in 1983 – the author explains that Switzerland is twice the size of New Jersey and much less densely populated than that state. In order to underscore precisely what this might mean socially, McPhee adds: 'If you understand the New York Yacht Club, the Cosmos Club, the Metropolitan Club, the Century Club, the Piedmont Driving Club, you would understand the Swiss Army' (4). In other words, the Swiss Army is a private social club where all eligible men under fifty are members.

The point of McPhee's comparison between elite American social networks and the Swiss Army is not just to provide a condescending sense of scale in order to be able to position the Swiss as globally anachronistic in military terms, but to signal the thoroughgoing integration of business and military affairs in Switzerland. More broadly, and in the context of the first term of Ronald Reagan's presidency (1981–1984), during which time McPhee researched and published his book, *La Place de la Concorde Suisse* serves as an oblique commentary on the military definition of reality as it was being shaped by Reagan's massive increase in military spending and revived Cold War rhetoric. McPhee does not once mention American domestic or foreign policy because he does not have to; his account of a nation that has seamlessly knitted together financial and military power – the fortress-like security of Swiss banks, McPhee notes, 'is emphasized by the Swiss Army' (56) – to the point where it shapes the entire fabric of society is at once peculiarly foreign and disquietingly familiar. 'The army', observes McPhee of the Swiss example, in addition to all its other functions, has long been considered a first-rate school of business management. . . . The special training required for a position on the general staff is looked upon as the equivalent of two years at the Harvard Business School.

(57)

Swiss banks, he concludes with one last nod to his American readers, are the closest counterpart the country has to West Point (63). The business of war and war-as-business in Switzerland cannot fail, here, to resonate with contemporary US affairs.[6]

The name 'La Place de la Concorde Suisse' does not refer to a grand public square but to what McPhee calls a 'frozen intersection' of glacier streams – 'a world too bright for unshielded eyes' – beneath a range of towering mountain peaks 'where all horizons are violent' (11). In 1857, the British Alpine mountaineer Rev. J. F. Hardy called this spot 'the Place de la Concorde of Nature; wherever you look there is a grand road and a lofty dome' (Hardy, 1860: 204). For McPhee, it is here that the 'place that will never need defending represents what the Swiss defend' (11). McPhee's title, then, designates a fusion of the natural and cultural, a site of geomorphological splendour that borrows its name from opulent Parisian urban design. It is a crossroads and a paradox, folding, as does its French namesake, intimations of terror and conciliation (built for a king in 1755, what was originally called Place Louis XV later served as the site for public executions during the Revolution before being renamed Place de la Concorde in 1795). Hardy was no doubt looking for a human analogue for divine creation when he saw Paris in the Alps; what McPhee sees (or is dazzled by) is a collaboration of nature and nation that converges along the violent horizon of peaceful (non)aggression.

At 60% alpine, Switzerland is indeed a natural fortress, although it is a nature thoroughly managed over centuries in order to appear untouched; the Swiss, writes McPhee, 'have not embarrassed their terrain' (21). Instead, they have mastered the art of camouflage. The Rhone valley, for example:

> contains two paved airstrips – but no airport, no evident hangars, no evident airplanes, no fuelling trucks, not even a wind sock. One sees such airstrips in many mountain valleys. Near the older ones are hangars that are rises in the ground. They are painted in camouflage and covered with living grass. Other strips are more enigmatic, since no apparent structures exist. If one just happens to be looking, though, one might see a mountain open – might see something like an enormous mousehole appear chimerically at the base of an alp. Out of the mountain comes a supersonic aircraft – a Tiger, a Mirage – bearing on its wings the national white cross.

(14)

Offering some of the most picturesque and most heavily photographed landscapes in the world, Switzerland, for McPhee, is nature ironized. 'Thorn and rose', he writes,

> there is scarcely a scene in Switzerland that would not sell a calendar, and – valley after valley, mountain after mountain, village after village, page after page – there is scarcely a scene in Switzerland that is not ready to erupt in fire to repel an invasive war.

(21)

Pilots sit inside mountains waiting to scramble; the airstrips do not appear on maps although they are plain to see, 'like Band-Aids all over the Alps' (21); there are forest clearings that 'make no sense' (22). The military presence is curiously obvious and denied, enigmatic, chimerical, yet at the same time banal and beyond remark. Townspeople go about their business while soldiers conduct war games in their midst. 'As any tourist can testify', notes McPhee, 'the Swiss Army is probably the most visible army in the world' (130). The military is so visible that it disappears.

This is a hollowed-out country, rustic scenery theatrically concealing networks of tunnels, weapons caches, hospitals, a year's worth of fuel for the army and regularly replenished food supplies; some sites are secret even within the army (25). It is a nation of locked doors and dead ends, of 'weapons and soldiers under barns', 'cannons under pretty houses' (23). There are firing ranges everywhere – 'Shooting rifles is a national sport. It is also compulsory' (93) – and military equipment, including guns, are kept at home, ready to use at short notice. In addition to large public bomb shelters, all new houses are required to include a fallout shelter. Not only is Switzerland calmly duplicitous, it is also ready to blow: 'Every railroad and highway tunnel has been prepared to pinch shut explosively' (23), and often, 'the civilian engineer who created the bridge will, in his capacity as a military officer, be given the task of planning its destruction' (22). Here, as everywhere in McPhee's Switzerland, creation and destruction reside together.

Conclusion

War cannot be abolished, Freud concluded in 1915, but if this is the case, 'the question then arises whether we shall be the ones to yield and adapt ourselves to it' (1918: 71). The Swiss, more than most, have adapted themselves; there is no denial in McPhee's Switzerland, and the oddly sanguine embrace of permanent preparedness there is strange only in relation to the continued denial of militarization among other countries such as the US, where the externalization of aggression fails to account for the long history of conflict that underpins that nation's global position. McPhee knows that the exception proves the rule, which is why *La Place de la Concorde Suisse* is never just about Switzerland. One means of addressing the concealments of the present is to approach them through the past, as DeLillo does in *Underworld*, yet the danger is always that the past becomes sealed off as complete unto itself. McPhee's reportorial nonfiction relies upon the rhetoric of direct experience, but if *La Place de la Concorde Suisse* is a testimony, it is to the obliqueness of language as it approaches the logic of militarization. McPhee's text appears to describe plainly just what there is to tell of Switzerland, but it is always the absent context – the global situation which Switzerland has both internalized (permanent war preparedness) and externalized (war is always outside its fortified borders) – that gives *La Place de la Concorde Suisse* its power. The novelty of Switzerland, as McPhee portrays the country, is not that it is so heavily militarized but that it is so unabashedly so; it is an uncanny site of overt concealment that reveals, by indirection, the militarized condition of everywhere else.

McPhee's Switzerland, then, is like Jean Baudrillard's famous conception of Disneyland as the 'real' place that exposes everywhere else as fake (1988: 172). The lesson here is the lesson of twentieth-century war: that there is no outside, no neutrality, no territory that is not governed by the permanent threat of armed conflict. Peace is simply preparation for war, and as such the paraphernalia of everyday life, including its language and literature, is formed somehow out of that endless preparedness. Switzerland as McPhee describes it is so thoroughly militarized that it has become unremarkable; it takes the American writer to remark upon it. Literature is not a privileged site for the investigation of militarization and may not even be the best one. But the procedures for reading literature involve attentiveness to indirection, to ambiguity and the unsaid. Without the conviction that literature never quite means what it says, literary scholarship would be no more than bookkeeping. It is here, in what Roman Jakobson once called language's 'poetic function' – the aspect of language that acknowledges its own constructedness, its own artifice, and thus its own falsity but also its particular musicality and design – that the folding of peace into war during the twentieth century might be most properly addressed. The military definition of reality is unlikely to be apprehended directly, since modern war is about nothing if not dissimulation. As such, the presence of the military in literature is most likely to be apprehended, not in depictions of battlefields and barracks, but, like the guns and tanks and bunkers in Switzerland, in the streets, fields, flowerbeds and kitchens; in the ordinary nouns and verbs of everyday language as it continues to dissemble, as it continues to persuade us that there is no military definition of reality at all.

Notes

1   On the history of war in literature, see McLoughlin, 2011; for an extensive survey of twentieth-century war writing, see Piette and Rawlinson, 2012.
2   For a disquieting assessment of contemporary militarization, see Graham, 2010.
3   The work is also published in English under the title 'Thoughts for the Times on War and Death', in Strachey, James (Ed.), *The Standard Edition of the Complete Psychological Works of Sigmund Freud.* (London: Hogarth, 1953–1974), vol. 14: 273–300.
4   On the militarization of US society during the Cold War, see Oakes, 1994; McEnaney, 2000. The best analysis of American Cold War 'containment culture' remains Nadel, 1995.
5   For consideration of US militarization in the wake of the 9/11 attacks, see Lutz, 2002; Mariscal, 2003; Giroux, 2004; Orr, 2004; Bacevich, 2005.
6   For a discussion of Reagan's conception of the United States as a 'gigantic, integrated, self-sufficient military machine', see Edwards, 1996, Ch. 9 (281).

References

Adorno, Theodor W. (1983) *Prisms*, trans. Samuel and Shierry Weber. Cambridge, MA: MIT Press.
Bacevich, A. (2005) *The New American Militarism: How Americans are Seduced By War*. New York: Oxford University Press.
Baudrillard, Jean (1988) Simulacra and Simulations. In Poster, Mark (Ed.), *Selected Writings*. Stanford: Stanford University Press, pp. 166–184.
Chaloupka, William (1993) *Knowing Nukes: The Politics and Culture of the Atom*. Minneapolis: University of Minnesota Press.
DeLillo, Don ([1972] 2011) *End Zone*. London: Picador.
DeLillo, Don (1997) *Underworld*. London: Picador.
Derrida, Jacques (1984) No Apocalypse, Not Now (Full Speed Ahead, Seven Missiles, Seven Missives). *Diacritics* 14(2), 20–31.

Eby, Cecil (1988) *The Road to Armageddon: The Martial Spirit in English Popular Literature, 1870–1914*. Durham, NC: Duke University Press.

Edwards, Paul N. (1996) *The Closed World: Computers and the Politics of Discourse in Cold War America*. Cambridge, MA: MIT Press.

Enloe, Cynthia H. (2000) *Bananas, Beaches and Bases: Making Feminist Sense of International Politics*. Berkeley: University of California Press.

Freud, Sigmund (1918) *Reflections on War and Death*, trans. A. A. Brill and Alfred B. Kuttner. New York: Moffat, Yard.

Giroux. Henry A. (2004) War on Terror: The Militarising of Public Space and Culture in the United States. *Third Text* 18(4), 211–221.

Graham, Stephen (2010) *Cities Under Siege: The New Military Urbanism*. London: Verso.

Greenberg, Yitzhak (2013) The Swiss Armed Forces as a Model for the IDF Reserve System – Indeed? *Israel Studies* 18(3), 95–111.

Hardy, J. F. (1860) Ascent of the Finsteraar Horn. In Ball, John (Ed.), *Peaks, Passes, and Glaciers, Being Excursions by Members of the Alpine Club* (5th edition). London: Longman, Green, Longman, and Roberts, pp. 198–215.

Kim, Jodi (2014) Militarization. In Lee, Rachel (Ed.), *The Routledge Companion to Asian American and Pacific Islander Literature*. London: Routledge, pp. 154–166.

Lutz, Catherine (2002) Making War at Home in the United States: Militarization and the Current Crisis. *American Anthropology* 104(3), 723–735.

McEnaney, Laura (2000) *Civil Defense Begins at Home: Militarization Meets Everyday Life in the Fifties*. Princeton, NJ: Princeton University Press.

McLoughlin, Kate (2011) *Authoring War: The Literary Representation of War from the Iliad to Iraq*. Cambridge: Cambridge University Press.

McPhee, John (1984) *La Place de la Concorde Suisse*. New York: Farrar, Straus and Giroux.

Mariscal, Jorge (2003) 'Lethal and Compassionate': The Militarization of US Culture. *CounterPunch*, 5 May. Available at: http://www.counterpunch.org/mariscal05052003.html

Meloy, Ellen (1999) *The Last Cheater's Waltz: Beauty and Violence in the Desert Southwest*. Tucson: University of Arizona Press.

Mills, C. Wright ([1956] 2000) *The Power Elite*. Oxford: Oxford University Press.

Morison, Samuel Eliot and Commager, Henry Steele (1956) *The Growth of the American Republic*, Volume 2. New York: Oxford University Press.

Nadel, Alan (1995) *Containment Culture: American Narratives, Postmodernism, and the Atomic Age*. Durham, NC: Duke University Press.

Oakes, Guy (1994) *The Imaginary War: Civil Defense and American Cold War Culture*. New York: Oxford University Press.

Orr, Jackie. 2004. The Militarization of Inner Space. *Critical Sociology* 30(2), 451–481.

Piette, Adam and Rawlinson, Mark (Eds.) (2012) *Edinburgh Companion to Twentieth-Century British and American War Literature*. Edinburgh: Edinburgh University Press.

Pynchon, Thomas ([1966] 1996) *The Crying of Lot 49*. London: Vintage.

Pynchon, Thomas (1973) *Gravity's Rainbow*. London: Picador.

Solnit, Rebecca ([1994] 1999) *Savage Dreams: A Journey into the Landscape Wars of the American West*. Berkeley: University of California Press.

Virilio, Paul ([1985] 2005) *Negative Horizon: An Essay on Dromoscopy*, trans. Michael Degener. London: Continuum.

Wallace, Molly (2001) 'Venerated emblems': DeLillo's *Underworld* and the History-Commodity. *Critique* 42(4), 367–383.

Williams, Terry Tempest (1991) *Refuge: An Unnatural History of Family and Place*. New York: Vintage.

8

ARCHAEOLOGICAL APPROACHES TO THE STUDY OF RECENT WARFARE

John Schofield and Wayne Cocroft

It is an unfortunate fact that warfare characterises the modern world. Yet archaeological evidence shows that warfare is not a new phenomenon, with suggestions that it existed at least from the Mesolithic period (e.g. Thorpe, 2003; Boehm, 2012), as hunting and gathering gave way to farming and more settled lifeways, involving larger communities and arguably more reliance upon territories and ownership. What is distinctive about warfare in the twentieth and now twenty-first centuries, however, is the scale and the intensity of conflict, the degree to which its influence reaches deep into society, and the extent of its close relationship with developments in industry and technology. But there is also common ground. Like earlier examples, these more recent conflicts leave significant trace, in the form of both *matériel* and other physical legacies such as sites and buildings spread across landscapes, as well as the more mundane objects which people associate with the personal experiences of a conflict, and which reside now in personal and museum collections. And just as with earlier remains, the study of these modern sites and objects can benefit from an archaeological approach. It is that approach that we address in this chapter, focusing not so much on the evidence and what it tells us (described in numerous other publications, such as Dobinson et al., 1997; Schofield et al., 2002; Cocroft and Thomas, 2003; Schofield, 2005, 2009; Schofield and Cocroft, 2007), but on the archaeological methods and approaches that can be used to document the sites and buildings that remain as physical traces of recent warfare, and how to interpret them (for which see also Schofield et al., 2006).

Frameworks

To provide context for this discussion, and by way of introduction to an archaeological approach that extends beyond the 'staple' of excavation, we first briefly examine some of the particular characteristics of recent warfare, and the various ways in which their archaeological signatures have been read and recorded. In particular we focus here on speed; techniques, technology and accuracy; and scale and alliances. With an interest in warfare, expressed most clearly in *Bunker Archaeology* (1994), all of these themes have been the subject of review by the theorist and philosopher Paul Virilio, whose ideas form the basis of this brief overview (and see Schofield, 2005 for an extended version of this summary).

Speed

Virilio has written much on the significance of speed in understanding modern warfare (Virilio and Lotringer, 1997; Virilio, 2002). He has noted how the speed of decision making reflects technological capability: in the nineteenth century battles unfolded over days, and decision making and response times could be measured in hours. The Napoleonic Wars of the early nineteenth century were the last major conflicts fought at the pace of man, horse and sail. A generation later, during the Crimean War (1854–56), the British field army was able to establish direct telegraph links with London. Later, in 1870, the combination of the telegraph and railway technology gave the Prussians a decisive advantage over the French. By the end of the century wireless and the internal combustion engine foretold of unimagined speed of communications and the mastery of land and air (Cocroft, 2013: 65–66). In the Second World War warning and response times came down to the few minutes between enemy aircraft appearing on early warning radar and being engaged by anti-aircraft artillery and intercept aircraft. In the Cold War the significance of the three-minute warning is well documented. Now response times are measured in nanoseconds.

The material record – artefacts and places – provides the physical manifestation of these rapid developments, while thematic studies of explosives (Cocroft, 2000) and Cold War technology (Cocroft and Thomas, 2003) include examples of the impact of speed on the material culture of warfare and combat.

Techniques, Technology and Accuracy

Mirroring the archaeological tendency to categorise and create order, Virilio identified three major 'epochs' of war: the 'tactical and prehistorical' epoch, consisting of limited violence and confrontations; the 'strategic epoch', historical and purely political; and the 'contemporary and transpolitical logistical epoch', where 'science and industry play a determining role in the destructive power of opposing forces' (Virilio, 2002: 6–7). Within this framework can be seen the development of weaponry, and its increasing significance alongside a specific 'mode of deterrence'. In the first period, weapons of obstruction predominated (ditches and ramparts; armour), linked closely to the practice of siege warfare; then came weapons of destruction (lances, bows, artillery and machine guns) which represented a war of movement; and finally real-time weapons of communication (information and transport, wireless telephone, radar and satellites).

These developments can be traced into the twentieth and twenty-first centuries, with many of the weapons and delivery systems now well known through media reports and popular culture. The materials used to wage trench warfare in the First World War are also well documented, due to recent publicity and media coverage of the 1914 anniversary (see for example various papers in Saunders, 2004), as are those of the Second World War and the Cold War. Rocket technology emerged in the Second World War through the development of

Figure 8.1 Fylingdales, North Yorkshire, the once familiar 'golf balls' of the Ballistic Missile Early
Warning System. Initially the system might have provided the United States with a 15- to
30-minute warning of an attack, while the United Kingdom and Western Europe would
receive a 2.5- to 17-minute warning.

Source: © English Heritage BB97/09913.

the V-1 and V-2 unmanned weapons, used to attack targets in Britain and liberated Europe. After the war some of the same scientists put this experience to use in developing British and American rocket technology (e.g. Cocroft, 2000: 248). Blue Streak was Britain's most ambitious Cold War missile programme, but in 1960 the programme was cancelled. And that is often the way with developing technology, and with research and development: programmes will be realigned, intensified or cancelled dependent upon their success, the promise shown in early stages of work, on developments within science and technology more generally, and the wider political agenda. Some archaeological work has been undertaken on Blue Streak (e.g. Wilson, 2007) and the archaeological approach is often the only scientific approach available, given that archives and oral historical evidence for these aspects of research and development may be lacking.

With time, weapons generally continue to become smarter, quicker and more accurate, inevitably reducing the scope for reaction time. Accuracy has great significance, as it allows an attack to be more strategic, more focused. It can also reduce the chance of civilian casualties (although errors still occur). The first and second Gulf Wars demonstrated how targets can be sought out precisely and then hit with virtually no prior warning. Improvements in technology and the accuracy of weapons systems also impacts on the sophistication of decoy and deception. The use of decoys in the Second World War, in the form of dummy targets and camouflage, is now well documented (e.g. Dobinson, 2000). But even decoy and deception have changed. Now, in the twenty-first century, with weapons technology having developed beyond first-hand observation, it is also necessary to:

> Camouflage the trajectories, to direct the enemy's attention away from the true trajectory to lure his surveillance towards false movements, towards illusory trajectories, thanks to decoys, electronic countermeasures that 'seduce' but do not 'requite' their weaponry.
>
> *(Virilio, 2002: 54–55)*

To date, experience has largely concentrated on the study of the physical traces of modern warfare prior to 1989. In the quarter century since the end of the Cold War, cyberwarfare has come to the fore with unmanned aerial vehicles (UAVs), or drones, and ground-based remote systems becoming increasingly common. Other trends include the mass monitoring of electronic communications and the ability to attack computer networks. Distinctive physical footprints are therefore being lost, and homogenised computer work stations and work rooms are the norm, providing limited insight into the activities undertaken. Although, as the investigation of the Teufelsberg signals intelligence station in Berlin has shown, analysis of spatial arrangements of such facilities can reveal something of work patterns and practices (Schofield and Cocroft, 2012). In contrast, battlefields in Iraq and Afghanistan are characterised by relatively crude field fortifications, training simulacra of which might be expected in the home countries of the participating nations. With an increased threat of terrorist attack, city centres are being made more resilient by increased surveillance, redesigned road layouts and the use of concrete barriers to protect buildings. Other more subtle design features, for example the use of armoured glass, may be less obvious.

Scale and Alliances

Recent warfare has typically extended beyond the confines of discrete battlefields, first to the entire landscape (the Western Front, for example), but extending ultimately to a global scale and incorporating sea, air and landscape, and impacting on everybody, however far from the front line they may have been. The Cold War involved the risk of mutually assured destruction, or 'massive retaliation' (Eric Groves, pers. comm.), introduced the reality of environmental pollution (e.g. Kuletz, 1998) and extended the physical limits to an infinite degree into space (with the Strategic Defense Initiative, or 'Star Wars' programme, for example) (Gorman and O'Leary, 2013). This development and the increase in geographical scale and reach was driven largely by technological capability, with the desire to win the Space Race.

Capability is one thing, but the potential impact of weapons at this scale is quite another, and the threat of global meltdown in the Cold War dominated many people's experiences of this period. The Cold War reached into everyone's homes around much of the world, as it had done within combatant countries of the world wars, represented now by the ubiquitous war memorials in virtually every community in the United Kingdom. Again Virilio's progression can be seen, from hand-to-hand combat and warfare at the scale of one's own personal space, to weapons that delivered munitions from a distance targeting troops and the places that contain or protect them, to those devices (now known to all as weapons of mass destruction) which have the potential to be remotely triggered, and could destroy entire regions, with wider global and environmental impact. The effects of such weapons are known through testing programmes, for example in the Pacific and the Nevada Test Site in the United States, and their use at Hiroshima and Nagasaki in 1945. The study of these remains has a strong international dimension, and one that relates also to the significance of military alliances. One obvious example concerns the infrastructure resulting from the United Kingdom's membership of NATO which was based on the operational requirements of the Alliance and not necessarily those of Britain. Furthermore, in Britain structures were built to NATO and not necessarily British standards and specifications, points that should be born in mind when recording and interpreting the buildings and sites that remain (Cocroft and Thomas, 2003). This is true also for the Warsaw Pact.

Archaeological Approaches

While excavation has long been the mainstay of archaeological investigations, it is by no means the only archaeological approach. Indeed for archaeological investigations of the recent past (Harrison and Schofield, 2010), excavation is rarely employed, with archaeologists instead reverting to making field observations on many sites over larger areas, or more detailed field or building surveys of particular sites, using aerial photographs or (where they are available) documentary sources to create the site distributions from which landscape-scale analyses can be undertaken. As this latter approach often provides the basis for the more detailed studies listed here, we begin at this wider scale, and with the use of primary documentation as a basis for further archaeological field investigation.

Documentation

By convention, archaeological studies of earlier periods involve and result in the production of site distribution maps, with sites often discriminated by type and/or date range, representing a visual impression of the historical process of human settlement and land use within a prescribed region. These maps are generated by the collection over time of archaeological materials, either from the discovery and excavation of sites, their visibility on aerial photographs (as a site plan distinctive to a particular type or period), or from the collection of artefacts of particular date from field surfaces where they have been exposed, for example by ploughing. From all of this accumulated evidence a distribution map can be created. As further discoveries are made, the maps are revised alongside, perhaps, their interpretation. In the 1960s and 1970s, amateur archaeologists took a similar approach to documenting anti-invasion defences of the Second World War in Britain, using field observations to generate maps, resulting in overview and interpretation (Wills, 1985).

What those amateur archaeologists did not realise, however, was that detailed documentation existed in the National Archives (formerly the Public Record Office), highlighting and explaining the extent, depth and diversity of Britain's anti-invasion defences, and the political, practical and logistical decisions and strategies behind their deployment. Archaeologists were not at that time used to working with such detailed documentation, which in this case (and for many areas of the country) highlighted precisely what was built, where, why and when. The problem with the anti-invasion defences was that documentation was so extensive as to make its assessment and analysis virtually impossible. For other categories of site, however, involving smaller numbers (such as coastal artillery sites, anti-aircraft gunsites and bombing decoys built to mimic real targets and confuse enemy bomb-aimers), the documentation was manageable. In the knowledge that this documentation existed, and the detail and accuracy of the information it contained (for a summary see Dobinson et al., 1997), two research strategies were developed. One strategy (for the more numerous anti-invasion defences) would begin in the field, using documentary sources where possible as a control on the results obtained, and for context. The other strategy would begin in the archives, generating full lists of sites, and their locations, which could then be field-checked. Historic England, created on 1 April 2015 after it separated from English Heritage, is the government's independent expert advisory service on the historic environment. It has statutory duties in England to promote the understanding and protection of England's historic environment. To justify a protection regime, a building must be deemed of 'historic interest', or a site as of 'national importance', and for this a strong case must be made. It was primarily for this reason that, in the 1990s and into the 2000s, English Heritage undertook and commissioned significant research to use fieldwork and documentation for these purposes. Alongside designation decisions, outputs included a series of books (e.g. Dobinson, 2000) for public, academic and professional audiences, and numerous well-founded recommendations for protection.

To take the example of Second World War anti-aircraft artillery sites in England, documentary sources revealed the locations of 981 sites. These were then investigated on recent aerial photographs to determine which of these originally established sites survived and how well preserved they were (Schofield, 2002: 274–276). Well-preserved sites might prove suitable for long-term protection, or for research. The results of this survey demonstrated just how many of these substantial and extensive sites had been levelled or even destroyed in the intervening fifty-five years (the survey was completed in circa 2000). Of the 981 sites originally built, only fifty-seven (5.8%) were revealed as being either complete (with a legible plan form) or near complete (with some building foundations). A further 119 (12.1%) had partial yet fragmentary remains, while in a very small number of cases the photographs proved inconclusive. In all therefore, since 1945, 81.7% of these heavy anti-aircraft gunsites have been removed, largely a result of urban development or agriculture.

This is a reflection of a wider trend, both in the United Kingdom and elsewhere. Many of these sites of recent conflict either find a new use after the conflict is over, or are removed in advance of new development or land use. The number of sites that survive by accident or by default appears very small. It is only through this close interrogation of documentary sources and field study that this trend can be recognised and quantified. But having looked at the value of documentary sources for determining total populations of sites, and creating the baseline information for management decision making, we now turn to the more detailed investigation of surviving sites, through field investigation.

Field Investigation

Survey

Field survey is regularly used to document recent military sites, often in situations where documentary sources are not available, or where archaeologists gain physical access to recently abandoned or decommissioned military sites before the supporting documentation is released. This may be up to twenty years, although freedom of information requests may now shorten this time. Furthermore, and even where documentary sources are available, they may provide an imperfect record of a site's construction history and use. Management of the site's infrastructure may have been carried out from remote locations and increasingly by private contractors, and at each change of administration documents may be mislaid; routine official weeding will destroy most working files. If a site has been sold, files and drawings may pass to the new owner or may be destroyed. It is also commonly found that files on a site's infrastructure will be held separately to that of the technical equipment, to the extent that the two will rarely appear on the same plan. If building drawings do survive, they rarely record former functions, or only at a particular date. During the investigation of an intelligence processing centre at Bletchley Park in Buckinghamshire, rooms were colour-coded to indicate their degree of originality, information that is crucial for the site's future management (Lake et al., 2006: 49–58), or for answering research questions, such as understanding social organisation and work routines.

The techniques used to survey modern military sites are no different to those used to record earlier sites. Practically, features may be surveyed using electronic theodolites and now more usually with global positioning systems, or a combination of the two. In common with all such surveys, the aim is to understand the evolution of the area being investigated. At Spadeadam in Cumbria, previously unrecorded traces of its construction phase including a navvy camp and concrete mixing area were found, as well as modifications for its later use as an electronic warfare range (Tuck and Cocroft, 2004: 29–32, 91–92). All the buildings

Figure 8.2   Orford Ness, Suffolk, global positioning satellite survey equipment being used to map the remains of the former Atomic Weapons Research Establishment.

Source: © English Heritage DP30364.

and other significant features were recorded on standard forms and cross-referenced to the plan. This created a basic record of the site in a format that could readily be used in producing protection documentation, and which will in the future provide estate managers with information to identify and assess features.

Since the National Buildings Record (now subsumed within the Historic England Archives) was founded in the midst of war in 1941 to hold images of England's buildings, high-quality photography has formed an important part of any field recording project. During the 1990s, faced with the post–Cold War drawdown of the defence estate for the majority of threatened sites, it was decided to concentrate on securing a photographic record and site descriptions based on field visits with a minimal amount of documentary research. For most sites, few if any photographs were available in the public domain, and any hesitation could mean the loss of buildings without record, or at best a record only after the building was severely damaged by vandalism. From the 1990s, photography followed contemporary historic building recording styles of three quarters views of buildings and close-ups of significant features; great care was taken to exclude people and vehicles. A decade later attitudes had changed, and when the opportunity arose to photograph RAF Coltishall, Norfolk, in its last months of operation, less emphasis was placed on the architecture and more on how buildings were used and how the spaces between them were occupied (Cocroft and Cole, 2007; Schofield et al., 2012).

A further challenge with investigating recent military sites, and especially those associated with complex technologies, is the paucity of historical studies and the relative lack of interest in the history of technology in the United Kingdom. This understanding is critical for assessing the historical significance of sites thus making the case for their protection. The former

Figure 8.3   Former RAF Coltishall, Norfolk, Propulsion and Aircraft Components Flight Adour Engine Assembly Facility. Recording prior to closure is able to document the use of space, evidence that is lost on the point of closure.

Source: © English Heritage AA054464.

Atomic Weapons Research Establishment, Orford Ness, Suffolk, was acquired by the National Trust in the mid-1990s. At that time there was little information in the public domain about Britain's nuclear weapons and the activities at Orford Ness. To further understand the site in advance of formal protection, a measured survey of the site and detailed descriptions of the buildings were prepared (Cocroft and Alexander, 2009). In the meantime, research by historians has emphasised the importance of the 1958 Mutual Defence Agreement in restoring nuclear cooperation between the United States and United Kingdom, a high-level political accord secured by the efforts of research establishments such as Orford Ness (Arnold and Smith, 2006: 13). The 1950s atomic bomb store at Barnham, Suffolk, was first documented by photography, a field visit and reference to secondary sources. Subsequently, it was given legal protection ('scheduled' as a 'monument'). By combining knowledge about the types of bombs held on this site with evidence from the buildings, it was possible to reconstruct something of the procedures involved in the handling and maintenance of these weapons (Cocroft and Thomas, 2003: 29–34). Recently, during the development and implementation of a Conservation Management Plan for the site, working with the owner, a loose consortium of independent researchers and the University of East Anglia have made contact with former personnel (Allistone, 2005: 54–61), leading to a deeper understanding of how the site operated.

The scale and complexity of many military sites can appear overwhelming. Airfields and munitions factories may include hundreds of buildings spread over a wide area. The Central Government War Headquarters at Corsham, Wiltshire, housed in the manmade caverns of a former stone quarry covered a vast area, both above and below ground (Phimester and Tait, 2014: 3–22). For such sites, characterisation is often applied. Characterisation is about more than academic understanding and was used at former RAF Upper Heyford, Oxfordshire, an archetypical late Cold War fast jet airbase, to inform how this place could change, while retaining the settings of its most significant Cold War elements (ACTA et al., 2005). Here considerable research was also undertaken to appreciate how the airfield functioned. How did aircraft move around the airfield? Which routes were used to move munitions? What were the functional relationships between buildings? How did they change through time, reflecting changing technologies and strategies? All were crucial questions that needed answers in the arguments for the retention of structures and their intelligibility to future visitors. In reality, more detailed investigation will sit alongside the broader understanding brought by characterization, and this may include in-depth studies of individual structures or recording of vulnerable features, such as wall art.

The physical investigation of military sites complementing archival sources is a technique that distinguishes an archaeological approach from a historical (entirely document-focused) approach. Field survey promotes a particularly intimate relationship with a place. The act of making record drawings is an important part of the investigation process, allowing time for deeper reflection and discussion on the phasing and functioning of buildings and sites. In contrast to a historical approach, which often involves lone working, fieldwork involves dialogue between team members leading to the shared interpretation of a site. Places investigated include active sites where access and the dissemination of reports has to be negotiated. At Spadeadam, an active RAF base, survey work of the derelict space-age relics was undertaken against a backdrop of low-level jet training and a landscape dotted with former Warsaw Pact vehicles. After closure, sites can quickly pass from high-tech, restricted places to everyday places used as scrapyards and ruined buildings used as farm animal shelters. The resulting archaeological survey records will nonetheless be very similar.

Aerial Photography

In common with field survey, the use of aerial and more recently satellite imagery owes its development to military reconnaissance techniques developed from the use of balloons in the nineteenth century, fixed wing aircraft from the First World War, and satellites from the 1960s. It was during the First World War that early practitioners, such as O.G.S. Crawford, both recognised the potential of aerial photography to reveal archaeological sites and landscapes, and learned the skills to acquire and interpret the information (Barber, 2011: 83–110). Today, one of the first steps in the investigation of any modern military site is to check online satellite images. This can rapidly confirm the location, condition and general form of a site. To go beyond this requires access to specialised collections, such as those held by the Historic England Archives. The benefits of such imagery include well-dated sequences that allow the phased development of sites to be understood, and critically acting as an independent check against cartographic and documentary sources. Specialist images also allow photographs to be viewed stereoscopically revealing far more detail than a 2D image.

Air photographs also provide an effective means of prospecting for undocumented military sites and activity and for plotting elements of military landscapes that might be unsurveyed. Sites may be discovered during general aerial survey programmes or targeted investigations in areas of interest. First World War practice trenches are a good example of sites that are rarely documented, but which are readily recognisable from the air.

Similarly, many elements of Second World War coastal defences, such as anti-tank scaffolding, barbed wire entanglements, and anti-glider poles, may have gone unsurveyed, but are clearly visible from the air (Hegarty and Newsome, 2007: 52–63). Aerial photography remains an important source for studies of Cold War–era sites. As we have seen, for many sites plans may be lost or inaccessible. During the investigation of the Atomic Weapons Research Establishment, Foulness, Essex, current Ordnance Survey mapping was used as the base plan and this was enhanced by air photographic transcription. This mapped traces of the earlier agricultural landscape and Second World War features that were either demolished or lost beneath the postwar structures. Within the research establishment, temporary features not

Figure 8.4   Gosport, Hampshire, First World War practice trenches. The trenches are closely modelled on contemporary field manuals and illustrate the care taken to prepare troops for life and fighting on the Western Front.

Source: © English Heritage RAF-540–453–5-Apr 1951.

Figure 8.5 Former RAF Coltishall, Norfolk. Local-level air photography is used to understand relationships between features and to illustrate the character of a place.

Source: © English Heritage NMR24369/037.

shown by the Ordnance Survey were mapped and the photographs used to produce a phased plan of the area (Cocroft and Newsome, 2009).

Most historic aerial photography was taken for mapping purposes and tends to be high level and black and white. To complement this, low-level oblique photography is regularly used to record large military sites. At RAF Coltishall, Norfolk, it was particularly valuable in demonstrating the imposition of the 1930s airfield on the surrounding rural landscape and the severing of ancient routes across it. Within the airfield it was effective in illustrating its strong symmetrical planning, the campus-like quality of its well-spaced buildings and tree-lined roads, and the interrelationships between buildings. To record the latter required excellent coordination between the pilot and photographer, and on the part of the photographer an understanding of the different functional areas with the airfield

Excavation

Excavation is traditionally considered a staple of archaeology, and for most periods it is one of the principal methodologies for data retrieval. But for periods of the recent past it often plays a less significant role. For sites of the modern period, the question is often asked whether excavation adds anything either to what we already know, or to information we can gather from surface inspection. But this assumes the only purpose of archaeology is data retrieval. Excavation is also increasingly a community endeavour, and one for which the process (the collective effort, and a physical engagement with the past) is as, if not more, important as the results deriving from it. A Ministry of Defence initiative, Operation Nightingale, led by professional archaeologists, has used archaeological projects to assist military personnel to recover after service in Iraq and Afghanistan. Work undertaken has included sites of clear military interest, such as the excavation of aircraft crash sites, but also themes of traditional archaeological attention, including prehistoric and Anglo-Saxon sites (Osgood, 2014).

Recent examples of excavations on twentieth-century military sites include First World War trench excavations in Belgium and France (Saunders, 2010), revealing not only the layout and construction of trench systems (which were already mapped or documented to large extent), but also details of life in the trenches, through the artefacts recovered. Excavations have also been conducted on trenches from the Spanish Civil War (Gonzalez-Ruibal, 2011), former concentration camp sites and at least one prisoner-of-war camp in Germany (Demuth, 2009; Theune, 2010). Excavations of a First World War camp in the Yorkshire Dales were undertaken to coincide with the centenary. On Cannock Chase, Staffordshire, a model of the First World War Messines Ridge model was excavated (Lee and Purvis, 2013), while excavations are often also conducted on crashed military aircraft, although rarely to professional standards (Osgood, 2014).

There is an obvious important distinction between excavations that address a specific research agenda, and those conducted (often by enthusiasts) to see what is there, and sometimes to 'stamp-collect' artefacts for personal gain. But this is true also for earlier periods. For any archaeological site, with a specific set of research questions and motivations, the methods and the techniques, the speed of the excavation and the infrastructure to support it will be determined by the social and economic conditions in which the excavation takes place. Every excavation, like every archaeological investigation, is unique.

Figure 8.6   Students of the University of York excavating at the site of a First World War camp in the Yorkshire Dales, at Breary Banks.

Source: © Jon Finch.

Conclusions

This chapter has outlined how archaeology, as a suite of distinctive methods and techniques designed to document and interpret material expressions (artefacts, places, buildings), can make unique scientific contributions to the study of recent warfare. The chapter has not sought to be definitive and to cover every method and approach, but rather to outline those commonly used and provide examples of each. As previously stated, archaeological approaches to the recent past are closely comparable to archaeological approaches to any period. One difference is that certain types of artefacts, such as organic artefacts, are better preserved for the recent past. Documents and perhaps oral testimonies may also exist, together or independently shedding further light on what those places or objects mean. But where documents and witnesses are absent, a scientific archaeological approach is often the only method available.

In terms of professional practice, surveys will continue to be made of less understood classes of sites leading to national and local protection and heritage decision making. As the significance of more recent military sites is acknowledged in local historic environment records, they will become better recognised within the planning system, ensuring mitigation against harm by protection, adaptation or, if threatened by loss, recording before demolition. Archaeologists, through scientific excavation standards, are increasingly exploring the new understandings that can be obtained through analysis of the material culture of recent conflict sites. They also enjoy the benefits of growing public interest. A large number of heritage and archaeological organisations combined to mark the centenary of the First World War in the United Kingdom. Like the Defence of Britain Project before it (funded by the national Heritage Lottery Fund, and involving the public in recording surviving sites from the Second World War), this is a democratisation of heritage and archaeological practice; a sharing of expertise. It may also prove an important stage in the evolution of archaeological practice – fieldwork not just in the hands of the 'qualified experts' who supposedly know what they are doing, and will always 'get it right'; but allowing others to have a go, knowing that the records they generate may be imperfect, or the site locations slightly inaccurate, or some small and potentially key finds may be missed. But perhaps this is a price worth paying – marshalling the enthusiasm of many at the possible cost of a few missed artefacts and the occasional 'poor' record. In this area at least, the archaeology of modern warfare is leading the way.

Acknowledgements

We are grateful to English Heritage and Jon Finch (University of York) for permission to use the illustrations in this chapter.

References

ACTA, Oxford Archaeology and the Tourism Company (2005) *Former RAF Upper Heyford Conservation Plan.* Typescript report.

Allistone, M. (2005) Nuclear Weapons and No 94 MU, RAF Barnham. *Royal Air Force Historical Society* 35, 54–61.

Arnold, L. and Smith, M. (2006) *Britain, Australia and the Bomb: The Nuclear Tests and Their Aftermath.* London: Palgrave Macmillan.

Barber, M. (2011) *A History of Aerial Photography and Archaeology: Mata Hari's Glass Eye and Other Stories.* Swindon: English Heritage.

Boehm, C. (2012) Ancestral Hierarchy and Conflict. *Science* 336, 844–847.

Cocroft, W. D. (2000) *Dangerous Energy: The Archaeology of Gunpowder and Military Explosives Manufacture.* London: English Heritage.

Cocroft, W. D. (2013) The Archaeology of Military Communications. *Industrial Archaeology Review* 35(1), 65–79.

Cocroft, W. D. and Alexander, M. (2009) *Atomic Weapons Research Establishment, Orford Ness, Suffolk, Cold War Research & Development Site.* English Heritage Research Department Report Series 10/2009.

Cocroft, W. D. and Cole, S. (2007) *RAF Coltishall, Norfolk a Photographic Characterisation.* English Heritage Research Department Report Series 68/2007.

Cocroft, W. D. and Newsome, S. (2009) *Atomic Weapons Research Establishment, Foulness, Essex*. English Heritage Research Department Report Series 13/2009.

Cocroft, W. D. and Thomas, R. J. C. (2003) *Cold War Building for Nuclear Confrontation 1946–1989*. Swindon: English Heritage.

Demuth, V. (2009) 'Those who survived the battlefields': Archaeological Investigations in a Prisoner of War Camp Near Quedlinburg (Harz/Germany) from the First World War. *Conflict Archaeology* 5(1), 163–181.

Dobinson, C. (2000) *Fields of Deception: Bombing Decoys of World War Two*. London: Methuen.

Dobinson, C., Lake, J. and Schofield, J. (1997) Monuments of War: Defining England's 20th-Century Defence Heritage. *Antiquity* 71(272), 288–299.

Gonzalez-Ruibal, A. (2011) Digging Franco's Trenches: An Archaeological Investigation of a Nationalist Position from the Spanish Civil War. *Journal of Conflict Archaeology* 6(2), 97–123.

Gorman, A. and O'Leary, B. (2013) The Archaeology of Space Exploration. In Graves-Brown, P., Harrison, R. and Piccini, A. (Eds.), *The Oxford Handbook of the Archaeology of the Contemporary World*. Oxford: Oxford University Press, pp. 409–424.

Harrison, R. and Schofield, J. (2010) *After Modernity: Archaeological Approaches to the Contemporary Past*. Oxford: Oxford University Press.

Hegarty, C. and Newsome, S. (2007) *Suffolk's Defended Shore Coastal Fortifications from the Air*. Swindon: English Heritage.

Klausmeier, A., Purbrick, L. and Schofield, J. (2006) Reflexivity and Record: Re-mapping Conflict Archaeology. In Schofield, J., Klausmeier A. and Purbrick, L. (Eds.), *Re-mapping the Field: New Approaches to Conflict Archaeology*. Berlin: Westkreuz-Verlag, pp. 5–8.

Kuletz, V. L. (1998) *The Tainted Desert: Environmental Ruin in the American West*. New York: Routledge.

Lake, J., Monckton, L. and Morrison, K. (2006) Interpreting Bletchley Park. In Schofield, J., Klausmeier, A. and Purbrick, L. (Eds.), *Re-Mapping the Field: New Approaches in Conflict Archaeology*. Berlin: Westkreuz-Verlag, pp. 49–58.

Lee, L. and Purvis, P. (2013) The Messines Model. *Britain at War*, September, 80–86.

Osgood, R. (2014) Recovering Spitfire P9503 Exercise Tally Ho! *British Archaeology*, May/June, 30–35.

Phimester, J. and Tait, J. (2014) Corsham's Hidden Landscape. *Landscapes* 15(1), 3–22.

Saunders, N. J. (2004) *Matters of Conflict: Material Culture, Memory and the First World War*. Abingdon: Routledge.

Saunders, N. J. (2010) *Killing Time: Archaeology and the First World War*. Stroud: History Press.

Schofield, J. (2002) The Role of Aerial Photographs in National Strategic Programmes: Assessing Recent Military Sites in England. In Bewley, B. and Raczkowski, W. (Eds.), *Aerial Archaeology – Developing Future Practice*. Amsterdam: NATO Science Series – Series 1. Life and Behavioural Sciences, Volume 337, pp. 269–282.

Schofield, J. (2005) *Combat Archaeology. Material Culture and Modern Conflict*. London: Duckworth.

Schofield, J. (2009) *Aftermath: Readings in the Archaeology of Recent Conflict*. New York: Springer.

Schofield, J. and Cocroft, W. D. (Eds) (2007) *A Fearsome Heritage: Diverse Legacies of the Cold War*. Walnut Creek, CA: Left Coast Press.

Schofield, J. and Cocroft, W. D. (2012) The Secret Hill: Cold War Archaeology of the Teufelsberg. *British Archaeology* 126, 38–43.

Schofield, J., Cocroft, W., Boulton, A., Dunlop, G. and Wilson, L. K. (2012) 'The aerodrome': Art, Heritage and Landscape at Former RAF Coltishall. *Journal of Social Archaeology* 12(1), 120–142.

Schofield, J., Johnson, W. G. and Beck, C. M. (Eds.) (2002) *Matériel Culture: The Archaeology of Twentieth Century Conflict*. London: Routledge (One World Archaeology).

Schofield, J., Klausmeier, A. and Purbrick, L. (Eds.) (2006) *Re-Mapping the Field: New Approaches in Conflict Archaeology*. Berlin: Westkreuz-Verlag.

Theune, C. (2010) Historical Archaeology in National Socialist Concentration Camps in Central Europe. *Historische Archäologie* 2, 1–13.

Thorpe, N. (2003) Anthropology, Archaeology and the Origins of Warfare. *World Archaeology* 35(1), 145–165.

Tuck, C. and Cocroft, W. D. (2004) *Spadeadam Rocket Establishment, Cumbria*. English Heritage Archaeological Investigation Series 20/2004.

Virilio, P. (1994) *Bunker Archaeology*. Paris: Les éditions du semi-cercle.

Virilio, P. (2002) *Desert Screen: War at the Speed of Light*, trans. from the French by Michael Degener. London: Continuum.

Virilio, P. and Lotringer, S. (1997) *Pure War* (revised edition). New York: Semiotext(e).

Wills, H. (1985) *Pillboxes: A Study of U.K. Defences 1940*. London: Leo Cooper.

Wilson, L. K. (2007) Out of the Waste: Spadeadam and the Cold War. In Schofield, J. and Cocroft, W. D. (Eds.), *A Fearsome Heritage: Diverse Legacies of the Cold War*. Walnut Creek, CA: Left Coast Press, pp. 155–180.

SECTION 2

Interactions

9

COMPARING MILITARIES

The Challenges of Datasets and Process-Tracing

Jocelyn Mawdsley

Why compare militaries? A great deal of the seminal social science literature on the military has been focused on nationally specific topics, and the nuances of national strategic cultures are generally held to be important in understanding why states and their militaries vary on questions of defence. However, it is undeniable that a reliance on what Williams (2007: 100) calls 'insular case studies' may mean that wider trends are missed. For some, the appeal of comparative research lies in the sense that militaries are changing rapidly: post–Cold War changes in the nature of warfare, the globalisation of security challenges, the increasing multi-national nature of missions and a growing homogeneity among advanced militaries mean that comparative work is needed to understand the changes and the implications they might have. Cross-national comparisons can help us to better understand the processes and mechanisms underlying change at the national level and thus identify what is important and what less so. Caforio (2007), for instance, claims that the extent of the recent changes to the functions and roles of the military, particularly in highly developed states, makes cross-national military research increasingly vital. For others, the desire to generate generalisable theories makes hypothesis testing on multiple cases necessary.

The type of research questions that inform comparative social science research on the military can vary considerably. Researchers might wish to see how similar pressures for change have impacted on militaries of a similar type. Forster (2005), for example examined the extent of change in civil-military relations in European states following the end of the Cold War. Researchers might want to discover the best way to demobilise armed forces in a post-conflict society, so a comparative study of states that have undergone such a process might be instructive; as in the previous case, this might be best achieved by a series of detailed case studies. Or, researchers might want to discover how particular variables affect the way in which militaries are managed in comparable states. Born et al. (2003), for example carried out a comparative study of democratic control of the armed forces, stressing the need for variable-driven comparative research to understand the issue properly rather than relying on highly nationally specific case studies.

In short, military researchers, like other comparative researchers, may find that their research questions drive the choice between what Della Porta (2008) calls case-oriented and variable-oriented research. Other factors that may influence this choice are researchers' epistemological preferences, their methodological skillsets (e.g. statistical or linguistic abilities) and practical considerations such as data availability or the feasibility of fieldwork. While mixed methods studies are growing in popularity, many military researchers wanting to do comparative work will find themselves choosing between large-N and small-N comparative research designs. Both pose particular challenges to the military researcher, beyond the more general difficulties of comparative research. This chapter aims to give an overview of some of the military-specific issues that large-N and small-N comparative research pose.

Large-N Comparisons

Quantitative methods have long played an important role in military research. The influence of the RAND Corporation's quantitative research on US defence policy during the Cold War has had a substantial impact on how social scientists engage with military research (Barnes, 2008). While the RAND style of statistical modelling has been criticised for divorcing military analysis from its historical and social context and thus lacking substance (Gray, 2002), it is nevertheless quantitative analysis that often appears to give the clear answers to research questions that the military establishment is seeking. Statistical analysis is therefore a popular form of research methodology for those carrying out research for the military or hoping to have impact on military thinking. This institutional preference, which as Müller-Wille (2014) points out is often based on only a rudimentary methodological understanding, can lead to the military commissioning and using poorly designed studies, which have involved methodological compromises in, for example, survey sampling techniques that render the findings unreliable, particularly in conflict environments, where reliable baseline data are absent. The clear answers that quantitative research appears to offer may not in fact be the case.

As Barnes (2008) argues, the bias towards quantitative work in military-commissioned research has been matched by a general trend towards the quantification of the social sciences, particularly in the United States. Indeed, some fields of military research such as civil war research have become predominantly quantitative in orientation (Florea, 2012). For comparative researchers, carrying out large-N cross-national studies on military topics entails an initial choice as to whether the researcher will collect data themselves or draw on existing datasets. The feasibility of data collection is obviously driven by the research question(s) being investigated. It is much easier for example to devise a survey instrument, and to expect reasonable response rates, if like Born et al. (2003) you are studying military democratic accountability issues in advanced democracies, than if you are studying causes of internal conflicts in failed states. Data collection may also involve the gatekeeper and access problems common to all military research, with the additional challenge for comparativists that unless the study is officially sanctioned/commissioned, it requires multiple national gatekeepers to grant the same access to the requisite number of people, and the infrastructure being in place to support effective data collection. This is a particularly challenging hurdle in underdeveloped or conflict/postconflict regions (Müller-Wille, 2014). Given this, it is perhaps unsurprising that many studies draw on existing datasets. Alongside official data compiled by international and national authorities, there are long-standing and freely available datasets compiled by researchers such as the Uppsala Conflict Data Program, which records instances of armed conflict. Other popular sources include the Correlates of War project, now based at Penn State University, which compiles datasets on variables that may influence the outbreak of war, or the data on military expenditure and arms exports at the Stockholm International Peace Research Institute (SIPRI).

Using existing datasets for comparative research is not, however, entirely straightforward. Schrodt (2014: 291), for example questions the value of continual 'reanalysis of a small number of canonical datasets, even when those have well-known problems'. In particular, Schrodt (2014) questions the value of multiple analyses of a dataset when the researcher only makes minor specification, operationalisation or methodological variations on the original. Moreover, reusing existing datasets means that you are tacitly accepting the data collection techniques, the coding decisions and the accuracy of the data. In a similar vein to Schrodt, Florea (2012) suggests that uncritical use of datasets can lead researchers into a conceptual morass. He gives the example of civil war datasets, which pinpoint the start and endpoint of civil wars by a casualty-threshold metric, which he argues leads to arbitrary and problematic coding and the conflation of civil war with violence. For the rest of this section, the chapter will consider the military expenditure (MILEX) data to show the difficulties facing a comparative researcher who wishes to use this data in a large-N comparative study. MILEX data matters as it is not only used directly, but also as an indicator in other key military-related datasets such as the Correlates of War dataset.

Military Expenditure (MILEX) Data

There are some basic questions about data that any comparative researcher needs to ask. How reliable are the data? Can the states in question be reasonably compared using this measure, or has it been subject to what Sartori (1970) described as 'conceptual stretching'? Do the data actually answer my research question? These are all issues that arise with MILEX data but which are complicated further by its military nature.

First, let us turn to data reliability. MILEX data are compiled by a number of national, international and independent bodies. The International Monetary Fund (IMF) and the UN Office for Disarmament Affairs collect this data. The US Bureau of Arms Control, Verification and Compliance issues annual 'World Military Expenditures and Arms Transfers' reports. For those states that are members, the Organisation for Economic Co-operation and Development (OECD), the North Atlantic Treaty Organization (NATO) and the European Defence Agency (EDA) also issue annual reports. Finally, independent research institutes such as the International Institute for Strategic Studies (IISS) and SIPRI compile *The Military Balance* and the *SIPRI Yearbook* respectively on an annual basis. For the researcher, who may be at first excited by this plethora of publically available data, it is confusing to note that the figures vary from body to body.

Brzoska (1981) argued that this variance is due to two separate problems: data origin and data preparation. First, definitions of military expenditure vary, and all sources are reliant to some degree on governments reporting in good faith their expenditure. There are multiple reasons why a government may choose not to report or to exaggerate or underreport spending. A government may want to heighten or alleviate regional geopolitical tensions; it might be responding to domestic concerns about levels of defence spending; or it might just consider such information best kept secret. Second, if these data are to be presented in a comparative fashion, there will need to be adjustments to account for different budgetary cycles, inflation and the conversion into a single currency (usually the US dollar). Particularly when overall military expenditure is broken down into subcategories such as defence procurement or military research spending, differences in the way governments report accumulate. Lebovic (1999) additionally points out that MILEX reports are heavily reliant on estimates, where governments have failed to or only partially reported (or are assumed to be unreliable), which may be revised subsequently as new data become available, and that this practice leads to substantial divergence between the various sources. Lebovic (1999) argues that the level of error in the data may be responsible for the degree of disagreement between researchers on a variety of hypotheses linked to MILEX data, such as arms race models or the hypothesis that military spending inhibits economic growth.

Let us examine one case as an example. As has been argued already, governments may deliberately choose to underreport or overreport military expenditure for political reasons, which can also skew analysis, where identification of deviant cases or outliers is the researcher's aim in a comparative study. Here, perhaps, the example of West Germany during the 1980s is instructive. In general, West German defence data from this period were thought to be of low quality (Cowen, 1986). Brzoska (1981) claimed that West Germany produced different figures on defence spending for domestic and international (NATO) consumption to suit national distaste for any hint of militarism and the foreign policy need to be seen by its allies as a reliable NATO member. But the issue was particularly problematic when it came to analysis of the role that expenditure on defence research played in creating economic growth. West Germany, like Japan, had often been used by economists as an example to support the 'crowding out' hypothesis (i.e. that government investment in defence research and development [R&D] crowds out civilian R&D, which would have resulted in higher rates of growth), given that until the early 1990s, despite a much lower level of defence R&D spending than comparable states, it enjoyed considerably higher rates of growth (Kaldor et al., 1986). This challenged the widespread view that spending on defence research correlated positively with economic growth.

West German peace activists like Rilling (1988) pointed out that this picture of West German research suited the actors involved and was therefore rarely challenged, even if it seemed unlikely – the research community prefer their civilian image, while it had suited successive West German governments to claim that defence production is in private hands and has little to do with them. Rilling also claimed that by 1988 'the military research budget in the FRG requires a far larger percentage of the national science resources than is officially declared' (Rilling, 1988: 317). His argument was that the visibility of such research was minimised by a combination of statistical secrecy, a large percentage of such work being carried out in the private sector rather than in national research facilities, and a conscious effort to marginalise the work by subsuming it under nonmilitary spending categories. Rilling was making a political point. However, parliamentary questions from the politician Edelgard Bulmahn led to an admission by the West German government that the figures on military research supplied to the OECD were too low; for 1990, the year in question, revised figures were estimated to be almost double the amount that stood in the statistics (Liebert, 1998). Had the economists been aware of the inaccuracy of the reported data, would West Germany have been such a deviant case? As it happens, subsequent meta-analyses of multiple studies on military research expenditure and economic growth suggest that the positive correlation was much weaker than was thought at the time, and that the crowding out hypothesis may well be correct (Dunne and Braddon, 2008). Nevertheless, the misleading West German data may have contributed to what Lebovic (1999) sees as the inconsistent results in these studies. While all sources relying on government reporting are subject to similar problems, the particular sensitivity of military data makes it likely that these problems occur more frequently.

Our next problem is whether states can be reasonably compared by such a measure or whether it is subject to conceptual stretching (Sartori, 1970). It should be noted that SIPRI, at least, is of the opinion that despite both governments and researchers using MILEX data to closely compare states, they are in fact more appropriately used for 'comparisons over time and as an approximate measure of the economic resources devoted to military activities' (Omitoogun and Sköns, 2006: 270–271). They point to one well-known case during the Cold War as an example of how this can be misleading. There was a lack of credible official statistics on the Soviet Union's military spending and so estimates were generated by the 'building-brick' method. As Omitoogun and Sköns (2006) point out, this was critiqued at the time as being methodologically unsound, as it used US costs and relative prices to estimate costs in the Soviet Union. Despite the fact that the data were known to be problematic, this approach led politicians and media in the West to uncritically accept that the Soviet Union had higher military expenditure than the United States, and used this to argue for defence spending to increase in the West. A similar worry might apply to the way in which Chinese MILEX data are being used politically at present, given they too are largely based on estimates in the absence of reliable official data.

The final consideration is whether the data actually answer the research question. Here MILEX data offer a good example of what Florea (2012) sees as a current problem with quantitative civil war research, namely that data continue to be collected, under classifications developed during the Cold War. But are these classifications still relevant to the way we conceptually now understand civil wars? Omitoogun and Sköns (2006) identify two potential problems with the use of MILEX data for contemporary security research: the impact of the war on terrorism and the broadening of the concept of security. The blurring of the dividing line between internal and external security means that increasingly governments are including items in MILEX reporting that they would not have done previously. Unless MILEX reporting is sufficiently disaggregated, this may mean a confusing picture for those seeking to use MILEX data as an indicator of increased hostility in a region. Moreover, MILEX data do not give a picture of the impact of nonmilitary security concerns on government spending (e.g. rises in climate change protection spending). As military force may not always be the response to newer security concerns, its use as an indicator of the likelihood of conflict may diminish. In short, while large-N comparative studies offer the tempting possibility of testing and yielding generalisable hypotheses, and for military researchers there are multiple existing datasets, care needs to be taken not to misuse the data.

Small-N Comparisons

Small-N comparative research generally relies on the development of small number of detailed case studies. Vennesson (2008) argues that there are three types: the descriptive case study, the interpretive case study and the hypothesis generating or refining case study. Case study research on military matters has often been of a historical nature, but has been used to go beyond description and to test theories on subjects like military doctrine. This type of research became more prevalent in security studies from the 1980s onwards, when more archival material was made available. This led to various studies questioning established inter-pretations of security events and the theories that had been largely unquestioned during the Cold War, in particular challenging deterrence theory (Walt, 1991).

Vennesson (2008) argues that researchers working with case studies can take a positivist or an interpretivist perspective. For a positivist the main purpose of process-tracing in a case study is to establish or evaluate whether the causal process of the proposed theoretical framework can be observed. The case study allows the investigation of links between different factors and the evaluation of the relative importance of potential causal factors. For an inter-pretivist, Vennesson (2008) argues that the focus is not just on what happened but on how it happened. In other words, 'interpretive approaches to political science focus on the meanings that shape actions and institutions, and the ways in which they do so' (Bevir and Rhodes, 2003: 17). Here the importance of the case study is to investigate the relationship between actors' beliefs and their behaviour. This type of approach for example might be valuable for a researcher interested in change in the armed forces. These different perspectives will influence how the researcher designs and carries out case study research. Military researchers come from both perspectives.

Case study research generally employs a process-tracing methodology. Usually this would entail an appropriate combination of open-ended interviewing, participant observation and document analysis (Vennesson, 2008). As most military researchers are unlikely to be able to engage in participant observation, particularly in a comparative context, this section will concentrate on the particular issues for comparative researchers in documentary analysis and interviewing.

Process-Tracing: Documentary Analysis

For any military researcher working on recent or contemporary topics, access to relevant government documents is likely to be heavily restricted. Military documents are more likely to be classified and for longer than other government documentation. The problem is mag-nified for a researcher carrying out a cross-national study. Not only does the researcher have to learn to navigate different national archives (each with their own peculiarities in terms of procedures, cataloguing and regulations), but the researcher also has to accept that decisions to classify/declassify documents are unlikely to be the same in each state. If one of the states being researched has strict rules about what may be classified in the first place as well as a robust freedom of information system, a detailed case study can be developed, only to find that none of the equivalent documentation is available in the other cases. Perhaps even more dispiritingly, but not unexpectedly, Deschaux-Beaume (2012) reports lengthy freedom of information procedures with eventual rejections from both her case study states (France and Germany).

This problem is also prevalent in the archives of international military organisations, which might be expected to be a helpful source for the comparative researcher as their documents are applicable to more than one state. Mastny (2002), for example bemoans the lack of a comprehensive history of NATO due to the declassification problem. Although in theory NATO documents are released after thirty years, objection by any member state can block their release. Deschaux-Beaume (2012) also complains about the lack of availability of European Union documents due to its Common Security and Defence Policy (CSDP). Political events of course may mean that archival material becomes available unexpectedly in one state that reveals information on its allies. For example, Heuser (1993) was able to draw on material from the opened East German military archives to trace the evolution of Soviet military doctrine through Warsaw Pact training documents.

The comparative military researcher needs to be resourceful and flexible in seeking documentation to help build up their case studies. There are various sources that can be helpful. Military researchers have the good fortune that military matters are the subject of multiple well-informed specialist magazines both in paper format and online. While the Jane's Information Group, with titles such as *Jane's Defence Weekly* and *Jane's Intelligence Review*, is perhaps the best-known military publisher (and has been publishing in the field since the nineteenth century), other useful publications include the online *Defense News* and *Flight International*. Most states also have journals that cut across the worlds of academia, practitioners and think tanks, such as the British *Royal United Services Institute (RUSI) Journal* or the French *Revue Défense Nationale*. Military training establishments may also have their own publication series which can offer insider insight. Depending on the states studied, if there is an active and institutionalised peace movement, their publications and collections of documents can also be helpful. Finally, of course, there are the defence-related think tanks and research institutes, which whether independent or closely associated to a national ministry of defence offer both publications but also libraries, with often well-catalogued and extensive grey literature collections including official documentation, which can be easier to navigate than official archives. The most helpful sources are likely to vary between the case studies, so it is worth investing time in conversations with more experienced researchers in each state to get a sense of where the most useful material is likely to be found, so that fieldwork can be as efficient as possible. Clearly as well, linguistic proficiency or otherwise will limit how much access a researcher can get, and varying language skills may lead to imbalance between case studies. In sum, documentary analysis on its own is unlikely to produce a full picture of the topic investigated, and so many comparative military researchers make interviewing a key part of their research design.

Process-Tracing: Interviewing

Interviewing can be a highly valuable information gathering resource. As Deschaux-Beaume (2012) argues, it is perhaps of particular value for researchers without existing connections to the military, as it brings them into contact with a highly specific and distinct social field. The degree of difficulty in setting up a series of interviews is highly dependent on who you wish to interview, the sensitivity of the topic and the states that are the focus of the case studies. While there are always access issues, the gatekeeper problem is reduced if, for instance, you seek to interview people working within ministries (even if they are serving officers) rather than multiple junior members of a regiment on their base. A basic problem for comparative military researchers is equality of access between states. A topic might be more sensitive in one state than another, or the time between fieldwork trips might mean that a topic was not sensitive in the first country, but a subsequent security incident has made it sensitive before the fieldwork is carried out in the other states. The gatekeeper in one system might be more supportive of researchers than in another. This might involve methodological compromises, whereby fewer interviews are carried out in one country than another. Again, while organigrammes of defence ministries vary substantially, if you are interviewing in ministries asking an official in state A who their opposite number in state B is, enables a form of snowball sampling to take place (i.e. the gathering of new interview partners from the suggestions of current ones; for example, who in practice does interviewer in State A work with closely in state B even if formal job titles are different?). Deschaux-Beaume (2012) also makes the point that personal relationships between officers from different militaries can be a significant enabler for comparative research. Finally, some states are more accustomed to researchers carrying out military research than others; this may be beneficial or disadvantageous to the researcher, depending on whether they are viewed as a novelty or a threat.

Interviewing soldiers or veterans in a conflict or postconflict setting clearly increases access and sensitivity problems. This may mean that some methodological flexibility is required. For example, Eriksson Baaz and Stern (2009) report that in their research on sexual violence carried out by soldiers in the Democratic Republic of the Congo, semis-tructured group interviews were more fruitful than individual interviews – which were their initial intent – as the former were less intimidating for the interviewees. In conflict zones, the principle of informed consent from interview partners is particularly relevant, and the researcher needs to be highly sensitive to cross-cultural differences and power dynamics.

Carrying out interviews across several nation-states also requires cross-cultural sensi-tivity in research design, concerning the conditions under which serving military person-nel may agree to be interviewed. In particular, when interviewing military and defence ministry personnel, their willingness to be interviewed on the record may vary between countries. It is important to recognise that the patterns emerging from fieldwork in one country may not be duplicated in another. Deschaux-Beaume (2012) offers the example of the different legal protection for freedom of speech for French and German soldiers, and the problems this caused for her own comparative study of the French and German mili-taries' experiences of the European Union's security institutions and networks. The French military statutes state that they may only give their own opinions when off duty and then only with the reserve expected of military personnel. In contrast, the German armed forces are officially 'citizens in uniform' and therefore protected by the right to freedom of speech contained in the German Basic Law if they criticise German defence policy on the record in an interview. For Deschaux-Beaume's (2012) research, as a practical matter this meant that some of her interviewees would not allow interviews to be recorded or to have their names or positions identified. Such factors need consideration in the initial research design, for as Deschaux-Beaume (2012: 110) points out, there will need to be trade-offs between 'research deontology and methodological rigour', and an inability to record a substantial number of interviews obviously means that some forms of analysis of interview data are consequently ruled out.

The question of the need for reflexivity on the part of the military researcher is also more complex when the researcher is carrying out comparative work. Higate and Cameron (2006) rightly point out that a reflexive military researcher needs to be aware of the potential of co-option, or being militarised, of the issues raised by interviewing in highly masculine environments, of the power relations between the researcher and the interviewee, and the question of whether or not the researcher is viewed as an insider. In addition to this already daunting list, the comparative military researcher needs to be aware of their own relations to each military organisation. For example, does shared citizenship, language or experience lead to unconscious bias in favour of one set of interviewees over another without that sense of familiarity? How much have the researcher's or interviewee's linguistic abilities helped or hindered effective communication?

Carrying out case study–based comparative research on militaries can be a fruitful enterprise. It is however likely to involve a degree of methodological flexibility, and an acceptance that the ways in which the case studies are developed will depend on local factors and will vary. The researcher has to be reflexive about the decisions being made, and also ready to be resourceful and resilient.

Conclusion

Caforio (2007) is correct to argue that our understanding of the nature and extent of the changes to militaries, following the end of the Cold War, can be greatly enhanced by comparative studies. Both large-N and small-N comparative studies can help to provide a richer and more detailed picture, and help researchers to generate generalisable theories, which may enhance our understanding of international security. There is a need to know what remains nationally specific and what broader trends are emerging.

Nevertheless, it is necessary to accept that the difficulties in carrying out military research in general are increased in much comparative work. This chapter has reviewed some of the challenges involved in conducting large-N and small-N comparative research in this difficult research terrain. It is certainly far from an exhaustive account of the potential issues involved, which will vary substantially between projects.

Questions of access, gatekeeping, reliability and availability of information and negotiating an unusual environment pose methodological challenges to military researchers working on a nationally focused study. The additional needs for cross-cultural awareness and sensitivity, the difficulties of consistently comparing different national militaries, and in many cases multiple periods of fieldwork mean that a comparative military researcher may need to be even more reflexive and flexible about the methodological choices made during and after the period of research design. This should not, however, deter the researcher: comparative military research can make both a substantial contribution to knowledge and be enriching for the researcher.

References

Barnes, Trevor (2008) Geography's Underworld: The Military-Industrial Complex, Mathematical Modelling and the Quantitative Revolution. *Geoforum* 39, 3–16.

Bevir, Mark and Rhodes, Rod (2003) *Interpreting British Governance*. Abingdon: Routledge.

Born, Hans, Haltiner, Karl and Malesic, Marjan (2003) *Renaissance of Democratic Control of Armed Forces in Contemporary Societies*. Baden-Baden: Nomos-Verlag.

Brzoska, Michael (1981) The Reporting of Military Expenditure. *Journal of Peace Research* 18(3), 261–275.

Caforio, Guiseppe (2007) Introduction: The Interdisciplinary and Cross-National Character of Social Studies on the Military – The Need for Such an Approach. In Caforio, Guiseppe (Ed.), *Social Sciences and the Military: An Interdisciplinary Overview*, Abingdon: Routledge, pp. 1–20.

Cowen, Regina (1986) *Defence Procurement in the Federal Republic of Germany*. Boulder: Westview Press.

Della Porta, Donatella (2008) Comparative Analysis: Case-Oriented Versus Variable-Oriented Research. In Della Porta, Donatella and Keating, Michael (Eds.), *Approaches and Methodologies in the Social Sciences: A Pluralist Perspective*. Cambridge: Cambridge University Press, pp. 198–222.

Deschaux-Beaume, Delphine (2012) Investigating the Military Field: Qualitative Research Strategy and Interviewing in the Defence Networks. *Current Sociology* 60(1), 101–117.

Dunne, J. Paul and Braddon, Derek (2008) *Economic Impact of Military R&D*. Brussels: Flemish Peace Institute.

Eriksson Baaz, Maria and Stern, Maria (2009) Why Do Soldiers Rape? Masculinity, Violence, and Sexuality in the Armed Forces in the Congo (DRC). *International Studies Quarterly* 53(2), 495–518.

Florea, Adrian (2012) Where Do We Go from Here? Conceptual, Theoretical and Methodological Gaps in the Large-N Civil War Research Program. *International Studies Review* 14(1), 78–98.

Forster, Anthony (2005) *Armed Forces and Society in Europe*. Basingstoke: Palgrave Macmillan.

Gray, Colin (2002) *Strategy for Chaos: Revolutions in Military Affairs and the Evidence of History*. London: Frank Cass.

Heuser, Beatrice (1993) Warsaw Pact Military Doctrines in the 1970s and 1980s: Findings in the East German Archives. *Comparative Strategy* 12(4), 437–457.

Higate, Paul and Cameron, Ailsa (2006) Reflexivity and Researching the Military. *Armed Forces and Society* 32(2), 219–233.

Kaldor, Mary, Sharp, Margaret and Walker, William (1986) Industrial Competitiveness and Britain's Defence. In *Lloyds Bank Review* 162(October), pp. 31–49.

Lebovic, James (1999) Using Military Spending Data: The Complexity of Simple Inference. *Journal of Peace Research* 36(6), 681–697.

Liebert, Wolfgang (1998) Dual-Use und Ambivalenz von Forschung und Technik – Neue Herausforderungen für die Rüstungskontrolle. In Martin, Grundmann and Hummel, Hartwig (Eds.), *Militär und Politik – Ende der Eindeutigkeiten*. Baden-Baden: Nomos, pp. 202–213.

Mastny, Vojtech (2002) The New History of Cold War Alliances. *Journal of Cold War Studies* 4(2), 55–84.

Müller-Wille, Björn (2014) Doing Military Research in Conflict Environments. In Soeters, Joseph, Shields, Patricia and Rietjens, Sebastiaan (Eds.), *Routledge Handbook of Research Methods in Military Studies*. Abingdon: Routledge, pp. 40–52.

Omitoogun, Wuyi and Sköns, Elisabeth (2006) Military Expenditure Data: A 40-Year Overview. In *SIPRI Yearbook 2006: Armaments, Disarmament, and International Security*. Oxford: Oxford University Press, pp. 269–294.

Rilling, Rainer (1988) Military R&D in the Federal Republic of Germany. *Bulletin of Peace Proposals* 19(3–4), 317–342.

Sartori, Giovanni (1970) Concept Misformation in Comparative Politics. *American Political Science Review* 64(4), 1033–1053.

Schrodt, Philip (2014) Seven Deadly Sins of Contemporary Quantitative Political Analysis. *Journal of Peace Research* 51(2), 287–300.

Vennesson, Pascal (2008) Case Studies and Process Tracing: Theories and Practices. In Della Porta, Donatella and Keating, Michael (Eds.), *Approaches and Methodologies in the Social Sciences: A Pluralist Perspective*. Cambridge: Cambridge University Press, pp. 223–239.

Walt, Stephen (1991) The Renaissance of Security Studies. *International Studies Quarterly* 35(2), 211–239.

Williams, John Allen (2007) Political Science Perspectives on the Military and Civil-Military Relations. In Caforio, Guiseppe (Ed.), *Social Sciences and the Military: An Interdisciplinary Overview*. Abingdon: Routledge, pp. 89–104.

10

CONDUCTING 'COMMUNITY-ORIENTATED' MILITARY RESEARCH

Ross McGarry

Upon their return to the UK, the bodies of British service personnel who have been killed in operations overseas are repatriated to an airhead at Royal Air Force (RAF) base Brize Norton. Once received by the tailgate procession, they are transported by hearse to the John Radcliffe hospital in Oxfordshire for postmortem by a military coroner. During 2007 a runway closure at RAF Brize Norton meant that the landing site for aircraft arriving from Afghanistan and Iraq moved to nearby RAF Lyneham. This caused the repatriated bodies of military personnel to be diverted en route to the John Radcliffe hospital via a small town in Wiltshire, the town of Wootton Bassett. On 1 and 2 April 2007, two British soldiers – Kingsman Danny Wilson and Rifleman Aaron Lincoln – were killed on duty while serving in Iraq. Days later, on 5 April 2007, the bodies of both men were driven by hearse along the high street of Wootton Bassett on their way to be received by the coroner. However unlike the deaths of the 188 British service personnel who perished in the years before them (52 in Afghanistan, 136 in Iraq), this time was different. Townspeople noticed the cortège and stopped by the roadside to pay their respects. For the next four years, the subsequent 167 military repatriations were attended by increasing numbers – at some times thousands – of townsfolk, passers-by, military veterans, bereaved families and military personnel. With these events came national and global media attention, causing Wootton Bassett to become both geographically and symbolically associated with the British war dead. Several years later, on 18 August 2011, Lieutenant Daniel Clack became the last British soldier to be repatriated through Wootton Bassett's high street. Although the occurrence of these events were never intended to last, the 'duty' of publicly receiving the deceased in Wootton Bassett was officially handed over to the nearby town of Carterton on 31 August 2011, following the reopening of the airhead at RAF Brize Norton. For its efforts in hosting 167 military repatriations during this period, Wootton Bassett was awarded royal assent – the first English town to receive this title in over 100 years. This chapter documents the processes that have served to facilitate a prolonged methodological engagement with the military repatriations in Royal Wootton Bassett (locally referred to as Bassett). First, the chapter introduces some of the attendant war literature relating to the study of military remembrance to find a place to situate this research within the human experiences of those who receive the war dead back into their communities. From here a brief biographical note is observed to indicate the gateway for studying military repatriations, followed by a methodological rationale centred upon ethnographic methods conducted at 'long' and 'short' range; a form of community-orientated research, engaging with those who lived and worked in the town who experienced first-hand the military repatriations in Bassett. To finish, the chapter reconnects with the theoretical contexts of military repatriation before providing a brief conclusion indicating the uniqueness of this approach for sociological military research.

Thinking About the Deceased

What we know of the war dead and how we, as a public, have been afforded ways to imagine and engage with them has undergone notable change during the last century. This change is well depicted by King (2010) who, in a discussion of the newly 'dead of Helmand', reminds us that public engagement with those killed at war overseas is one dictated by scale (large/small), autonomy (unknown/known), distance (far/near) and the capacities of mass communication (restricted/pervasive). As Winter (1998) notes, the significant numbers of British service personnel killed across Europe during the First World War meant that publicly reporting each death was difficult to facilitate with any speed or accuracy, a process hindered all the more through the limited capability of communication to either the families of the dead or the public. During this period the British government had ruled out returning the war dead from where they were killed "on grounds of expense and equality"; bringing the dead home was not only costly to the state and logistically difficult, but some bodies could be identified while others could not (Winter, 1998: 27). The parsimonious nature of the state and absence of the war dead in public life created by these circumstances brought with it a demand to reclaim the war dead by bereaved families and the public (Winter, 1998). This reclamation took shape in what King (2010: 7) has termed "lapidary conventions": the memorialisation of the war dead using fixed stone structures (i.e. war graves and cenotaphs) encompassed within remembrance practices. In the events preceding the wars in Afghanistan and Iraq we can note a similar conflict between state tenure and public reclamation of the dead. Edkins (2003) asserts that in the aftermath of 9/11 the US government quickly gained 'ownership' of the *civilian* dead, appropriating a politics of grief to help facilitate the wars in Afghanistan and Iraq. In response, the public are said to have demonstrated a 'rush to commemorate' as a means of reclaiming the dead from political ownership (qua Winter, 1998); remembering the dead in obituaries centred on the banality of everyday life rather than the catastrophic events that had brought the dead into the public imagination (Edkins, 2003). In doing so, the obituaries of the dead stripped away the political controversy relating to 9/11 and unwittingly served to help depoliticise the events that followed in Afghanistan and Iraq (Edkins, 2003). By contrast to the First and Second World Wars, the size and scale of these contemporary wars are considerably smaller. The number of service personnel deployed to fight is less and those subsequently killed serving the *military* have become fewer.[1] In spite of this, the deaths of British military personnel have become a consistent and pervasive feature of public life for over a decade since the wars in Afghanistan and Iraq began. Although

King (2010) observes that the deaths of military personnel killed in these wars have been absorbed into the lapidary practices of remembrance events which provide continuity with the past, like Edkins (2003), King (2010) also notes that the contemporary war dead are not only documented and reported to the public, but also that their deaths are personalised and their existences intimately eulogised. As such, contemporary memorialisation serves to 'domesticate' military deaths in ways that differ from simply individualising the sacrifices of military personnel (King, 2010).

The military repatriations in Bassett exemplify this shift in military commemoration, and academic commentary related to these events has continued to reflect similar critical observations. Although costly, the wars in Afghanistan and Iraq are smaller in scale. Military deaths are quickly made known to bereaved families, and later the public (see BBC News, 2014); the bodies of *all* British military personnel killed are repatriated back to the UK, a process supported by the Joint Casualty and Compassionate Centre.[2] Their proximity to us publicly was placed at close quarters through the events in Bassett, made visible and immediately recognisable via British media coverage of each event (see e.g. Gillan, 2010) and international recognition further afield[3] (see BBC News, 2010). Making use of this extensive media exposure Jenkings et al. (2012) provide a detailed analysis of print media coverage of the military repatriations that demonstrates their pervasiveness within the public domain. But there is a critical message here too. They consider Bassett to have become a militarised civil space, politically appropriated as part of a broader 'rehabilitation' of the relationship between the military and the public, fractured by aggressive British foreign policy in Iraq and Afghanistan (Jenkings et al., 2012). Other theoretical work has been similarly concerned with a conceptual analysis of the war dead (see Drake, 2011, 2013). With particular reference to Bassett, Drake (2011) suggests that these events have seen the war dead 'diverted' into 'extrapolitical spaces'. Apropos of Edkins's (2003) interpretation of the *civilian* dead of 9/11, Drake (2011) similarly considers the repatriated *military* dead from Afghanistan and Iraq to have facilitated a militarisation of everyday life and created a depoliticisation (and thus legitimatising) of war and the military. As Freeden (2011: 6) avers, the most proliferate elements of military repatriation events "are the perceived gulf between their proclaimed apolitical nature and the very strong political components they exhibit."

These observations provide valuable starting points for asking broader sociological questions of military repatriation. However, without visiting Bassett or experiencing a military repatriation, what can we really know of the town and these events? Recounting his attendance at the repatriation of one of his soldiers in 2008, former infantry officer Patrick Bury (2010: 286) notes,

> What we hadn't rehearsed is the procession through the small town of Wootton Bassett, whose high street runs between Lyneham and the road to the coroner's, and of which I know nothing. We arrive before the hearse. The shops are closing. Old veterans, in blue blazers adorned with gold medals, ribbons of yellows, whites, reds, khaki, green and black berets sitting loosely on their greying heads, line the road. As do women and children. Three deep in places. Old colours dip in unison and shake slightly as old hands proudly hold them. *At least they appreciate us. At least they are proud of us.* It pulls at my emotions. I fight, fight the tears. I lose.

This brief extract offers a different interpretation of military repatriation. It depicts Bassett as a popular public space, well populated with existential rather than lapidary conventions, a commemoration event that connects the past (i.e. *old* military veterans) with the present (i.e. *young* children). It indicates a perceived disjuncture between the military and the public that is suggestive of a more nuanced rather than apolitical relationship between Bassett and the state. In addition, it portrays an event that is charged with emotion, illustrating that even for those who may have experienced Bassett as a member of the military receiving the war dead, this does not mean that the repatriation events, or the Bassett community, are known or instantly knowable.

Selecting Methodology: Biography, Ethnography and Community-Orientated Study

Of course this is only one particular reading of associated literature. But what if we consider moving beyond such theoretical observations of commemoration in an attempt to get closer to the experiences depicted by Patrick Bury that are "both allegorical and real" (Winter, 1998: 28)? To do so first requires situating the researcher methodologically to help consider their own starting point and direction they wish to pursue. At this point it is useful for my own research voice to offer an entry point into this discussion, to provide my own reflections on the methodological processes employed during a sociological study[4] of Bassett that started sometime during 2010.

A Biographical Point of Entry: The Vulnerable Body of the Soldier

My initial attention was drawn to Bassett from a previous career in the British military and subsequent academic work relating to British soldiers (see e.g. Higate and Cameron, 2006; McGarry, 2012). Other sociological work with similar starting points has researched the military institution using autoethnography (see Hockey, 1986; Kirke, 2010, 2013), but the focus of my research has differed in several respects. It has been inopportune to write myself in as either an 'insider' or 'outsider' (qua Higate and Cameron, 2006) to previous published work, and I have always observed the military from a distance or via contact with military veterans rather than as an integrated member of the institutional setting. But the main departure of my research has been its focus on the vulnerability of military personnel through either perpetrating or witnessing violence at war (see McGarry and Walklate, 2011), being poorly supported and equipped by the state (see McGarry et al., 2012), or represented as 'victims' of war in the public domain following death and injury (see McGarry and Ferguson, 2012). This particular focus – underscored by victimology – allowed my sociological research interests to segue from the vulnerabilities of British soldiers at war to the receipt of the war dead via military repatriations in Bassett. The frequent but low numbers of deaths of British military personnel in Afghanistan and Iraq repatriated through Bassett brought the vulnerabilities of soldiering I had previously researched squarely into the public imagination. The visibility of the war dead in Bassett and their systematic documentation by the military (see Ministry of Defence, 2014) greatly increased their accessibility and connectivity to the public (Martinsen, 2013). But it is a focus on the *public* that attended the repatriations in

Bassett that provided a point of departure from my previous work while still paying attention to the military institution.

Directing the Research From Theory to Methodology

Making a departure from a theoretical understanding of Bassett to one that is empirically focused on those who engaged with the military repatriations required careful consideration. To help centre this thought process, it is first possible to consider this project as resembling ethnography, given that the main thrust of ethnographic work is to research – for extended periods – groups or social environments about which little is known (Henn et al., 2006). As Cosgrove and Francis (2011: 201) note:

> Ethnography . . . combines cultural interpretation – that is eliciting an understanding of the shared meaning of the group so as to develop understanding of their actions – with prolonged participant engagement in the natural settings within which the group operate.

While it is acknowledged that the traditions of ethnographic work reside in the observation of social environments for prolonged periods, ethnographies are open to a wide variety of approaches to *method*, including but not limited to observations, field notes, interviews, focus groups, document and visual analysis. This indicates that ethnography is open to some adaptation in relation to the methods employed for researching social groups and environments (Bryman, 2012). As the accommodating advice of Henn et al. (2006: 198) suggests:

> The ethnographic researcher is a fieldwork pragmatist, flexible and resourceful in his or her approach to the use of whichever methods and sources of data are at her or his disposal . . . the key is to ensure that the research approach is one that is appropriate for the research question and also maximises *Verstehen* [understanding].

Although these observations provide some methodological context there is something more nuanced to say about *who* and *where* this research was engaging. As the purpose of this study looked to the 'liminal' space (qua Kellaher and Warpole, 2010) within which military repatriations took place to engage the townsfolk who facilitated and sustained them locally, it became evident that this research was concerned with both the military *and* Bassett as a community. Despite the various problems noted by Bell and Newby (1971) regarding the definition and treatment of community as an object of study, they suggest that community study can be employed as a *method* to understand elements of the social world. Although this research does not claim to be a community study proper, some of the methods employed do share elements with what is termed as "community-orientated" study, research conducted on "social life as affected by the community" (Simpson, 1974: 312). Some of these characteristics include the research tool of the sociologist to be him or herself; this means that what is recorded and when, and how it is interpreted is a subjective and creative judgement made through observation "to seek information without asking questions" (Bell and Newby, 1971: 65). In addition, sociologists are required to live in the communities they research for short or long periods of time; it is particularly important to place researchers in physical and emotional proximity to those they are studying (Bell and Newby, 1971, 1974). Finally, sociologists studying within a community require a 'sponsor' to establish entry into the environment; this is followed by a process of being 'placed' at an appropriate entry point in the community, then having to 'fit in' and be 'fitted in' through an ongoing interaction between the researcher, sponsor and other community members (Bell and Newby, 1971). These latter processes of sponsorship, placement and fitting in were achieved over a prolonged period of several years by establishing a working relationship with the Bassett Town Council.[5] These unseen methodological practices should be considered as the undergirding to the fieldwork described in the following section. Finally, as Bell and Newby (1974: 309) suggest, "The best community studies, however, are eclectic, utilizing whole batteries of techniques over and above the central one of observation." In the following section I describe the battery of techniques employed to conduct an ethnographic study of the military repatriations in Bassett from what I have termed distances at long and short range.

Ethnography at Long and Short Range

During the course of this study, since 2010 the events occurring in Bassett have been continually monitored and frequently discussed with colleagues during research meetings and in classroom activities. This process of teaching, collegiate discussion and continual engagement with Bassett as a site of sociological interest resulted in a written publication about the military repatriations as they had been understood at *long range*. Building a more comprehensive ethnographic study of Bassett was achieved by devising methods to seek empirically informed knowledge at *short range*. Through a network of student interaction, collegiate discussion, publication and fieldwork this prolonged engagement with Bassett demonstrates an ethnographic commitment (qua Deegan, 2001) to better understand the military repatriations as sociological events, and the role of the community in facilitation of the social processes that produced and sustained them. I will briefly discussion each of the methods employed before returning to make some reflections on the military repatriations re-engaging with some of the literature outlined earlier.

Long Range: Asking "What Is Wootton Bassett About?"

This ethnographic research began in the classroom at the University of Liverpool, whereby a simple question was posed as the basis for a seminar discussion: "What is Wootton Bassett about?" To facilitate this discussion, several photographs taken of a military repatriation at Bassett by an independent professional photographer (and colleague)[6] were provided to students, accompanied by a range of sociological journal articles offering different ways of interpreting each image. Using these materials the students were asked to document their own ideas of what they thought 'Wootton Bassett was about'. A typical response included: 'Displays of peace and solidarity to bring calm back to the community and country after soldiers have lost their lives in the war. It is a time of reflection.' Student responses generally reflected a normative understanding of what the military repatriations represented, and the interpretation of these images was perhaps understandably more influenced by common sense rather than sociological reflection. Nevertheless, the discussion emerging from this seminar became the catalyst to analyse Bassett using visual methods for publication.

Long Range: Visual Ethnography

When employing photographs in ethnographic work the images should be understood as a means of analysis more than just information (Ball and Smith, 2001). Commenting on the value of employing images as an ethnographic method, Ball and Smith (2001: 308) suggest that doing so "together with a descriptively precise and theoretically informed commentating text, can serve to illuminate and further ethnographic understanding." In an attempt to frame a cultural understanding of the military repatriations from the same distance the world's onlookers had received them (e.g. via media reporting and imagery), my colleagues and I presented a visual analysis of photographs that did not feature as part of the mainstream or global news media (see Walklate, Mythen and McGarry, 2011). In doing so we proposed that the events in Bassett could be understood to represent different frames of meaning: as a public outlet for privatised grief, as evidence of 'dark tourism' (Foley and Lennon, 2000) or as displays of apolitical resistance against the wars in Afghanistan and Iraq. Rather than language being the driving force behind ethnography (as it is in most cases) this stage of the research permitted our 'eyes to do the work' (Ball and Smith, 2001). This further assists in recognising the biographical and theoretical position of the researcher (as noted in the previous section) as an imperative to how Bassett was perceived and what preconceived ideas were held in regards to it. However, following Rock (2001: 29),

> it does not do to presume too much in advance. Knowledge, it is held, is not won in the library but in the field, and it is for that very reason that ethnographers conduct fieldwork.

These initial long range, a priori observations set an agenda for research interest in Bassett to be pursued through fieldwork at short range.

Short Range: Observation

Observations are a more commonplace means of conducting ethnographic work, and are intended to capture descriptive accounts, personal experiences, dialogue and interactions on a day-to-day basis (Emerson et al., 2001). This juncture of the research took a more traditional ethnographic approach, entering the 'field' for the first time to move beyond a priori assumptions of Bassett to observe how repatriations were experienced. Two military repatriations were attended, one in Bassett and one in the new host site of Carterton to understand how each were experienced. They were documented using field notes to capture "an expression of the ethnographer's deepening local knowledge, emerging sensitivities and evolving substantive concerns and theoretical insights" (Emerson et al., 2001: 355), allowing for later reflection during the research process. Developing on from some of the characteristics of ethnography noted by Noaks and Wincup (2004), beyond observations ethnographic work also requires researchers to follow up their findings with interviews. Once military repatriations had been experienced personally, it became pertinent to speak to the townsfolk of Bassett to see how the military repatriations were experienced locally.

Short Range: Focus Groups and Interviews

Following the processes of entering and fitting in to the community as noted earlier, the next stage of this research involved conducting focus groups and semistructured interviews.[7] These took place over a series of months with members of the Bassett community who were consistently in attendance at the repatriations. A number of visits were paid to Bassett on several occasions, visits which included living in the town for short periods at a time. These experiences were also supplemented with field notes. Prior to conducting the focus groups and interviews, a research schedule was designed to address three key areas of academic interest: place, repatriation, and soldiers and soldiering. A standardised range of questions relating to these themes was asked during the focus groups and interviews, each of which lasted up to one hour. Once completed, the focus groups and interviews were transcribed and analysed thematically using the qualitative data analysis package NVivo 10.

Initial Research Findings

Initial findings from this research illustrated military repatriations to be highly complex events, fronted with a politics of respect for the dead and bereaved families, mitigated against tacit rules of conduct for mourning appropriately, behaviour which has ineffable consequences for those attending (Walklate et al., 2015). The broader interpretation of this fieldwork remains ongoing.

Reflecting on Theory

So what has been learned about military repatriations throughout this research process to help reconnect with the theoretical starting points of this chapter? Much like a roadside shrine, the military repatriations in Bassett occupied a physical space that expanded and collapsed with each event. During each repatriation Bassett high street *temporarily* represented an 'end point' for the dead through the demonstration of private cemetery practices in a public place (Kellaher and Warpole, 2010). Although we must remember that this was not the final resting place for the British war dead, the prevalence and consistency of the repatriation events over time provided Bassett with *symbolic* permanence. The geographical space of Bassett signified fixity, denoting a stable place associated with the receipt and reintegration of the war dead back to the UK. Rather than pertaining to lapidary conventions as described by King (2010), the military repatriations in Bassett instead occupied a liminal space (Kellaher and Warpole, 2010), a space attended by crowds of people. In order to transcend a fixation on the body of the war dead, Winter (1998) urges us to consider the social networks that emerge following the deaths of individual soldiers, often developed following poor support from the state. Within the literature described in the early sections of this chapter, and indeed depicted within the imagery of repatriation events in Bassett, it is assumed that the dead rather than the living are the central figures for our attention. But as Winter (1998: 29) reminds us, when receiving the war dead "many of these moments are lived within families supported by social networks," requiring a shift in focus to view military repatriations from the perspective of the public who receive them. Housed in coffins draped in the Union Flag, the deceased gave affordance to the repatriation events and demonstrations of public mourning. But those visible were the social networks of families, townsfolk and military personnel demonstrating individual and collective grief and bereavement articulated through respectful acts, gestures and symbols of mourning (Winter, 1998). The public were very much at the centre rather than the periphery of each military repatriation. The body of the soldier drew our attention to Bassett, the procession sequestered the geographical space, but it was the people in Bassett who simultaneously created and filled a "rupture . . . in the social world" (Kellaher and Warpole, 2010: 162) during each repatriation event. As such this research provides an insight into military repatriations that has some connections with previous repatriation literature. But it also offers a new focus that shifts attention from the symbolism and politics elicited from the body of the soldier to the experiences of the public who received them.

Conclusion

To conclude, it is worth noting that although unique to Bassett, and despite being replicated at Carterton, similar displays of public mourning for repatriated military personnel on this scale have also been experienced in at least one other location. Since 2002, Highway 401 in Canada has witnessed mass gatherings of the public, military and emergency service personnel, lining a route of motorway overpasses and roadsides known officially as the Highway of Heroes and the Route of Heroes (see Fisher, 2011). This route charted the transportation of deceased military personnel from CFB Trenton to the Forensic Services and Coroner's Complex in downtown Toronto. Recognising this as a rare sociological phenomenon, Fletcher and Hove (2012) attended a number of repatriations, conducting brief interviews to gauge the ways in which the repatriations had influenced Canadian public support for the war in Afghanistan. Although employing a less diverse range of methods than the methodological outline described in this chapter, both this and the current study perhaps set precedents for military research given the inimitability of these militarised events occurring in the public domain. In doing so they not only provide the platforms for interesting community based studies, but they also indicate something important about researching the military institution from the outside in. As Jenkings et al. (2011) have previously noted, studies of the military, as in other areas of social science, are stratified across qualitative and quantitative approaches, with a penchant for functionalist and orthodox research, frequently conducted through close linkages between military and educational establishments. For the most part the proximity of sociological research with the military renders both qualitative research and critical appraisals of the military estate underused and underexplored (Jenkings et al., 2011). But the uniqueness of military repatriation research offers a further rupture to the methodological fabric upon which military sociology has established itself. The current study of Bassett crossed the civil-military divide to directly learn something about the vulnerable occupation of soldiering and the destructive end product of state-legitimised violence, conducted without the structural constraints of the military environment. Military personnel had found a presence within the public domain, meaning that the methodological approach adopted during this study was permitted to be creative and wholly qualitative – as Jenkings et al. (2011) advocate for military research – without having to buttress against the methodological influences of orthodox military interests. In having this freedom the research has been able to adopt 'fieldwork pragmatism' (qua Henn et al., 2006): employing an attentive methodology to pursue an absence of data using the most appropriate methods available. But principally this project has always been anchored to methodology. That anchor has dropped in different types of ethnographic methods, and although presented in a logical order throughout this chapter, constructing this research strategy was frequently a reactive, creative and iterative process. It required a considerable amount of methodological flexibility and artistic license to help balance the demands of the 'research-teaching nexus' of higher education (Simons and Elen, 2007), with the momentum needed to conduct research requiring prolonged sociological engagement with a specific place and population. By interpreting military repatriations through the eyes of a community rather than by way of a military institution, this study has had the opportunity to embrace the nuances of "methodological pluralism," whereby "no longer can there be only one style of social research with one method that is to be the method. Rather there are many" (Bell and Newby, 1977: 11). The end result is a bottom-up study of the military (Jenkings et al., 2011) conducted from the perspective of community-orientated research.

Notes

1 Although as Ruggerio (2015) notes the most notable change in the conduct and style of warfare over the past century is that those who suffer and are killed in extremis are civilians, not military personnel.

2 See https://www.gov.uk/joint-casualty-and-compassionate-centre-jccc for more information.

3 The poignancy of the repatriations reached further afield than Wiltshire, with Bassett's global recognition acknowledged by US President Barack Obama during a public address in 2010.

4 Although written from a single author's perspective, it is to be noted that this research has been conducted collaboratively with Professor Sandra Walklate and Professor Gabe Mythen (both University of Liverpool). Dr Lindsey Metcalf (Liverpool John Moores University) provided support for the preliminary focus group data analysis.

5 Our thanks and gratitude to Bassett Town Council who helped to support and facilitate this research.

6 Griffiths, Stuart, see: http://stuartgriffiths.net.

7 Although it would be easy to associate this research with ethnographic interviewing, this is not what has been undertaken during this project. Qualifying this sort of interviewing as a method (over using focus groups and semistructured interviews, for example) would mean a focus on the context, time and place of the interview, having repeated interviews with the same participants to build up a bigger picture, and with an element of reflexivity present in the interviews (Rock, 2001). While there was an awareness of the latter in the interviews and focus groups conducted, this research did not undertake ethnographic interviewing.

References

Ball, M. and Smith, G. (2001) Technologies of Realism? Ethnographic Uses of Photography and Film. In Atkinson, P., Coffey, A., Delamont, S., Loftland, J. and Loftland, L. (Eds.), *Handbook of Ethnography*. London: Sage, pp. 302–319.

BBC News (2010) Barak Obama Praises People of Wootton Bassett. Available at: http://www.bbc.co.uk/news/uk-england-wiltshire-10709266

BBC News (2014) UK Military Deaths in Afghanistan. Available at: http://www.bbc.co.uk/news/uk-10629358

Bell, C. and Newby, H. (1971) *Community Studies: An Introduction to the Sociology of the Local Community*. London: George Allen and Unwin.

Bell, C. and Newby, H. (1974) The Sociology of Community – Appraisals of the field. In Bell, C. and Newby, H. (Eds.), *The Sociology of Community: A Selection of Readings*. London: Frank Cass, pp. 309–311.

Bell, C. and Newby, H. (1977) Introduction: The Rise of Methodological Pluralism. In Bell, C. and Newby, H. (Eds.), *Doing Sociological Research*. London: George Allen and Unwin, pp. 9–29.

Bryman, A. (2012) *Social Research Methods* (4th edition). Oxford: Oxford University Press.

Bury, P. (2010) *Callsign Hades*. London: Simon and Schuster.

Cosgrove, F. and Francis, P. (2011) Ethnographic Research in the Context of Policing. In Davies, P., Francis, P. and Jupp, V. (Eds.), *Doing Criminological Research* (2nd edition). London: Sage, pp. 199–222.

Deegan, M. J. (2001) The Chicago School of Ethnography. In Atkinson, P., Coffey, A., Delamont, S., Loftland, J. and Loftland, L. (Eds.), *Handbook of Ethnography*. London: Sage, pp. 11–25.

Drake, M. (2011) The Returns of War: Repatriating the War Dead from Iraq and Afghanistan. In Karatzogianni, A. (Ed.), *Violence and War in Culture and the Media*. Oxon: Routledge, pp. 131–147.

Drake, M. (2013) The War Dead and the Body Politic: Rendering the Dead Soldier's Body in the New Global (Dis)order. In McSorley, K. (Ed.), *War and the Body: Militarisation, Practice and Experience*. Oxon: Routledge, pp. 210–224.

Edkins, J. (2003) The Rush to Memory and the Rhetoric of War. *Journal of Political and Military Sociology* 31(2), 231–250.

Emerson, R. M., Fretz, R. I. and Shaw, L. L. (2001) Participant Observation and Fieldnotes. In Atkinson, P., Coffey, A., Delamont, S., Loftland, J. and Loftland, L. (Eds.), *Handbook of Ethnography*. London: Sage, pp. 352–368.

Fisher, P. (2011) *Highway of Heroes: True Patriot Love*. Toronto: Dunduran.

Fletcher, J. F. and Hove, J. (2012) Emotional Determinants of Support for the Canadian Mission in Afghanistan: A View from the Bridge. *Canadian Journal of Political Science* 45(1), 33–62.

Foley, M. and Lennon, J. (2000) *Dark Tourism: the Attraction of Death Disaster*. London: Cengage Learning EMEA.

Freeden, M. (2011) Editorial: The Politics of Ceremony: The Wootton Bassett Phenomenon. *Journal of Political Ideologies* 16(1), 1–10.

Gillan, A. (2010) How Wootton Bassett Became the Town That Cried. *Guardian*. Available at: http://www.theguardian.com/uk/2010/feb/25/wootton-bassett-audrey-gillan

Henn, M., Weinstein, M. and Foard, N. (2006) *A Critical Introduction to Social Research*. London: Sage.

Higate, P. and Cameron, A. (2006) Reflexivity and Researching the Military. *Armed Forces & Society* 32(2), 219–233.

Hockey, J. (1986) *Squaddies: Portrait of a Subculture*. Exeter: Exeter University Press.

Jenkings, K. N., Megoran, N., Woodward, R. and Bos, D. (2012) Wootton Bassett and the Political Spaces of Remembrance and Mourning. *Area* 44(3), 356–363.

Jenkings, K. N., Woodward, R., Williams, A. J., Rech, M. F., Murphy, A. and Bos, D. (2011) Military Occupations: Methodological Approaches and the Military–Academy Research Nexus. *Sociology Compass* 5(1), 37–51.

Kellaher, L. and Warpole, K. (2010) Bringing the Dead Back Home: Urban Public Spaces as Sites for New Patterns of Mourning and Memorialisation. In Maddrell, A. and Sidaway, J. D. (Eds.), *Deathscapes: Spaces for Death, Dying, Mourning and Remembrance*. Farnham: Ashgate, pp. 161–180.

King, A. (2010) The Afghan War and 'postmodern' Memory: Commemoration and the Dead of Helmand. *British Journal of Sociology* 61(1), 1–25.

Kirke, C. (2010) Orders Is Orders . . . Aren't They? Rule Bending and Rule Breaking in the British Army. *Ethnography* 11(3), 359–380.

Kirke, C. (2013) Insider Anthropology: Theoretical and Empirical Issues for the Researcher. In Carreiras, H. and Castro, C. (Eds.), *Qualitative Methods in Military Studies: Research Experiences and Challenges*. Oxon: Routledge, pp. 17–30.

Martinsen, K. D. (2013) *Soldier Repatriation: Popular and Political Responses*. Surrey: Ashgate.

McGarry, R. and Ferguson, N. (2012) Exploring Representations of the Soldier as Victim: From Northern Ireland to Iraq. In Gibson, S. and Mollan, S. (Eds.), *Representations of Peace and Conflict*. Basingstoke: Palgrave, pp. 120–142.

McGarry, R., Mythen, G. and Walklate, S. (2012) The Soldier, Human Rights and the Military Covenant: A Permissible State of Exception? *International Journal of Human Rights. Special Issue: New Directions in the Sociology of Human Rights* 16(8), 1183–1195.

McGarry, R. and Walklate, S. (2011) The Soldier as Victim: Peering Through the Looking Glass. *British Journal of Criminology* 51(6), 900–917.

McGarry, S. R. (2012). *Developing a Victimological Imagination: An Auto/biographical Study of British Military Veterans*. Unpublished PhD Thesis, Liverpool Hope University.

Ministry of Defence (2014) Op Herrick Casualty and fatality tables. Available at: https://www.gov.uk/government/publications/op-herrick-casualty-and-fatality-tables-released-in-2014

Noaks, L. and Wincup, E. (2004) *Criminological Research Understanding Qualitative Methods*. London: Sage.

Rock, P. (2001) Symbolic Interactionism and Ethnography. In Atkinson, P., Coffey, A., Delamont, S., Loftland, J. and Loftland, L. (Eds.), *Handbook of Ethnography*. London: Sage, pp. 26–38.

Ruggerio, V. (2015) War and the Death of Achilles. In Walklate, S. and McGarry, R. (Eds.), *Criminology and War: Transgressing the Borders*. Oxon: Routledge, pp. 21–37.

Simons, M. and Elen, J. (2007) The 'research–teaching nexus' and 'education through research': An Exploration of Ambivalences. *Studies in Higher Education* 32(5), 617–631.

Simpson, R. L. (1974) Sociology of the Community: Current Status and Prospects. In Bell, C. and Newby, H. (Eds.), *The Sociology of Community: A Selection of Readings*. London: Frank Cass, pp. 312–334.

Walklate, S., Mythen, G. and McGarry, R. (2011) Witnessing Wootton Bassett: An Exploration in Cultural Victimology. *Crime Media Culture* 7(2), 149–166.

Walklate, S., Mythen, G. and McGarry, R. (2015) "When you see the lipstick kisses . . ." Military Repatriation, Public Mourning and the Politics of Respect. *Palgrave Communications* 1(15009), open access. doi:10.1057/palcomms.2015.9.

Winter, J. (1998) *Sites of Memory, Sites of Mourning: The Great War in European Cultural History*. Cambridge: Cambridge University Press.

11

ETHNOGRAPHY IN CONFLICT ZONES

The Perils of Researching Private Security Contractors

Amanda Chisholm

Introducing My Field Community

Ethnography, as a methodology, has deep roots in anthropology and only recently has been seen by political scientists and international relations (IR) scholars as an important methodology. As a research method, ethnography advocates long-term immersion in the field, situating the researcher among the community she is researching, either as an active participant or an observer or a combination of the two. Such a method offers researchers an important opportunity to produce empirically rich research on militarised communities. Ethnographic research has proven vital in connecting the global with the local in the production of security. However, ethnography remains a challenging method in so much as it requires researchers to immerse themselves in the security community. Such immersion brings about issues regarding gaining and maintaining access and how the investigator then represents the dynamic and conflicting nature of these communities. Gaining access is key to sustained ethnography, and yet such access depends heavily on the perception of the research and researcher within the community, in which cultural awareness and identity politics play a key role.

This chapter discusses some benefits and challenges in doing ethnographic fieldwork in particularly dangerous and militarised contexts. To do this, I draw upon my experiences as a researcher exploring on-the-ground security practices of private security contractors in Afghanistan and Nepal between January 2008 and May 2010. Specifically, my research focused on the experiences of Gurkhas, men from Nepal with a 200-year military history with the British, working as Global South contractors. Because my research was interested in the ways in which race, gender and colonial histories are employed in producing and sustaining hierarchies among security contractors, ethnography allowed me to explore how these practices materialise in the field, how they are resisted and how they are sustained.

My fieldwork consisted of two visits to Kabul and one visit to Nepal. The first visit to Kabul was January 2008 to July 2008. While there I observed how Gurkhas interacted with each other and with their white security company manager, directors and Western clients.

I participated in the registration process of private military and security companies (PMSCs) with the Afghan government, attended informal meetings with other company country managers and with Afghan government representatives. I also interviewed PMSC country managers and socialised with them in various settings, from evening drinks and barbecues to Friday brunches at the Serena Hotel and Kabul Koffee House. Such socialising allowed me to both observe and interview my research community from an insider position. My second visit to Kabul was from September 2009 to April 2010. During this visit, I was living in a different compound, where Gurkhas and security contractors did not live. I thus relied upon interviews rather than ethnography, based upon direct participation and observation with Gurkhas, to inform my research. In May 2010, I travelled to Nepal to interview more Gurkhas and five local recruitment and marketing agencies that contract Gurkha services internationally. While in Nepal, I lived with a Gurkha family; spent time doing leisure, work, and community activities with them; and interviewed their friends and others within larger Gurkha communities in both Kathmandu and Pokhara (two areas of Nepal with large Gurkha settlements).

This fieldwork was complemented with interviews with four white British national former Gurkha officers: two formerly and two currently in director positions for private security companies that solely contracted Gurkhas. While the interviews were certainly important to my research agenda, ethnography remained a key method for my research in order to explore how constitutions of race and colonial histories continue to play out in everyday practices of security. Such a method allowed me to demonstrate how race and colonial histories are not only observed in oral accounts, located in interviews, but also social interactions, using common space and body language – all observations which remain difficult to reveal through other methodologies.

The remainder of this chapter proceeds with a discussion of the epistemological positioning of my research with regards to feminism and postcolonial research and the need for a reflexive awareness during the fieldwork process. This section is followed by a reflexive description of the contextual realities of doing ethnographic research in a war zone. I use reflexivity in this chapter to demonstrate that while the methodological tools I draw upon are used in most ethnographic examinations, the tools I adopted were context-specific, as the communities I sought to research were heavily militarised and very secretive towards outsiders. As I discuss in more detail later, my own background as a former military soldier and my ability to empathise (even as I did not completely agree) with military assumptions of what constituted security enabled me to have more access into the communities I researched. Conversely, and sharing with other ethnographic research, I was not always in a position to control how others perceived me and my credentials, and so my embodiments as a female, former soldier and now Western, white academic mattered differently depending on the context. The conclusion draws these themes together before offering some final reflections which new researchers to ethnography should consider.

Reflexivity, Ethnography and Resisting the Colonial Gaze

Anthropologists, feminists and postcolonial scholars have long argued that ethnography is "circumscribed by a myriad of power relations" (Eriksson Baaz and Stern, 2009: 504), whereby the relationship between the researcher and the researched has a direct bearing on the types of information produced. These power relations are also observed in the author's authority to represent the field through his or her writing (Rooke, 2009: 150). Ethnography also continues to be intimately connected to colonial enterprise (Gobo, 2008: 2). Weeden (2009: 75) traces ethnography and the use of fieldwork to Victorian-era history, and by the nineteenth century fieldwork was a part of "established standards" within natural science. In fact, this method's history was embedded, even with the best intentions of researchers, with the colonial enterprise and British imperial expansion (Gobo, 2008; Vrasti, 2008; Weeden, 2009). Military ethnography in particular was used to construct ideas of who Gurkhas were, which allowed the British to constitute and, therefore, govern them. As such, it created foundations for disseminating essentialising features of Gurkhas as martial race warriors (Des Chene, 1991; Caplan, 1995; Streets, 2004).

Contemporary colonial applications of ethnography are aptly demonstrated in current uses of military ethnography in Human Terrain Mapping (HTM),[1] which purport that we can know and understand (and thereby manipulate or control) the 'other' through ethnography (Gilbert, 2008; Gobo, 2008). Decolonising ethnography, that is disrupting existing power dynamics embedded in the method, then comes not by treating ethnography as a machine through which one can access knowledge, but in applying the method with a sensitivity regarding its potential to reproduce power practices when the researcher asserts authority to represent sole discovery of the other (Vrasti, 2008). As Basham reminds us, we as researchers create our stories through the stories we are told. We then abstract from those stories detailed in our interviews and observations what we find most relevant (Basham, 2013: 9). As such it is imperative for ethnographic researchers to write themselves into their research. Such writing improves transparency in research as it allows the reader to determine how the researcher came to the conclusions she did.

Feminists and critical ethnographers attempt to address some of the aforementioned power relations and challenges in fieldwork through the practice of reflexivity, a practice that places the researcher within his or her research and offers context within which the information was produced. Reflexivity is a strategy that exposes power in research, as it requires one to pay attention to "how the researcher is socially situated, and how the research agenda/process has been constituted" (Ramazanoglu and Holland, 2002: 118). Indeed, Weeden (2009: 76) describes reflexivity as the "ways in which concepts and styles of reasoning, as well as scholarly commitments, are historically situated and enmeshed in power relationships." Consequently, there is a lot to be learned about power by focusing on how research is produced through intersubjectivities and power practices observed through the interactions between the researcher and the researched (Ramazanoglu and Holland, 2002: 118; Gunaratnam, 2003; Pintchman, 2009). For my own work, the practice of reflexivity allows me to acknowledge how certain power practices within my own fieldwork mediated what could, and perhaps would, be said in interviews, whom I could talk to and what topics I could engage with.

Reflexivity also offers a trustworthy and dependable account of how research is produced because it openly details the researcher's understandings, positionalities and biases within his or her own analysis. By placing themselves in their own research, researchers ensure that readers are able to understand the unique set of social conditions and contexts that have led the researchers to their particular conclusions and analysis (Higate and Cameron, 2006: 223).

Through this understanding, we then all "become answerable for what we learn and how we see" (Haraway, 1988: 583, cited in Gunaratnam, 2003: 33). Not only does reflexivity therefore offer a more transparent account of our research, but it also highlights both the power practices and the privilege of the researcher that are embedded in all research projects (Skeggs, 2009). The research is made more transparent so the reader can make up their own mind on the conclusions reached. Reflexivity not only makes us, as researchers, accountable to the stories we produce – thereby working to decolonise the method – but it also enables us to explore our own emotional connections to our research community, as well as the ways we can attempt to find a balance between keeping distance and expressing empathy (Behar, 1996: 3; Rooke, 2009: 152). These positionalities, the stories we tell and the connections we make all impact on the type of research produced. Reflexivity then, as a tool, allows us to remain open and transparent about these research practices.

Bearing all of this in mind, I introduce myself here as a white, English-speaking, graduate-level educated Canadian woman. I have former military experience, having served for five years in the Canadian Forces as a medical assistant. As will be discussed in more detail in the following sections, I believe my background afforded me as many opportunities as it presented challenges. From the onset of my research, I was an outsider to this Gurkha military family, a military family deep-seated in colonial history. With only four years to get to *know* the men constituted as Gurkhas, in practice I could only begin to forge relationships and an understanding of them and their long histories. Unlike the British Gurkha officers I interviewed, my accounts and understandings of Gurkhas are mediated not through military regiments but through my experience as a white Canadian woman with a military background, intellectually informed by postcolonial perspectives of race and gender, and self-identified to the men I interviewed as an academic researcher. My representations as an outsider impacted on how I could access these research communities and the knowledge I could produce. But so did the ways in which I made sense of the field and in particular, how my need to feel physically and emotionally secure impacted on how I understood security and the questions I asked in interviews.

Ethnography in Dangerous Places: A Reflection on Limitations of Practice

Being in a militarised environment for an extended period also impacts upon the research questions one asks and the research one can conduct (Cohn, 1985); it is also the context in which the research is situated and impacts upon that whole research process. Before expanding upon the research process, I want to put the research into a general context. Feminist anthropologists have long argued that conducting ethnographic research in a place perceived as dangerous and in a militarised environment is something that many feminist political researchers have embarked upon. Nordstrom and Robben (1995), Henry et al. (2009), Higate and Henry (2004) and Cohn (2007), among other social ethnography researchers, have conducted ethnographic research in conflict zones and other 'dangerous places'.[2] Importantly, these scholars have illustrated that the field is never neutral and production of knowledge is not a straightforward *textbook* methodological application. The aforementioned scholars have raised considerations over how research about violence and security gets produced and the challenges researchers have in the field – challenges that inform their research process and that have a direct bearing on the type of knowledge produced. These feminist applications of ethnography and the researchers' subsequent experiences in dangerous and militarised fields resonate within my own research.

Kabul is an insecure city, and perceptions of immediate danger to physical life were everywhere. Granted, these perceptions were not always supported by empirical evidence, as insecurity (at least for the private security contractors I was working with and living among in Kabul) was determined through the concept of risk mitigation. This latter point was made clear during my two encounters with being protected. When I first got to Kabul I was working and living at a local research institute. My job there was to develop and deliver research-training programs to the Afghan researchers employed at the institute. The institute was small, employing eight Afghan nationals and three international staff. My protection at the institute came in the form of a local Afghan who stood static guard with a rifle. I was aware of the potential insecurity of my living and working conditions given the limited physical protection of the compound, but I reassured myself that because I was the only westerner located at this institute and it was in the middle of a quiet residential suburb of Kabul, I was not likely to be a target. I also maintained a lower profile, dressing like the local Afghan women, to blend in and not draw attention to who I was.

After a few months of working in Kabul I moved to a PMSC compound and took on work as a research consultant and bid preparation coordinator. This gave me the opportunity to conduct participatory observations on the security contractors and companies I was researching. Security and insecurity were articulated much differently in the PMSC compound. Unlike the research institute which had one armed guard at the front gate, security took on a more organised and hierarchal form – a form where I ceased to have any authority on my own security. The security contractors were regarded as the experts, and my role was to adhere to their advice and follow their guidance. The compound was also heavily guarded and I continually saw men with guns coming and going from the compound. For the PMSC security contractors who took on the role of protecting me, insecurity came not through actual acts of violence but the potential for violence in its varying forms. Violence and insecurities were always understood to come from the outside. Outside, in these communities, meant someone who was local and who was almost certainly nonwhite and not a westerner. Mitigating risk, then, involves understanding who the potential victim is – that is, the person's particular subjectivities and the relational vulnerabilities they could potentially encounter. Within private security, the victim was someone inside; that is someone who as a client, working for the particular company and almost always white and western. For me, being white, western, and female, I was told, increased my level of certain potential insecurities such as those relating to sexual violence, or kidnapping for ransom or political motives. The potential perpetrator was never identified, but again was assumed to be outside the security community and the people the community protected. This perpetrator could potentially embody both genders – as a female or male suicide bomber or as an orchestrated kidnapping ring. Importantly, while the perpetrator was never identified in specifics, s/he was almost certainly a nonwhite, non-western outsider within the security companies who predominately consisted of white, western men with western military training. Before moving to this compound I never thought of all the variations of my own insecurity – nor did I think the only way to assure my security was to stay in the compound and under the protection of these men.

Unlike my time at the research institute in the PMSC compound, the threat of violence informed my daily routines, how I saw Kabul, and how I understood the people living and working there. I began to situate violence as occurring outside the compound and my protection from this by adhering to the advice and guidance of the western security contactors surrounding me. There was a banality to the violence and the threat or potential to experience or witness devastating bodily harm that animated me throughout my fieldwork in Kabul. In the beginning I was alarmed at the overt militarised space observed in the abundant show of guns, men in military uniforms, static security guards in front of doorways, and fortified compounds adorned with razor wire. While at first all of this appeared alarming and perhaps theatrical, as I immersed myself more in the field I started to empathise with the people I was researching and even began using their terminology, talking in their security speak and ignoring the overt posturing that these security performances highlight.[3] I began to emotionally and intellectually invest in the understandings of security I was supposed to be there to be critical of.

As such, contradictions in security practices did not seem so obvious to me the more time I spent in Kabul. The same contradictions of security and the associated gendered tropes that underpin security subjectivities – discussed in detail by scholars, such as in Stiehm's (1982) protected/protector gender binaries, Higate and Henry's (2009) *Insecure Spaces* and their (2004) work on engendering insecurity – all demonstrate the ways in which state-led security is gendered and is performed in a multitude of often conflicting ways in the field. Additionally, Higate's (forthcoming) protected/protector analysis in his ethnographic work on clients of PMSCs, his discussions of intersubjective security performances of PMSC contractors through the study of personal memoirs (2011a) and his participatory observations in security training courses (2015) further demonstrate the ways in which these security practices extend to the private security communities. Such gendered security tropes and the contradictions of what constitutes security were also present in my fieldwork. For example, in one place where I lived, our security manager offered daily intelligence briefings where going out alone at night to a restaurant or social gathering was prohibited. Yet this same manager often regaled us with stories of how he stumbled home drunk from one of the local bars with a loaded handgun, at the same time drawing the caveat of *do as I say, not as I do*. This contradiction created in me feelings of both security and insecurity.

While recognising these security contradictions, at the time I was complicit in my investment in militarised ways of understanding security and in the performativities of militarised security, as I required them to feel safe. Militarised security is security that is understood through the control of an outside threat. Security, in this view, is best achieved through adhering to hierarchies of power and privileging the expert male voice. We are all seduced into adopting militarised ways of behaving (Enloe, 2007: 82–83). While in the field, I adopted a military understanding of security that was achieved through control of space and behaviour. I not only entrusted my safety to these contractors. They largely controlled my movements around Kabul through their regular monitoring of where I went and my seeking their approval prior to leaving the compound. However, I was constantly reminded through intelligence briefings and security incidents, including bombings and kidnappings, that the allusive threat could not be completely controlled. The threat, which was heavily racialised as a nonwhite, non-English-speaking embodiment, was always looming, and it was often repeated that many times attacks would occur when the victim was in the wrong place at the wrong time.

These potential insecurities were always beyond my immediate control, and I was repeatedly informed by the security experts as to my own inability to properly address them should they arise. In retrospect, the insecurities that lurked around our daily living and rituals were surmountable if one took the time to think through them all – a practice I attempted to avoid. As a way of coping with these perceptions of insecurity and security and my lack of control in addressing any of them, I focused my control inwards. Like many other expat women and men, I socialised and worked with while in Kabul, and I became fixated on controlling my diet and my consumption practices. Maintaining my diet allowed me to find some control over the decisions I made in my life. Security contractors – because they managed my security and because insecurity, I was told, was everywhere – also managed my work routines (when I went out of the compound and when I came back) and were to accompany me on research trips. This control over the day-to-day living of the protected expats was common. My regular conversations with other expat men and women about what was the best restrictive calorie-controlled diet to try indicated that a controlled diet was not just a coping mechanism for myself, but for many other men and women who were also protected by security contractors. Various diets were practiced, such as a cabbage soup diet, a vegetable broth detox diet and a lemon juice detox diet. These diets were coupled with fairly excessive exercise programs and continual communal conversations over which workout program produced the quickest change in body shape. As a collective we became immersed in how to control our calorie intake and the most effective way to change the appearance of our bodies. For me, the calorie-controlled diet and excessive exercise was a manifestation of the emotional and physical investments I had with the militarised understandings of security, a security I was not an authority on and could not control, and yet I was compelled to still be in control of *something*.

While I was able to critique the different forms of violence and how security gets articulated and performed, the longer I stayed in Kabul the more militarised (Enloe, 2007: 5) I became (as reflected in my need for control as way to increase security). This made conducting critical research on the very security I relied upon to keep me safe a complicated task. Over time, and in line with Cohn's (1985: 704) reflections of her research, I began to adhere to the gendered protected/protector security performances (Stiehm, 1982), because they made me less afraid. I was emotionally invested in the security-speak of risk mitigation through limited and varied movement and constant vigilance. Local intelligence suggested that mitigation of threats is performed through daily security checks by the armed guards (who are men) at your compound perimeter, regular security intelligence briefings (administered by white men) and restrictions on outside movements (controlled and monitored by white men and Gurkhas). Observing these security practices disconnected me from the unnamed and unknown security threat *out there*. This all made me feel safer in my compound, and as if the cement walls, razor wire, and static security guards were all I needed to *be* safe. The actual threats remained ambiguous, and thus it was clear to me that there is certainly room for resistance by those like me who are the protected (Higate, forthcoming). However, in my experience, the male security experts always defined the insecurity and security. Adhering to the role of the protected, as much as it frustrated me, allowed me to feel safe at the time.[4] Due to this, much of the critical reflection into the ways I invested in the field practices I was there to critique came upon me leaving the field and reexamining my field notes and observations. Researchers need to be aware that their personal security is important and that it also necessitates concessions on data collection – this a necessary compromise, but one that needs to be recognised in one's reflexive scrutiny of the data and how the researcher's own biases and experience mediate her analyses.

Beyond shaping how I understood social concepts of security and violence, being militarised and protected also mediated the various places I went to, how long I could spend there and with whom I could associate. I only ended up talking with Gurkhas when they came back from their security work, because they often worked outside Kabul or in more insecure parts of the city, where it was deemed – by their assessments of my personal security – too dangerous for me to go. Because of this, I was unable to see or observe a lot of their security performances. What I was able to observe was the Gurkhas and western expats (a term given to westerners working/living in Afghanistan) performing security in and around my own compound as well as when I was in transit from one location in the city to another. However, just because I emotionally invested in such practices and conceptualisations of militarised security does not mean I am unable to look critically at these practices. My critical analysis of these practices occurred once I left the field and was able to more openly reflect on how I was seduced by those practices. The point of this section has thus been to offer some insights into the ways in which the field shaped my own understandings of security and how these understandings mediated how I conducted interviews with Gurkhas while in Kabul. The next section details how perceptions held of me by my research community also had an immediate bearing the research process.

'You're Like Joanna Lumley, You're Such a Brave Woman': Gendered Perceptions of the Researcher

I recall sitting in the dining room of my compound in Wazir Akbar Khan, a wealthier and predominantly international populated area of Kabul. I was anxiously awaiting the arrival of my first focus group with Gurkhas. Gurkhas, men from Nepal with an over 200 year military history with the British military, were working as TCNs [third country nationals – a label given to men from the Global South who work in security] security contractors in Kabul after serving with the Indian army and Singaporean Police as Gurkhas. There were five of them coming to talk to me in this particular focus group. The meeting, arranged by my then fiancé Steve, a former manager of Gurkhas in commercial security, had been set up only a few hours ago and I suspect they were talking to me because they were friends with Steve.

The men all arrived at the same time and after Steve introduced us all he left the room. After introducing myself and what my research was to these men, Jitendra, a former Singaporean Police Gurkha, remarked "Oh, so you're like Joanna Lumley." The others nodded in agreement and smiled. I was immediately taken aback at this comment. I laughed off the comment and politely responded that no I'm not like Lumley. Their comparison of me to Lumley, a woman reproduced in UK media as the daughter of a white British national Gurkha Officer, who almost single-handedly fought for Gurkhas' settlement rights in the UK was unsettling to me. It struck me as overly colonial and riddled with privilege and power relations I was keen not to reproduce in my own research. Taking postcolonial perspectives seriously in my work, I had certainly not intended to be represented as a white female colonial saviour of the Gurkha. I'm not entirely sure if this is what Jitendra meant in his offhanded comment. On retrospect, I did embody the white western female and my interest in Gurkhas on the surface did appear to be similar to Lumley's. Did my similarities with Lumley mean that I was only seen as a colonising saviour? Could I be seen as something else to these men? What did this representation mean to my own future research with Gurkhas?

(Extract from Fieldwork Diary, Kabul, September 2009)

This field note entry illuminates the ways in which representations of self in the field are conflicting, nuanced, situated and negotiated and beyond the complete control of the researcher. These representations matter in gaining and maintaining access to the research community. As much as ethnography offers important and empirically rich research, it is intimately shaped by how the researcher relates to her research community. In order to really understand how research is shaped and produced within the field, an account of how the researcher is complicit in this knowledge production is important. This account offers the reader the context to understand the specific ways that representations of the researcher matter in gaining access to the community, the questions one can ask and the answers that can be offered.

Gaining access to these communities was as much about me expressing an emotional understanding of their positions as it was about how they understood me as a researcher. As such, Rooke's (2010: 154) call for researchers to be vulnerable and to open their own life up to those they are researching as a necessary part of ethnography is poignant in my own work. I believe that presenting myself with situated knowledge and empathy, in that I was a former serving soldier and thus empathised with the challenges of *doing* security, opened access for me. These men saw me as an ally: whether as a former serving soldier, as an interested white female western academic and as a surrogate daughter, I was able to make an emotional connection with each of the men I interviewed. I also believe that my personal relationship with the man who was a security manager of Gurkhas, and who was now my husband, also endeared me to the Gurkha community, bringing our relationship beyond researcher and researched. These men saw me as intimately involved with a man they respected and whom they believed was invested in their economic and emotional welfare. However, I was not always able to control how others saw me or felt about my research and me, as the extract from my field notes illustrates.

Throughout my interviews and fieldwork I was situated in a highly masculinised space, predominantly containing men with former military experience working in private security. This made it difficult for me, as a civilian woman with no combat experience, to integrate into this militarized community. At first I believe this community looked at me with suspicion and wonder. Why would a woman who was not associated with a media outlet, United Nations or a non-governmental organisation be in Afghanistan? For the men within this community who did get to know me, I believe they took pity on me. They saw me as naive and in need of their protection. In order to gain initial access, I had to accept the role that I was to be protected. This meant I could not make particular claims about knowing about security because I lacked the military/masculine experience. To be taken more seriously by this group, I strategised by drawing upon my own former military service with the Canadian Armed Forces. For some of the men in the community this was enough to allow them to share a common military history and background, but the more I immersed myself within the security community, the more I realised they were more accurately understood as communities.

Distinctions were not only drawn between civilians and men with military training; further differences were drawn between those of different nationalities and with different former military training. When conversing with men from different military backgrounds and nationalities, my gendered, raced and military-associated background mattered differently. For example, with British contractors, being Canadian and associated with the Canadian military was seen positively – although I was the brunt of a lot of jokes about the size of our military, I was commended for the overt patriotism Canadians showed – as if I were responsible for this patriotism. I was told repeatedly that the British could learn a lot from Canada's support of their military. With the American contractors I was assumed to be 'one of them'. American security contractors repeatedly told me that we are all North Americans. Gurkhas and local national security contractors were more difficult to get to know, and my integration with these men – with the exception of a few Gurkhas living in the same compound as me during my first field trip – remained minimal.

Initially, I shared Cohn's (2007: 97) observations and reflections, as described in her research. Like Cohn, my gendered subjectivity as a young civilian woman was seen as non-threatening and allowed me access to certain security contractor communities. Additionally, because I was a female who had formal military training who was *out there* in Afghanistan researching, I was privy to conversations that other researchers may not have had access to. Specifically, many of the men I interviewed (either formally or informally) were very forthcoming with their own personal experiences and insights about the security industry, their own positions and their understandings of the role Gurkhas play in the industry.

Being *out there* in Afghanistan allowed me to garner a lot of credibility with my research community, more so than if I had just undertaken interviews back in the UK. One director of a Gurkha security company openly expressed my credibility. Admiration for my ethnography in Afghanistan was also expressed by Gurkhas I interviewed in Nepal, and by a local agent from Nepal who recruited and marketed Gurkhas internationally. These men commented, "you're such a brave woman," and said that my efforts for going out into the field should be commended; their admiration resulted in fairly substantial, generous and forthcoming responses. While general praise was given, for each group of men I embodied subtle differences in gendered traits that, for the most part, allowed me to connect with them in different ways.

For the white British national Gurkha managers, my gender and former military experience allowed for candid discussions in interviews. My previous military experience as a medic offered me an important insider perspective on the military culture that largely informs private security culture. My feminist, political and intellectual foundations were also rather novel to many of the western men. During more casual conversations with western expats, when political discussions came up where I identified as being a feminist, many men such as Forster and other security company managers in Kabul would comment on how they believed in gender equality and that women are just as capable as men. This point, I believe, was said more for my benefit and to develop friendly relations with me rather than as a declaration of their political commitment to gender equality. Yet as the conversations continued over a period of time, one security company director commented that our discussions made him rethink certain beliefs he had towards gender and security. For example, the director of the Gurkha security company and another British national former Gurkha officer were both curious when I proclaimed that my feminist research was not about women's equality, but that it was focused on the ways in which men are differentiated in the security industry; on how men, like women, are gendered; and how their experiences as gendered beings warranted examination. While they were indeed curious, my clarification of my research appeared to be warmly received and these men were fairly open in their reflections upon their own masculinities – specifically a security company director, who talked at length of what it means to be a 'good' British national Gurkha officer – discussions which were underpinned through masculinities rooted in the caring liberal subject, who views everyone as equal regardless of their class, race or gender.

Overall, my being a feminist did not explicitly come up in conversations with Gurkhas, with the exception being my discussions with a Gurkha man and his family I was living with in Nepal. When my political beliefs did come up in my extended conversations with some of my Gurkha contacts, they tended not to find them threatening, in part because they associated such beliefs with women, and also because the fact that I was not explicitly questioning the absence of women in my research, an absence which I think perplexed them. In general conversations with security directors and managers, the introduction of my research was almost immediately followed by their response that they believed in gender equality but that women do not belong in combat arms. These conversations indicated to me how important it was for them to be understood as liberal men who believed in gender equality – equating to liberal equality. It also indicated how they automatically assumed that because I was a woman studying gender that I must also have been studying women. I wanted to disrupt this association by stating that I was interested in studying men, and that men too have genders that warranted examination. This, however, as the field note entry at the beginning of this chapter indicates, led to other problematic representations about what my fieldwork and I were about.

Importantly, because I was paying particular attention to Gurkhas in security, these managers thought my work would eventually lead to the promotion of Gurkhas' struggles and challenges and their capabilities as security contractors. Other Gurkhas, Gurkha agents, and security company directors who employed their services were clear that they believed that Gurkhas are undervalued or underrecognised. As such, all these men, in turn, were very generous with their time and devoted a lot of space to talk about the Gurkha martial race and how their particular company harnessed this talent, mentored these men, and provided a high quality of security as a result. Of course, this generosity through relationship building is not unique to research militarised communities. What is particular to relationship building within militarised communities is how the researcher attempts to gain access to this secretive masculine community. I appealed to legitimacy through my own personal connections to the Canadian military alongside with me being willing to be protected by their security services, adhering to their security authority. I sought to empathise with their personal perspectives and beliefs as to what constituted good security and who were good security contractors. I engaged positively with their political views even if I did not share them. I think, with specific reference to the some Gurkhas referring to me akin to Lumley, was that I was often seen as an ally and perhaps a champion of Gurkhas. I believe these particular techniques and my different embodiments as a former military, western academic *out there* in the field and an empathiser with Gurkhas helped foster these relationships.

Throughout my fieldwork, my representation as a female academic did not impede my actual interviews, but it did produce barriers to accessing Gurkhas in the field. One way this was broken down was through discussion of educational opportunities. Gurkhas see education as key to overcoming structural inequalities. They were impressed with my educational background and often drew comparisons between my educational pursuits and those of their sons and daughters – many of whom were in postgraduate programs in Australia, the US, and the UK. These familial associations allowed for a commonality between the men I interviewed and myself, which turned the interviews into informal conversations where a variety of themes were explored with ease. As such, my problem was not in getting Gurkhas to talk to me, but rather in getting access to *them* through the gatekeepers (security managers/company owners). In one particular case, an Afghan-based security director in Kabul granted me access to interview Gurkhas he had contracted. However, without offering any explanation to me, he then grew suspicious of my research and subsequently denied me access to those interviewees. The denial of access coincided with a verbal altercation I had with him over what I felt was aggressive posturing he was taking with one of his Gurkha employees. Without any other explanation, I can only assume that my access was denied because I spoke up regarding how he was treating his contractors. According to some anecdotal information fed back to me, he further threatened his contractors with termination of their contract if they spoke to me. Despite this one incident with this local security director, my research is heavily indebted to the time, as well as the open and candid responses, that the Gurkhas and white British national Gurkha officers afforded me.

My time in the field and regular contact with the field community enabled me to move beyond the position of researcher to embody other subjectivities such as surrogate daughter and friend, and I have remained in regular contact with many of these men today. Embodying a daughter and/or friend subjectivity was demonstrated either with explicit statements during interviews or by including me in familial activities and communal domestic tasks in Nepal. This allowed me to position myself in a more intimate relationship with the men (and women) I was interviewing, which resulted in more candid responses to my questions. In the beginning being an outsider allowed me to observe the things I found both strange, particular and novel (Mann and Stewart, 2000: 385). Over time, I was able to reflect upon why I found these initial interactions and ideas interesting and novel, and how my own understandings changed throughout the fieldwork process as I integrated into the community. While reflecting upon the novel moved my fieldwork in different directions, it also allowed me to engage in how my own understandings of what we mean when we talk about security informed the research process. Over time, I was both seduced by the military understanding of security (with the need to feel physically safe) while remaining intellectually at odds with it – continually feeling it lacked in its ability to encompass multitudes of insecurities and how security is contested and negotiated. Conversely, however, the time spent in the field produced a certain familiarity that being an outsider could not obtain. While remaining personally at odds with how militarised communities understand and perform security, the immersion allowed me to garner insights into why they understand security in particular ways and how logics and performances are ambiguous, contested and negotiated. Over time I was able to observe how military logistics defining security also were contested in everyday practices of security and how, beyond just myself as a researcher, practitioners were also conflicted with how to perform security and how one understands security.

Conclusion

This chapter discussed some of the opportunities and challenges associated with applying ethnographic methods in research on militaries and militarised communities. Ethnography has been a key method for anthropologists, and more recently political science scholars, who examine questions pertaining to on-the-ground experiences of military and security communities (Basham, 2013: 10–13; Leander, 2015). For me, ethnography was a key method in answering my research questions that focused on how constitutions of race and gender mediated the experiences of Gurkhas and the men they worked with in Kabul. Specifically ethnography allowed me to go into the field to see how local articulations of race and gender were practiced and how these articulations related to global articulations. It also enabled analysis into how resistance to these articulations is practiced. Overall, researching in Afghanistan offered richness to the various ways in which the representations of Gurkhas are understood and applied in different settings. In particular, using ethnography allowed me to observe how Gurkhas are recruited, vetted and then work as third-country national labourers in the private security industry, and how whiteness remained normalised and privileged in constituting the ideal security contractor.

Ethnography remains an important method for revealing richness in research that interviews or surveys alone cannot achieve. It has also remained vital in connecting local security practices with the global. However, there is messiness in ethnographic fieldwork for which one cannot completely prepare. Furthermore, research produced in the field is laden with power dynamics that do not always favour the researcher. Certainly, the growing number of articles highlighting the challenges individual researchers face while in the field is comforting, but at the same time that number highlights the various problems and how each researcher navigates them differently (Cohn, 1985; Schatz, 2009). These publications assert that practising a certain 'common sense' in the field and having the ability to navigate potential problems have a direct bearing on one's access to research groups, on gaining the trust of gatekeepers to maintain that access, and on being able to foster a 'safe' space within which to interview. This common sense takes on different forms depending on the context and community the research seeks to engage with. Problems that occur in the field are at times beyond the researcher's control and can potentially restrict the type of research that can be done, as demonstrated in my experience with one particular gatekeeper. The intent of this chapter has been not only to highlight the usefulness of ethnography, but also to reveal how this type of fieldwork is conditioned through the ways in which the researcher herself relates to her community and how she works to maintain such access in dynamic and shifting political and social environments.

Notes

1   The Human Terrain Mapping (HTM) website can be found at http://humanterrainsystem.army.mil (Accessed 23 August 2013).

2 See Mertus (2009) for the detailed and varied challenges and opportunities involved in the production of research through fieldwork.

3 Carol Cohn (1985) discusses how, after immersion in the field with nuclear defence experts and her time in their summer program, she began to adopt their language and think in the way they did. This was in part because adopting the language allowed her to feel in control, be a part of the group, and make the serious implications of nuclear warfare less scary, even almost funny.

4 Higate (2011) develops Stiehm's protector/protected gendered binaries to explore how security is embodied and performed in Afghanistan. Higate asserts that, in fieldwork experiences, there are certain ambiguities in the client/security protector relationship, which allow the protected to subvert his/her passive and protected status. While this is certainly the case on many occasions, overall my experience involved adhering to their protective status, even if I felt frustrated over the lack of control I had over my own security.

References

Basham, V. M. (2013) *War, Identity and the Liberal State: Everyday Experiences of the Geopolitical in the Armed Forces.* London: Routledge.

Behar, R. (1996) *The Vulnerable Observer: Anthropology That Breaks Your Heart.* Boston: Beacon Press.

Caplan, L. (1995) *Warrior Gentlemen: 'Gurkhas' in the Western Imagination.* Oxford: Berghahn Books.

Cohn, C. (1985) Sex and Death in the Rational World of Defense Intellectuals. *Signs* 12(4), 687–718.

Cohn, C. (2007) Motives and Methods: Using Multi-Sited Ethnography to Study US National Security Discourses. In Ackerly, B. A., Stern, M. and True, J. (Eds.), *Feminist Methodologies for International Relations.* Cambridge: Cambridge University Press, pp. 91–107.

Des Chene, M. (1991, June) *Relics of Empire: A Cultural History of Gurkhas, 1815–1987.* Stanford: Stanford University.

Enloe, C. (2007) *Globalization and Militarization; Feminists Make the Link.* Maryland: Roman and Littlefield.

Eriksson Baaz, M. and Stern, M. (2009) Why Do Soldiers Rape? Masculinity, Violence and Sexuality in the Armed Forces in the Congo (DRC). *International Studies Quarterly* 53, 495–518.

Gilbert, N. (2008) *Researching Social Life*, Volume 3. Los Angeles: Sage.

Gobo, G. (2008) *Doing Ethnography*, trans. A. Belton. Los Angeles: Sage.

Gunaratnam, Y. (2003) *Researching 'Race' and Ethnicity. Methods, Knowledge and Power.* London: Sage.

Henry, M., Higate, P. and Sanghera, G. S. (2009) Positionality and Power: The Politics of Peacekeeping Research. *International Peacekeeping* 16(4), 467–482.

Higate, P. (2011) 'Cowboys and Professionals' The Politics of Identity Work in the Private and Military Security Company. *Journal of International Studies* 40(2), 321–341.

Higate, P. (2015) Aversions to Masculine Excess in the Private Military and Security Company and Their Effects: Don't Be a 'Billy Big Bollocks' and Beware of the 'Ninja!'. In Eicher, Maya (Ed.), *Gender and Private Security in Global Politics.* Oxford. Oxford University Press, pp. 131–145.

Higate, P. (forthcoming). Cat Food and Clients: Gendering the Politics of Protection in Private Militarised Security Company. In Mathers, Jenny (Ed.), *Handbook on Gender and War.* Cheltenham. Edward Elgar.

Higate, P. and Cameron, A. (2006) Reflexivity and Researching the Military. *Armed Forces and Society* 32(2), 219–233.

Higate, P. and Henry, M. (2004) Engendering (In)security in Peacekeeping Operations. *Security Dialogue* 35(4), 481–498.

Higate, P. and Henry, M. (2009) *Insecure Spaces. Peacekeeping, Power and Performance in Haiti, Kosovo and Liberia.* New York: Zed Books.

Leander, A. (2015) Ethnographic Contributions to Method Development: 'Strong Objectivity' in Security Studies. *International Studies Perspectives*. Advance online publication, doi:10.1093/isp/ekv021 (Accessed 30 January 2016).

Mann, C. and Stewart, F. (2000) *Internet Communication and Qualitative Research. A Handbook for Researching Online*. London: Sage.

Mertus, J. (2009) Introduction: Surviving Field Research. In Lekha Sriram, C., King, J. C., Mertus, J. A., Martin-Ortega, O. and Herman, J. (Eds.), *Surviving Field Research. Working in Violent and Difficult Situations*. Oxon: Routledge, pp. 1–7.

Nordstrom, C. and Robben, A. C. (1995) *Fieldwork Under Fire. Contemporary Studies of Violence and Culture*. Berkeley: University of California Press.

Pintchman, T. (2009) Reflections on Power and the Postcolonial Context. Tales from the Field. *Method and Theory in the Study of Religion* 21, 66–72.

Ramazanoglu, C. and Holland, J. (2002) *Feminist Methodology. Challenges and Choices*. London: Sage.

Rooke, A. (2009) Queer in the Field: On Emotions Temporality and Performativity in Ethnography. In Kath, B. and Nash, K. J. (Eds.), *Queer Methods and Methodologies: Intersecting Queer Theories and Social Science Research*. Farnham: Ashgate, pp. 25–41.

Schatz, E. (2009) *Political Ethnography. What Immersion Contributes to the Study of Power*. Chicago: University of Chicago Press.

Skeggs, B. (2009) Feminist Ethnography. In Atkinson, P., Coffey, A., Delamont, S., Lofland, J. and Lofland, l. (Eds.), *Handbook of Ethnography*. Los Angeles: Sage, pp. 426–442.

Stiehm Hicks, J. (1982) The Protector, The Protected, The Defender. *Women Studies International Forum* 5(4), 367–376.

Streets, H. (2004) *Martial Races: The Military, Race and Masculinity in British Imperial Culture, 1857–1914*. Manchester: Manchester University Press.

Vrasti, W. (2008) The Strange Case of Ethnography and International Relations. *Millennium: Journal of International Studies* 37(2), 279–301.

Weeden, L. (2009) Ethnography as Interpretive Enterprise. In Schatz, E. (Ed.), *Political Ethnography*. Chicago: University of Chicago Press, pp. 75–94.

12

RESEARCHING PROSCRIBED ARMED GROUPS

Interviewing Loyalist and Republican Paramilitaries in Northern Ireland

Neil Ferguson

This chapter will describe the use of biographical-narrative methods, interpretive phenomenological analysis (IPA) and thematic analysis to delve into the Northern Irish conflict. The chapter is divided into two main sections. I begin with a brief discussion of social science research into the Northern Irish conflict and engagement in political violence, before describing our research. This discussion of our research describes the stages involved, but instead of focusing on the methodology per se I illustrate how the data was employed to build up an understanding of the phenomenon of participation with armed groups in Northern Ireland, focusing on the Irish Republican Army (IRA), Ulster Volunteer Force (UVF) and Red Hand Commando (RHC). Following this discussion, I describe the research methodology and methods used in these studies of politically motivated violence. Finally, I reflect on experiences researching in a postconflict environment with people who have engaged in violence or been victimized by political violence. The theoretical approaches taken in this chapter are taken from social and political psychology.

Historical Context

While the causes of conflict in Northern Ireland have been contested, with some pointing to the Norman Conquest or growing English interference in Ireland during the seventeenth century as the start of the sectarian conflict in Ireland, no generation has been spared conflict since the Plantation of Ulster. The latest and most sustained period of violence, euphemistically called the 'Troubles', began in the late 1960s when agitation aimed at securing equal rights to housing and votes for Northern Irish Catholics led to widespread disorder and the arrival of British troops in August of 1969 as part of Operation Banner. This seemingly intractable conflict continued for decades until the paramilitary ceasefires in 1994 and the signing of the Belfast (or Good Friday) Agreement in 1998 (The Agreement, 1998), which moved Northern Ireland towards peace.

During thirty years of sustained political conflict over 3,600 people were killed (Fay et al., 1998). Over half of the people killed (53%) were civilians, 16% were British soldiers, 15%

were local members of the Royal Ulster Constabulary (RUC) and Ulster Defence Regiment/ Royal Irish Regiment Home Battalions (UDR/RIR), 10% were republican paramilitaries and 3% were loyalist paramilitaries. The vast majority of these victims were killed by paramilitary groups (87% of the total; 59% by republicans, 28% by loyalists) with approximately 11% of the fatalities attributed to the security forces.

In addition to the fatalities between 40,000 and 50,000 people were also physically injured, many severely, suffering blindness and loss of limbs. Although this figure is probably an underestimate, it demonstrates the impact this conflict had on a small (1.7 million), closely knit population (Breen-Smyth, 2008). In terms of the psychological casualties there are no reliable data, but much of the psychological distress went unreported and untreated (Ferguson et al., 2010). Indeed this widespread exposure to violence is viewed as a significant factor in the intractability of the conflict (Hayes and McAllister, 2002) and certainly is a major factor in the holding back of reconciliation today.

Since 1998 the level of violence has significantly reduced, and a power-sharing devolved Assembly has been established, with an executive drawn from the main political parties across Northern Ireland's political spectrum (Mac Ginty et al., 2007). The agreement led to the release of prisoners who were part of paramilitary groups on ceasefire, and while there is still a threat to life from small dissident paramilitary groups, all the main republican and loyalist paramilitary groups have decommissioned weapons, and Sinn Fein have accepted the legitimacy of the Police Service for Northern Ireland (PSNI) to police Northern Ireland. These seismic political changes have demonstrated that the peace process in Northern Ireland can be viewed as successful, and many lessons are being drawn from the situation in Northern Ireland for application to conflicts across the globe (Mac Ginty et al., 2007; White, 2013; Ferguson et al., 2014).

Research With Armed Groups in Northern Ireland

Northern Ireland is probably the most researched conflict in the world (Mac Ginty et al., 2007). For the past decade my colleagues and I have been adding to this comprehensive body of research into political conflict by interviewing members of different armed factions in Northern Ireland, from both republican and loyalist paramilitary groups, in addition to members of the British security forces. These interviews have tended to focus on three themes, initially exploring engagement in politically motivated violence (Burgess et al., 2005a, 2005b; Ferguson and Burgess, 2008, 2009; Ferguson et al., 2008); then the impact of engaging in armed actions, and perspectives on victimizing experiences (Burgess et al., 2007; Ferguson et al., 2010; McGarry and Ferguson, 2012), and finally disengagement from politically motivated violence and the demobilization of paramilitary groups (Ferguson, 2010a, 2010b, 2014; Ferguson et al., 2015).

Engaging in Politically Motivated Violence

Since 9/11 there has been a fascination with the radicalization of 'normal' young men into fanatics motivated to kill to bring political change. Indeed Horgan (2005) documented that over 800 books focusing on terrorism were published in the year after 9/11. However, much of this work is misinformed, short-sighted and based on weak data, with a scarcity of primary, first-hand research (Horgan, 2014), with over 80% of studies into terrorism reliant on a secondary analysis of books, journals or reports, and only 13% based on actual interviews with terrorists (Silke, 2001). The need to attend to these methodological weaknesses and develop research on engagement in politically motivated violence based on quality data, coupled with experience of researching different psychological aspects of the Troubles, prompted us to explore the processes involved in joining an armed paramilitary group.

The research findings challenged the commonly expounded myth that people who join armed insurgent groups or engage in acts of terrorism are 'mad' or sociopathic (Silke, 1998; Horgan, 2014) and resonated with research from other conflicts and with other armed groups from across the globe (Victoroff, 2005; Horgan, 2014). This suggests that a more comprehensive analysis incorporating intra-individual factors, social factors and the nature of the conflict is needed to understand this radicalization process.

The research demonstrated that there was a range of antecedent push and pull factors which combined to increase the likelihood of participation in politically motivated violence. These factors included, but were not limited to being subject to perceived injustice; the importance of family involvement with the cause or movement; membership of the armed group being respected and supported by the wider community; strong in-group identity; and having witnessed or been traumatised by violence against the self, family, friends or the wider community (for a more exhaustive account see Ferguson and Burgess, 2009).

In addition to these push and pull factors, our research also demonstrated the importance of *critical incidents* or the experience of some catalyst event which triggered future involvement in action. For our participants these critical incidents were attacks on themselves, members of their family or the wider community they strongly identified with. For example, a former member of the UVF succinctly illustrates this process when he reflects on the factors which led to his engagement in politically motivated violence. In the interview he discusses Bloody Friday[1] as the critical incident which led him to join an armed group and reflects on the impact Bloody Friday had on him:

> And I thought, 'That's my fence sitting days over,' and I joined the UVF. And there's so many stories like that where you talk to republicans or loyalists and you find out there was a moment. There was a moment when they crossed the Rubicon.
>
> *(Burgess et al., 2005b: 26)*

There were many stories from both loyalists and republicans which pointed to a particular critical or trigger incident which caused them to turn their back on their 'normal' life and seek out options to engage in political violence. These incidents are usually followed by a *period of reflection* during which the participants consider their options and weigh up how to act to bring about change. A member of the IRA explores this process in his account of the critical incident which fuelled his engagement in political violence, an event which centred on the killing of a teenage boy by a British Army sniper. After the young man died in his arms the interviewee went to a derelict house and reflected on his options:

> So, quite honestly, within, within maybe an hour I'd been sat there on my own on me hunkers, and I said, 'Right, that's it. Ahh, the gloves are off as far as I'm concerned.' So, I went in and I seen other people and to cut a very, very, long story short, I got stuck into the Brits every time I got a chance.
>
> *(Burgess et al., 2005b: 26)*

In these cases, as in others, the participants had previous opportunities to join armed groups; both had been invited to join the UVF and IRA prior to these events but declined. Indeed, many of the participants who suffered from indirect and direct violent experiences did not join paramilitary groups. Instead they became involved in peace work or civil protest, or simply did nothing. It was during this period of reflection that the participants consciously determined the parameters of their protest, considering factors such as age, career aspiration, family commitments, education prospects and so forth as they decided on how to react to these critical moments and the perceived assault on their family, community and way of life.

The importance of these critical incidents has been reported and discussed elsewhere within criminology, philosophy and psychology (Jaspers, 1970; Denzin, 1989; Goodey, 2000; Brett and Specht, 2004; Horgan, 2014), and when coupled with the push and pull factors discussed earlier they assist in our understanding of the complex multidimensional character of engaging in terrorism or political violence.

Being Involved: Trauma and Victimhood

The second round of studies explored the impact of engaging in violence for the perpetrators of violence, the victims of political violence and those who were both perpetrator and victim. Again, we employed the same research approach within Northern Ireland, but this time focusing on what happens once you put your head above the parapet, stand up and engage in political violence, and what the consequences of that violence are for victims and the perpetrators themselves. The research team interviewed members and former members of republican and loyalist paramilitary groups, the British security forces and civilian victims of the Troubles.

The studies illustrated how being actively engaged in protest movements brought both positive and negative psychological consequences. For example, the interviewees reported how they felt a greater sense of agency and efficacy through their actions and a greater sense of collective identity. However, there was also a greater awareness of the personal costs of activism and the impact it had on the individual and their family.

Many of the interviews also explored issues around being a paramilitary and perpetrator of political violence, with interviewees discussing feelings of guilt and regret about the violence they had committed and the psychological pain this caused them. These feelings of guilt and self-induced trauma are reflected in this quote from a former member of the IRA:

> I'm someone that's living that lived in the past, that went through it and is able to recount and tell them [the young today] the horrors of it. And how much it can take lumps out of your head. Because it has taken lumps out of mine, there's no doubt

about it. I have the rest of my life to live thinking on things that I've done and maybe hurt people. And I'm very, very, sorry for it. I never wanted to do it.

(Burgess et al., 2007: 79)

Indeed many participants reflected on how they found it difficult to live with having perpetrated immoral acts in their attempts to bring about violent political change, with many talking about something changing in them or how they unlocked their 'dark side' through their actions and through the traumatic experiences endured as an active paramilitary. This is visibly illustrated by another former IRA volunteer:

You see, they let the genie out of the bottle. They turned us into monsters, if you like, that got out of control, that they couldn't beat. I can describe myself as that. And I feel awful about that. That's what they did to me. They brought this instinct out in me, the violent part that wasn't me.

(Ferguson et al., 2010: 877)

Many of the paramilitary interviewees discussed their activities with detachment, demonstrating dehumanization of the enemy, or members of the 'other' community, and engaged in moral disengagement. These can be viewed as symptoms of perpetration-induced traumatic stress (PITS; MacNair, 2002, 2005), which are commonly displayed among combat veterans, as are the earlier reported feelings of guilt and the high levels of alcoholism and suicide which were common themes for discussion among former combatants.

Being both perpetrator and victim is difficult terrain to navigate, and many of the interviewees struggled to reconcile these conflicting experiences, on the one hand being brutalized and harmed by violence from 'them', and on the other hand brutalizing and harming the 'other' to defend the status quo or bring about political change. This challenge is illustrated by this former combatant:

Very emotionally wounded and sad, almost a victim, but I was forced into doing that, I can't really explain, it is too difficult to explain.

(Ferguson et al., 2010: 877)

This challenge was also reflected in wider discussions about the hierarchy of Troubles victims, and how being a paramilitary – or to a degree, a victim who served in the security forces – meant that you were a lesser victim or not a 'real' victim in the same way the vast majority of 'innocent' victims were due to their agency and ability to choose their fate:

They (the paramilitaries) are not victims of circumstances. They chose to become involved. The ones that's getting out, the ones who of course, use that cry the most are prisoners, the ones who have been found out, 'yeah the reason I done that is because I became involved in the Troubles at an early age, because of blah.' The same old rhetoric turned out again, and again and again. No, they chose to go that way.

(Ferguson et al., 2010: 867)

These studies have helped open up a better understanding of perpetrator victimhood and the psychological implications of engaging in armed violence and killing or attempting to kill people. Indeed the research demonstrates that the PITS caused by engaging in violent extremism has the capacity to impact on the perpetrator well beyond the duration of the conflict.

Leaving Political Violence Behind

In the third round of studies we explored issues around leaving political violence behind and explored narratives of disengagement from politically motivated violence. These interviews took place against a backdrop of UVF and IRA moves towards decommissioning of weapon stocks and explored individual narratives of disengagement within this wider organizational process of disarmament and demobilization. The interviewees discussed a number of themes related to incentives and barriers to disengagement. The factors which pushed or pulled par-amilitaries away from violence included spending time in prison and life changes driving the members of the armed groups towards peaceful disengagement from violence.

The main driving factor for the former combatants we interviewed was getting the space to think and reformulate strategy while serving long sentences in the Maze Prison. A loyalist paramilitary illustrates how prison provided the opportunity to sharpen political ideology and strategy away from the heat of the conflict raging on the streets outside:

> Some people say that the likes of the Maze [a prison in Northern Ireland] and places like that were universities of terrorism, they were terrorist training camps, but I actually believe that they were the university of peace, in terms of what we discussed in there, how we decided, how we came about in our discussions, how do we get out of this? How do we get, you know, where is this all going?
>
> *(Ferguson, 2010a: 113)*

Another former combatant further illustrates the importance of having a prison experience in guiding perpetrators of political violence towards utilizing peaceful political solutions and disengaging from violence:

> I've been involved for something like thirty-five years and the next stage obviously when you get involved in the conflict, the more operations you carry out the more you get involved and the bigger chance you've got of getting caught or killed. So I was caught, and put in prison, so I had those prison years where, and it should be no surprise to anybody, because some of the best leaders in the world developed their political thinking in prisons, Nelson Mandela. Some of the world's best leaders are ex-prisoners of the British particularly . . . so it should come as no surprise that people in prison do develop because you've been removed from the conflict.
>
> *(Ferguson, 2014: 277)*

The participants also discussed how getting older, getting an education – usually while in prison – having a family and thinking about the next generation were important drivers towards disengagement. There was a particular desire not to have the traumatic experiences they'd faced revisited on the next generation of Northern Ireland.

> The other thing is that I have children of my own and I certainly wouldn't like them to have the same life experience as I have.
>
> *(Ferguson et al., 2015: 205)*

The former combatants also discussed a number of barriers to both individual-level and group-level disengagement and disarmament. These barriers included the need to hold the groups together to guard against splits, and how the legacy of the conflict put pressure on people to accept violence as a solution to political problems and thus maintain arms and a readiness for re-engagement in political violence.

In a series of interviews with loyalist paramilitaries discussing disengagement and disarmament, all the participants discussed their fears that the peace process could lead to the organisation fracturing into smaller rival armed groups similar to the creation of the Loyalist Volunteer Force (LVF) in 1996, or that members would break away to form independent criminal gangs.

> And that's the problem. And I know there are these thoughts within the organisation, how do we address these kids? How do we basically get rid of them, you know what I mean, without them falling into ruin the way the LVF [the Loyalist Volunteer Force, a dissident faction of the UVF] went, you know. It would be a tragedy to see another LVF or another group in that guise, who are only interested in gangsterism and drugs, you know what I mean. So its how do you keep them on board and how you disarm them gracefully and, that's what you're toying with the whole time.
>
> *(Ferguson, 2010a: 118)*

To bring about stable and sustainable change the organization and the wider community need to provide opportunities to members which will allow them to rebuild their lives away from paramilitarism and meet external disarmament, demobilization and reintegration (DDR) agendas. This takes time, and as within any other organization undertaking a time of change management and repositioning, this can be a difficult process.

> You know, because they're talking about they are going to pension people off, that's fine, but how do you turn round and say to somebody's that's had 36 years service and who sees themselves as still having something to offer? How do you go to that guy and say, right, you are no longer in the organization? So we have to find a role for those people to say look you can come here, but what you've been doing in the past is no longer needed.
>
> *(Ferguson et al., 2015: 208)*

Another key factor in holding back full disengagement from violence was the culture of violence and widespread segregation which had been created by the Troubles. Decades of violence had created a culture where there were pressures to accept or support political violence, and violence more generally, as a solution to political and local social problems. This was compounded by high levels of residential segregation which fostered the development of negative stereotypes of the 'other'.

> I think that the emotional difficulty of dealing with the other side – let's look at governments for example. Governments spend billions to vilify their enemy, and then when the conflict ends, they have to talk to the enemy they've just vilified. Billions! You only realize what a damn good job you did vilifying them when you have to talk to them. And it's a bit like that for communities as well, that the other side could never be honourable, never be decent, never be genuine, never be real, and meanwhile back at the ranch having espoused that for as long as I can ever remember, then a society goes to talk to each other and we get into trouble because they've done such a good job of vilifying one another in the past. We don't trust each other and the failure of trust is not the issue, our not trusting each other, how could we? We don't know each other.
>
> *(Ferguson et al., 2015: 209)*

These interviews illustrate some of the factors involved in assisting or hindering disengagement from armed groups in Northern Ireland with many of the push and pull factors resonating with research focusing on other armed groups across the globe (Bjorgo, 2009; Horgan, 2009; Reinares, 2011), and how the challenges to disengagement are in part due to Northern Ireland's wider struggle to come to terms with its past and the legacy of decades of violence and division.

Research Methodology

This section will examine in detail the methodology we employed in conducting our research in Northern Ireland. Our approach draws on biographical-narrative methods, interpretative phenomenological analysis (IPA) and thematic analysis.

Interview Procedure

The structure of the interview and style of the initial questions were based on Wengraf's (2001) biographical-narrative methods. The interview itself consists of subsessions that encourage participants to give narrative-based personal accounts relevant to the specific topic under investigation. Initially, participants were asked to give a biographical account of events and personal experiences related to living in the Northern Ireland conflict. At this stage, the interviewer gave 'gentle nudges' for the participant to elaborate further rather than ask follow-up questions that were explicitly grounded in prior theoretical knowledge. This is in keeping with traditional phenomenological approaches.

Many IPA researchers have adopted this approach in order that the researchers' preexisting academic expertise does not drive the data collection in a way that minimises the degree to which participants give phenomenological descriptions of life from their own (as opposed to a theoretical) perspective (Smith and Eatough, 2007). Participants ordinarily give some kind of 'coda' to indicate that their personal phenomenological description has come to an end (Wengraf, 2001), and this marks the end of the first subsession (initial question and follow-up to induce further biographic narrative from the insider's perspective) of the interview. The final subsession of the interview is designed to elicit answers in response to questions derived from previous theoretical knowledge. This section is characteristic of IPA researchers who are also interested in integrating phenomenological descriptions of life with contemporary theoretical perspectives (Smith and Osborn, 2008). It should be noted that this section of the interview is also semistructured in nature, still allowing the interviewer to explore emerging areas of interest in real time (Smith and Eatough, 2007), while also maintaining a sense of participant agency (Patton, 1990; Parker, 2005), and recognizing the participant as an 'experiential expert' who is still capable of steering the direction of the interview through accounts of their personal experiences (Smith and Osborn, 2008).

It is this structure that enables the IPA researcher to be confident that the investigation is indeed privileging the experience of the participant and yet is still linked to the best available academic work. An alternative semistructured format could be adopted whereby questions were based exclusively on extant literature. However, this does not give as great an opportunity for discovering new information (i.e. material that goes beyond preexisting theoretical knowledge) regarding the way in which participants experience and make sense of their lives.

Participants were able to speak for as long as they wished, an approach that allows the researchers flexibility and increases the trust between the participants and the researchers (Finch, 1999). In order to maximize the degree to which participants felt at ease, each interview was conducted in an environment selected by the participant. These environments included the interviewee's home, public bars, offices and friend's homes. Three of us[2] were involved in interviewing the participants, and we attempted to insure that at least two authors were present for each interview. We tend to interview in pairs, as we are from different socioeconomic, national, political and religious backgrounds, thus at different times we are either 'insiders' or 'outsiders' depending on whom we are interviewing and where the interview is taking place. It also allowed us to cover the interview topics and keep each other on track. Most interviews lasted between 30 and 100 minutes, but some were considerably longer; all were digitally recorded.

Data Analysis, Procedure, and Issues of Validity

The qualitative analysis was based on principles common to interpretative phenomenological analysis and thematic analysis (Smith, 1995). The first author conducted a detailed analysis of each interview, annotating and coding each participant's transcript fully before starting the next one. Transcripts were analysed according to the IPA guidelines provided by Smith et al. (1999). The initial stages firmly emphasize the 'P' in IPA, the phenomenological aspect of the analysis. First, the idiographic focus starts by selecting a single transcript and reading it several times to become familiar with the participant's holistic subjective interpretation. The transcript is then analysed in a line-by-line fashion, with the researcher noting anything of psychological interest in the left-hand margin. These notes should be grounded in the participant's descriptions.

At the completion of this stage of analysis, the researcher goes through the transcript line by line again, but at this stage the original statements are transformed into emerging themes that reflect a broader level of meaning. These notes are made in the right-hand margin and often start to link to theoretical constructs to make maximum sense of the data from a psychological perspective (hence, the researcher begins to take a stronger interpretative stance regarding the data at this stage of the analysis). Despite the researcher taking a small step away from a participant's interview account during this stage of the analysis, the phenomenological nature of this method means that themes and interpretations must still always be grounded in what the participant has said (Smith and Eatough, 2007). At this point, the researcher resists the temptation to produce hierarchies of meaning by creating dominant themes and sub-themes. Instead, each participant statement is regarded as equally important. Each transcript was coded in the same manner, recognizing themes that had emerged in previous transcripts but allowing new themes to emerge from successive participants' data (Smith et al., 1999).

Now, finally, the different transcripts were compared to determine whether some themes could be reduced to a more inclusive category and to see whether superordinate master themes comprised subthemes. The researchers repeatedly checked original transcripts to ensure that each theme emerged from participants' accounts and not as a result of the researcher's preexisting theoretical knowledge.

One of the other members of the research team for each study independently conducted a mini-audit of the transcripts and summary documents and agreed with the coding and themes identified in line with accepted procedures (Yin, 1989). This mini-audit of interview transcripts was designed to gauge the extent to which the primary analyst's interpretations of data were grounded in participants' accounts, and to gauge the extent to which emerging themes adequately represented the data. During this audit, the research team could identify new potential themes that were not identified in the original analysis or modify existing codes to enhance the coherence of the analysis (Yardley, 2008). For detailed information on how IPA differs from descriptive phenomenological and grounded theory methods, see Smith (2008).

The conceptualization of validity in phenomenological research differs significantly from the conceptualization of validity in quantitative methods (Smith, 2008). The dominance of quantitative methods within psychology often leads to criteria for validity mistakenly being applied to qualitative methods from quantitative methods (see Yardley, 2008). As IPA researchers, we acknowledge that participants will have a unique perspective on and interpretation of reality which is informed and influenced by their personal experiences and their sociocultural environs. Similarly, we acknowledge that different researchers will have their own unique perspectives that they draw on when interpreting and evaluating participants' meanings. To enhance the qualitative validity of our research, we followed Yardley's (2000) guidelines to ensure the themes raised were sensitive to the context, demonstrated rigour and were coherent, and that the analytic steps were transparent and that the findings had an appropriate level of impact and importance.

Reflection and Discussion

Our interviews in Northern Ireland took place over a number of years, and the topics explored developed in relation to the knowledge we accumulated over that time, in reaction to changes on the ground in Northern Ireland and further afield and in relation to the development of research areas and questions being asked in psychology elsewhere. The research challenged some widely held assumptions around 'terrorism' and 'terrorists' and demonstrated that lessons could be learned in Northern Ireland which could be applied to armed groups in other conflict zones across the world. Indeed, challenging some of the common assumptions and suggesting that the processes of engagement in political violence occurring in Northern Ireland would be reflected in places like Iraq and Afghanistan, or even in the creation of home-grown terrorists from within the UK, made it difficult to publish some of our work, particularly prior to the events of 7/7 when there was a clear distinction between domestic and international terrorism and a belief that violence related to violent Islamic militants was unique and incomparable to violent groups of the past, or those based in other settings.

To some degree, when I reflect back on the ease of which we were able to interview members and former members of armed groups, I see those as almost halcyon days, first, because the interviewees believed that, having served sentences and been released from prison, they had served their time and were not at risk of rearrest for Troubles-era crimes and felt at ease to talk and discuss their past, present and future openly. Indeed, many participants were transitioning into political or community roles and felt it would be good experience. However, the requisition of the Boston College tapes[3] by the PSNI and the subsequent arrest of Gerry Adams and others related to the disappearance of Jean McConville have made former combatants nervous about talking to academics about their past just as openly. Second, the peace process made it possible for us to travel to parts of Northern Ireland and go into homes in areas which would have been closed off to us – or indeed dangerous for some of us to have spent any time in – just years previously. So you could say, we were in the right place at the right time.

Our methodology proved effective for us, and it developed as we developed as qualitative researchers. While working as a small team provided support and mentorship within a community of practice, it also allowed us the opportunity to debrief after interviews and deal with some of the vicarious disturbance that resulted from listening to and analysing stories of trauma and violence during the fieldwork. This is a practice that I would wholeheartedly recommend to others embarking on similar military-related research projects, because all military research has the potential to unearth issues related to violence, conflict and trauma. Therefore, it is important for researchers in this area to build in mechanisms to provide them with support and guidance on how to deal with these issues when they occur, both within the fieldwork setting and also within the research team when analysing potentially distressing data back at the office.

While these qualitative accounts of individuals' experiences do not easily allow for simple generalisation, it is important to note that these findings from Northern Ireland share many similarities with research conducted elsewhere in the world with armed groups (see Ferguson et al., 2015), which both lends confidence to these findings and demonstrates that there are parallels in routes into and out of armed groups with different structures, ideological perspectives and military strategies. This implies that further research with members of diverse armed groups, in different landscapes or battlefields, with different experiences of conflict, will assist in building general theories on how and why people engage in and disengage from political violence.

In terms of the wider field of military research or terrorism studies, the approach outlined here resonates with the development of a more interdisciplinary, critical and qualitative approach to both fields of study over the last few years. This increasing emphasis on methodological issues and the development of more sophisticated qualitative and mixed-method approaches can begin to tackle some of the perceived weaknesses inherent in both qualitative and quantitative research, which will benefit the areas of military and terrorism studies. With improved methodology and research practices future research will assist in unpacking and understanding the complexity of factors that underlie terrorism and the impact of political violence has on military and nonmilitary personnel with increased rigour. This will help us understand why politically motivated violence begins, its impact on both perpetrator and victim, and how we can lay the groundwork to facilitate individuals or armed groups to desist from political violence.

Notes

1   Bloody Friday, took place on 21 July 1972 in Belfast, when the IRA exploded twenty-six bombs in eighty minutes, killing nine people and injuring 130 more.
2   Mark Burgess and Ian Hollywood were also involved in the interviews.
3   Boston College conducted an oral history project on the Troubles in Northern Ireland. During the project they recorded interviews with loyalist and republican paramilitaries about their activities during the conflict, with the understanding that this material would not be released until after their death.

References

The Agreement: The Agreement Reached in the Multi-party Negotiations (1998) Belfast: HMSO.

Bjorgo, T. (2009) Processes of Disengagement from Violent Groups of the Extreme Right. In Bjorgo, T. and Horgan, J. (Eds.), *Leaving Terrorism Behind: Individual and Collective Disengagement*. London: Routledge, pp. 30–48.

Breen-Smyth, M. (2008) Dealing With the Past and the Politics of Victim-Hood. In O'Hagan, L. (Ed.), *Stories in Conflict: Towards Understanding and Healing*. Derry/Londonderry: Yes! Publications, pp. 71–96.

Brett, R. and Specht, I. (2004) *Young Soldiers: Why They Choose to Fight*. London: Lynne Rienner.

Burgess, M., Ferguson, N. and Hollywood, I. (2005a) Violence Begets Violence: Drawing Ordinary Civilians Into the Cycle of Military Intervention and Violent Resistance. *Australasian Journal of Human Security* 1, 41–52.

Burgess, M., Ferguson, N. and Hollywood, I. (2005b) A Social Psychology of Defiance: From Discontent to Action. In Sönser-Breen, M. (Ed.), *Minding Evil: Explorations of Human Iniquity*. Amsterdam/New York: Rodolpi, pp. 19–38.

Burgess, M., Ferguson, N. and Hollywood, I. (2007) Rebels' Perspectives of the Legacy of Past Violence and of the Current Peace in Post-Agreement Northern Ireland: An Interpretative Phenomenological Analysis. *Political Psychology* 28(1), 69–88.

Denzin, N. (1989) *Interpretive Biography*. London: Sage.

Fay, M.T., Morissey, M. and Smyth, M. (1998) *Northern Ireland's Troubles: The Human Costs*. London: Pluto.

Ferguson, N. (2010a). Disengaging from Terrorism. In Silke, A. (Ed.), *The Psychology of Counter-Terrorism*. London: Routledge, pp. 111–122.

Ferguson, N. (2010b) Disarmament, Demobilization, Reinsertion and Reintegration: The Northern Ireland Experience. In Ferguson, N. (Ed.), *Post Conflict Reconstruction*. Newcastle: CSP, pp. 151–164.

Ferguson, N. (2014) Northern Irish Ex-Prisoners: The Impact of Imprisonment on Prisoners and the Peace Process in Northern Ireland. In Silke, A. (Ed.), *Prisons, Terrorism and Extremism: Critical Issues in Management, Radicalisation and Reform*. London: Routledge, pp. 270–282.

Ferguson, N. and Burgess, M. (2008) The Road to Insurgency: Drawing Ordinary Civilians into the Cycle of Military Intervention and Violent Resistance. In Ulusoy, M. D. (Ed.), *Political Violence, Organized Crime, Terrorism and Youth*. Amsterdam: IOS, pp. 63–71.

Ferguson, N. and Burgess, M. (2009) From Naivety to Insurgency: The Causes and Consequences of Joining a Northern Irish Paramilitary Group. In Canter, D. (Ed.), *Faces of Terrorism: Cross-Disciplinary Explorations*. Chichester: Wiley, pp. 19–33.

Ferguson, N., Burgess, M. and Hollywood, I. (2008) Crossing the Rubicon: Deciding to Become a Paramilitary in Northern Ireland. *International Journal of Conflict and Violence* 2(1), 130–137.

Ferguson, N., Burgess, M. and Hollywood, I. (2010) Who are the Victims? Victimhood Experiences in Post Agreement Northern Ireland. *Political Psychology* 31(6), 857–886.

Ferguson, N., Burgess, M. and Hollywood, I. (2015) Leaving violence behind: Disengaging from politically motivated violence in Northern Ireland. *Political Psychology* 36(2), 199–214.

Ferguson, N., Muldoon, O. and McKeown, S. (2014) A Political Psychology of Conflict: The Case of Northern Ireland. In Kinvall, C., Nesbitt-Larking, P. and Capelos, T. (Eds.), *Palgrave Handbook of Global Political Psychology*. London: Palgrave, pp. 372–389.

Finch, K. (1999) Doing Data: The Local Organisation of a Sociological Review. *British Journal of Sociology* 45, 675–695.

Goodey, J. (2000) Biographical Lessons for Criminology. *Theoretical Criminology* 4(4), 473–498.

Hayes, B. C. and McAllister, I. (2002) Sowing Dragon's Teeth: Public Support for Political Violence and Paramilitarism in Northern Ireland. *Political Studies* 49, 901–922.

Horgan, J. (2005) *The Psychology of Terrorism*. London: Routledge.

Horgan, J. (2009) *Walking Away from Terrorism: Accounts of Disengagement from Radical and Extremist Movements*. London: Routledge.

Horgan, J. (2014) *The Psychology of Terrorism*. London: Routledge.

Jaspers, K. (1970) *Existential Elucidation of Philosophy*, Volume 2. Chicago: University of Chicago Press.

Mac Ginty, R., Muldoon, O. and Ferguson, N. (2007) No War, No Peace: Northern Ireland After the Agreement. *Political Psychology* 28(1), 1–12.

MacNair, R. M. (2002) *Perpetration-Induced Traumatic Stress: The Psychological Consequences of Killing*. Westport, CT: Praeger.

MacNair, R. M. (2005) Violence Begets Violence: The Consequences of Violence Become Causation. In Fitzduff, M. and Stout, C. E. (Eds.), *The Psychology of Resolving Global Conflicts: From War to Peace: Vol. 2. Group and Social Factors*. Westport, CT: Praeger, pp. 191–210.

McGarry, S. R. and Ferguson, N. (2012) Exploring Representations of the Soldier as Victim: from Northern Ireland and Iraq. In Gibson, S. and Mollan, S. (Eds.), *Representations of Peace and Conflict*. Basingstoke: Palgrave, pp. 120–142.

Parker, I. (2005) *Qualitative Psychology: Introducing Radical Research*. Maidenhead: Open University Press.

Patton, M. Q. (1990) *Qualitative Evaluation and Research Methods*. Newbury Park, CA: Sage.

Reinares, F. (2011) Exit from Terrorism: A Qualitative Empirical Study on Disengagement and Deradicalization Among Members of ETA. *Terrorism and Political Violence*, 23, 780–803.

Silke, A. (1998) Cheshire-Cat Logic: The Recurring Theme of Terrorist Abnormality in Psychological Research. *Psychology, Crime and Law* 4, 51–69.

Silke, A. (2001) The Devil You Know: Continuing Problems With Research on Terrorism. *Terrorism and Political Violence* 13(4), 1–14.

Smith, J.A. (1995) Semi-Structured Interviewing and Qualitative Analysis. In Smith, J.A., Harre, R. and Van Langenhove, L. (Eds.), *Rethinking Methods in Psychology*. London: Sage, pp. 9–26.

Smith, J.A. (2008) *Qualitative Psychology: A Practical Guide to Research Methods*. London: Sage.

Smith, J.A. and Eatough, V. (2007) Interpretative Phenomenological Analysis. In Lyons, E. and Coyle, A. (Eds.), *Analysing Qualitative Data in Psychology*. London: Sage, pp. 35–50.

Smith, J.A., Jarman, M. and Osborn, M. (1999) Doing Interpretative Phenomenological Analysis. In Murray, M. and Chamberlain, K. (Eds.), *Qualitative Health Psychology: Theories and Methods*. London: Sage, pp. 218–239.

Smith, J.A. and Osborn, M. (2008) Interpretive Phenomenological Analysis. In Smith, J.A. (Ed.), *Qualitative Psychology: A Practical Guide to Research Methods*. London: Sage, pp. 53–80.

Victoroff, J. (2005) The Mind of the Terrorist: A Review and Critique of Psychological Approaches. *Journal of Conflict Resolution* 49, 3–42.

Wengraf, T. (2001) *Qualitative Research Interviewing*. London: Sage.

White, T. (2013) *Lessons from the Northern Ireland Peace Process*. Madison: University of Wisconsin Press.

Yardley, L. (2000) Dilemmas in Qualitative Health Research. *Psychology and Health* 15, 215–228.

Yardley, L. (2008) Demonstrating Validity in Qualitative Research. In Smith, J.A. (Ed.), *Qualitative Psychology: A Practical Guide to Research Methods*. London: Sage, pp. 235–251.

Yin, R.K. (1989) *Case Study Research: Design and Methods*. Newbury Park: Sage.

13

PSYCHOANALYTICALLY INFORMED REFLEXIVE RESEARCH WITH SERVICE SPOUSES

Sue Jervis

In this chapter I discuss applying a psycho-social, interview-based methodology to military phenomena. In doing so, I aim to illustrate the intertwined nature of the research topic and the method used to study it, with the biographies of both the researcher and respondents. Thus, this will be a personal account of the research process and it will address the practicalities of undertaking such emotionally involved, reflexive research. The account will include a case study of one respondent, 'Sonia', illustrating the interactional context of the research method as well as the data and analysis it produces. In addition, the chapter describes my experience of some of the issues that are specific to doing psychoanalytically informed research in military environments, issues which other researchers might also encounter. I reflect on how my status as a researcher who was also a military wife may have affected gaining access to the military and its reception of the research findings. I reflect, too, on the problem of presenting qualitative data to the military. Finally, I suggest that the emotional responses of service families to the practical, everyday experience of military life require greater attention and further research.

Utilizing qualitative, interpretive methodologies remains unusual in military studies, where quantitative, positivist approaches have dominated (Higate and Cameron, 2006; Boëne, 2008; Jenkings et al., 2011). Higate and Cameron (2006) maintain that military researchers are particularly resistant to using reflexivity. I suggest that this resistance applies especially to reflexive research methods that seek to understand underlying emotions. Historically, military research has often been conducted by veterans habituated to the armed forces' ethos of stoicism and discipline, an ethos that eschews less pragmatic, more emotional attitudes. Such familiarity with what is taken for granted within the group being studied offers valuable insights, but it can also render researchers blind to how those dominant ideas influence group members (Winch, 1958; Finlay, 1998; Alvesson, 1999). Given that the military institution *itself* discourages any engagement with potentially messy and unruly feelings, it is hardly surprising that military research has tended to overlook whatever emotions exist beneath the surface of military communities.

My personal interest in such emotions first arose when I was living abroad as a frequently relocated navy wife. I was married to my husband, an officer in the Royal Navy, for over a decade before experiencing any moves at the military's behest. Thereafter, however, my husband was sent to several foreign postings and I accompanied him to each one. Although I enjoyed living in interesting new places, there was a price to pay. Moving abroad meant leaving behind family and friends, a rewarding career and a pleasantly located home. I anticipated that these personal costs might arouse various uncomfortable feelings. Nevertheless, some of my emotions took me completely by surprise, prompting me to question whether similar reactions are common among trailing service spouses. That questioning determined my research aim: to conduct an in-depth study of the emotional responses of British servicemen's wives to military relocation overseas. A full description of the entire research project is provided in *Relocation, Gender and Emotion: A Psycho-Social Perspective on the Experiences of Military Wives* (Jervis, 2011).

Other life experiences influenced my choice of research method. Prior to relocating overseas, I worked in a field of counselling informed by psychoanalytic thinking. There, I learned that people's emotions are complicated; personal feelings are constantly influenced by social situations, and vice versa, often in unrecognized ways. That is, psychological and sociological factors perpetually intertwine. So, for me, any attempt to understand emotional experiences must also consider the circumstances within which they occur. Consequently, I used an interdisciplinary methodology drawn from a still evolving group of approaches called 'psychosocial studies'.

Research Method

I have argued elsewhere (Jervis, 2014) that, notwithstanding the ethical and interpretive problems that can arise when psychoanalytic techniques and theories are used outside clinical settings, psychoanalysis has much to offer social research. What it offers may be especially helpful for military studies, as I will discuss shortly. Potentially useful to *any* social researcher, however, are three fundamental aspects of psychoanalysis: first, it helps people to talk freely, thereby producing a lot of data; second, it addresses feelings, ideas and processes that are usually outside people's conscious awareness, hence generally inaccessible; and third, it puts a spotlight on the underlying dynamics of relationships, allowing researchers to learn more from their research relationships than what is apparent on the surface. I chose to use the 'reflexive psychoanalytic research methodology' described by Clarke (2000, 2002), which aims to explore aspects of people that ordinarily remain hidden, within their social contexts, to reach deep understandings of their experiences. At the same time as Clarke was developing this research method, Hollway and Jefferson (2013) were developing a similar approach called the Free Association Narrative Interview method, so I draw from both methodologies.

The research practice that I have adopted involves

> qualitative unstructured interviews comprising a few open questions that invite respondents to talk, rather than inducing 'yes' or 'no' responses; the researcher keeping interventions to a minimum; following wherever respondents' ideas lead; encouraging the expression of free association; paying careful attention to the feelings that

respondents, or their material, evoke; and separating out the task of interpreting research data so that it does not take place during the interviews.

The aim of each interview is to elicit narratives through listening empathically to respondents' experiences. Clarke (2002) argues that if researchers avoid asking 'why' questions then respondents are less likely to respond with clichés. Moreover, by using respondents' terminology and word order, researchers minimize the risk of influencing their choice of material. . . . Afterwards: interviews are transcribed in detail and painstakingly analysed; the researcher reflects upon each interview, considering not only what was said but also, as in psychoanalysis, what might have been unconsciously imparted within it; and, theoretical knowledge about sociological and psychoanalytic processes is utilized to make sense of the data.

. . . Clarke (2002, 2006) and Hollway and Jefferson (2000) argue that researchers must become very familiar with their data before attempting to analyse it. Ideally, they should fully transcribe all research interviews personally so that they are immersed in, and thus completely saturated by, the material. This both elicits researchers' own potentially informative free associations and helps them to make appropriate theoretical connections. In addition to considering what respondents actually say about their experiences, researchers should also reflect upon: contradictory material; pauses and gaps in respondents' narratives; any themes that emerge; the order in which issues are raised; sudden changes of subject; what attracts, or discourages, the researcher's interest; and the feelings aroused in the researcher.

(Jervis, 2011: 112–113)

Applying such a comprehensive reflexivity to the minutiae of each research relationship can help researchers to identify underlying emotions and interpersonal dynamics that might inform the research as a whole.

Reflexivity is increasingly being used in social studies to address the inevitable influence of the researcher's personal history (ethnicity, gender, class, beliefs, etc.) on his/her research project. As Higate and Cameron (2006) argue, findings are more credible when researchers reflect transparently and honestly on their personal involvement in their studies. It is difficult, however, for researchers to be completely transparent because so much of what happens within and between people is not consciously recognized. Indeed, psychoanalytic theories suggest that everyone unwittingly defends themselves against engaging with unconscious ideas and emotions that might evoke anxiety (Klein, 1946).

Being 'defended subjects' (Hollway and Jefferson, 2013: 17–19) means that, alongside their biographies, both respondents and researchers bring various unrecognized aspects of themselves into their relationships. A major premise of psychoanalytically informed reflexive research is that such elements are sometimes expressed through nonverbal 'projective communications' (Clarke, 2000: 146). These largely unconscious interpersonal dynamics, which are similar to processes called transference and countertransference in psychoanalysis, can communicate an individual's unrecognized ideas or feelings by arousing similar states of mind in other people (Hollway and Jefferson, 2013: 158–164). Hence, researchers who remain receptive enough to empathically identify with anything that is unconsciously communicated by respondents, and then take time to reflect on their own emotional reactions to all research communications, might discover parts of respondents' experiences that, being outside conscious awareness, could not have been put into words (Clarke, 2008; Hollway, 2008). That is, by adopting an attentive open-mindedness, comparable to that used in psychoanalysis, researchers may be able to *feel* something similar to what respondents feel, or have previously felt. For example, occasionally during my research I experienced perplexing emotions such as sadness or anxiety. After reflecting on those apparently inexplicable feelings, however, I realized that they were not mine alone; they also mirrored certain aspects of respondents' experiences that had not been verbalized. So, by treating the unusual emotions evoked in me as additional data to be thought about, my attention was drawn to feelings that respondents had experienced but not acknowledged.

Various emotional experiences typically pass unnoticed within military families and their communities, which is precisely why they warrant close examination. After all, how service families feel affects not just their well-being but the military task, too. A major reason why I chose a psychoanalytically informed reflexive research method was that it provided a means of uncovering and exploring these potentially informative yet largely hidden emotions. For me, this capacity to expose and investigate feelings that would ordinarily remain unrecognized, or concealed, is just one of the important contributions that psychoanalysis can make to military research.

In addition, I suggest, psychoanalytic theories can help military researchers to actively address their own unavoidable involvement in the research process. Since military researchers so often have a personal connection with the armed forces, finding ways to consider how their experiences and feelings might influence their studies, not just consciously but also *unconsciously*, can help to ensure that research conclusions are as reliable as possible. As already intimated, however, the very nature of unconscious influences makes them difficult to spot. Even though I had spent several years attending to unconscious processes in my former work as a counsellor, I still reacted to some respondents in ways that I did not immediately notice, which could have had a detrimental impact on my research findings.

To explain: being a serviceman's wife myself, and hence an 'insider researcher', I knew that members of military communities seldom complain about their challenging lifestyle: they 'just get on with it'. I was not surprised, therefore, when several respondents were philosophical about significant personal losses that they had sustained through accompanying service personnel to foreign postings. I assumed that these matter-of-fact accounts might conceal some distress. What I initially failed to recognize, however, was the depth of that distress. As I have discussed elsewhere (Jervis, 2011, 2012a, 2014), one reason for my initial blindness to the extent of some respondents' suffering was that I was as influenced as they were by the stoic ethos and social norms of the military community to which we all belonged. Consequently, I occasionally (unwittingly) colluded with respondents to minimize their pain, unconsciously complying with what I later came to understand as the community's taboo against 'whingeing'.

It was only through feeling, for myself, something akin to the experiences of individual respondents that my understanding of their underlying emotions, and also an awareness of my own collusiveness in minimizing those painful feelings, really began to develop. Using a psychoanalytically informed reflexive research method was vital, then, in ameliorating unhelpful aspects of my personal involvement in my research while simultaneously facilitating an emotional connection with respondents which was to prove informative. Various unusual responses were evoked in me that ultimately helped me to recognize the deep extent of respondents' distress and some of the ways in which that distress was disregarded within the military community. Later, I will illustrate this dynamic by discussing one of the research relationships in which it occurred. First, I want to describe the context of my research and how I went about obtaining a representative sample of respondents.

Research Context

I conducted my research with civilian spouses of British Armed Forces personnel posted to a NATO establishment in northwestern Europe. Before starting my interviews I undertook a small pilot exercise, issuing questionnaires to ten spouses. Two questionnaires went to acquaintances and eight to randomly chosen occupants of Service Family Accommodation. I should explain that it remains common for British service families to reside in military-controlled housing; indeed, 61% of military spouses who responded to a recent official survey (MoD, 2013: 4) did so. Following an overseas posting, almost all service families live in such housing, or in locally rented substitutes known as hirings, often seen for the first time on the day that families move in. By visiting Service Family Housing estates to hand out the questionnaires personally, I was able to explain that I was a civilian researcher, acting independently of the military.

The questionnaires asked how many times, and over what period, respondents had moved to new locations with service personnel; then respondents were invited to describe their feelings about military relocation. I hoped that beginning my research with the questionnaires would help me in two ways: first, by identifying a few broad themes to investigate during my subsequent interviews; and second, by providing access to individuals who might be willing to participate in those interviews. The seven respondents who returned completed questionnaires had relocated between one and eleven times.

The questionnaire respondents reported a diverse range of both positive and negative emotional reactions to military relocation. One common theme to emerge, however, was loss. Respondents had left behind familiar environments, careers or other valued roles, and friends and family members, including some children who were sent to boarding school to protect their education from the discontinuity caused by repeated relocation. In addition to these losses, respondents reported losing some of their independence. One contributory factor to this undermining of respondents' personal autonomy might have been that they had no tenancy rights in relation to the military housing that all of them now occupied. Tenants' 'licences' are granted only to service personnel, not to their civilian spouses or partners (MoD, 2014). Consequently, respondents became reliant upon others to arrange things for them, or to grant authority for certain life choices that they would have been able to organize or authorize for themselves in a civilian environment. For instance, if they wished to run a business from home, or keep a pet, they would have to obtain official permission to do so (MoD, 2014).

Given the various losses that they had sustained, I was not surprised to find that several respondents had experienced upsetting post-relocation anxieties similar to the disturbing feelings typically found among the recently bereaved. I had anticipated that, since it necessarily involves loss, military relocation might arouse this sort of emotional upheaval. By confirming not just that such feelings existed, but also that they were at odds with how respondents usually felt, the questionnaires served to affirm my initial hypothesis.

Unfortunately, however, the questionnaires did not yield the hoped-for sample of potential interviewees; only two respondents were willing to be interviewed. Perhaps the others were reluctant to talk in any depth about their experiences for fear of becoming too distressed (Jervis, 2011: 125). Whatever the explanation, my use of questionnaires failed to produce a large enough sample for the interview phase of my research, and it was now necessary for me to seek more respondents.

Luckily, I was able to identify additional respondents by making several personal approaches to individuals who then connected me with other potential interviewees. This process of snowballing enabled me to obtain a sample of fifteen respondents: fourteen women and one man. The respondents – ranging in age from twenty-six to sixty, with partners holding various commissioned or other ranks in the Royal Navy, the Royal Air Force or the Army – provided a reasonably representative, albeit small sample. Between them, they had experienced from one to twenty-three military relocations.

My original intention was to interview each of the fifteen respondents just once. However, after noticing that my emotional equilibrium was less affected when I interviewed respondents who felt settled in the new location than it was while interviewing newcomers, I decided to interview ten respondents, some newcomers and some not, twice. The follow-up interviews allowed me not only to compare whether any of those ten respondents' feelings changed in the interval between our two meetings, but also to compare my personal reactions to respondents still experiencing the turbulence of adapting to their new environment with my reactions to those who had already negotiated that unsettling process. Through these comparisons, I hoped to learn to what extent levels of anxiety differed among respondents at various stages of post-relocation adjustment. So, I conducted a total of twenty-five interviews.

I started each initial interview by extending an invitation along the lines of 'tell me about your experience of military relocation'. This invitation elicited stories from most respondents, who moved naturally from one theme to the next as new associations occurred to them. A few respondents, however, said very little unless prompted, so I encouraged them to say more about their experiences by asking open questions. The average length of each interview was an hour, although two interviews lasted for approximately two and a half hours while another two ended after only about half an hour.

All fifteen respondents gave me permission to tape-record our interviews, and I transcribed each recording personally, producing 450 pages of research material. Other data that I collected included factual information, such as details of respondents' life histories and how often they had relocated; notes about my emotional reactions to respondents and personal associations to their narratives; any preoccupying questions relating to respondents, our relationships or the process of each interview; and anything else that caught my attention. Added to this data were two accounts of relocation experiences sent to me by servicemen's wives who learned about my research through an academic acquaintance, and also an account by another woman who wrote, anonymously, about her feelings on military moves in an article (Anon, 2005) that I happened to read. Although it wasn't possible for me to interview these three women, having their narratives – which bore strong similarities to those of several

interviewees – provided helpful 'triangulation' as extra data and validated the evidence produced during the interviews.

I carefully considered all of this material, beginning with a thorough examination of what
respondents had told me. By repeatedly shifting my focus between each individual interview
and the data as a whole, I discovered several recurring themes. Next, I looked for anything
that might indicate unconscious influences or hidden meanings, focusing as much on the presentation of respondents' stories as on their content. Uncovering the defences that respondents unconsciously use to guard against encountering unacceptable thoughts or feelings can
help researchers to understand respondents' narratives (Hollway and Jefferson, 2013). I hope
that the following extracts from my interviews with a respondent whom I have called Sonia (a
pseudonym) will illustrate how reflecting on the *process* of those interviews highlighted some
of the unconscious dynamics that influence servicemen's wives.

Sonia's Story

Accompanying her husband to an overseas military posting presented Sonia with various
problems that were common among respondents, including distress about leaving loved relatives in Britain; worries around children's disrupted education; and difficulty finding employment as a trailing spouse. Being allocated a 'terrible' flat increased Sonia's homesickness and
for several months after relocating she was 'so depressed'. My focus here, however, is less on
what Sonia said and more on how she conveyed her experiences, which, I suggest, ultimately
communicated something deeper.

When I arrived at Sonia's home her husband was there too. Declining my offer to return
another day, Sonia invited me into a room where her husband was working. He remained
throughout our interview and I became increasingly conscious of his presence. Although Sonia
seemed to describe her experiences openly, she sometimes glanced towards her husband or
asked him to confirm details; I speculated about whether Sonia might have said more had he
not been there. Interestingly, Sonia's husband was not present during our second interview,
which lasted almost twice as long as the first.

Like other respondents, Sonia occasionally laughed when describing painful experiences.
As my research progressed, I realized that this inappropriate laughter was an unconscious
defence that enabled respondents to comply with the stoicism expected of servicemen's
wives by laughing off the emotional significance of their painful losses. Similarly, unwitting
contradictions sometimes pointed to underlying distress. A particularly obvious example in
Sonia's story was that she said 'I feel fine, yeah, I'm happy,' only to confide moments later that
she still had 'a lot of bad days . . . I can be alright and then . . ., other days, I can be so unhappy.'

Sonia's presentation of herself included other contradictory elements, too. Describing
herself as 'strong' and able to 'say what I want to say', Sonia told me a lot about her personal
experiences, and related grievances, as a serviceman's wife. One such grievance was her
irritation with officers' wives who 'don't think they should speak to anybody lower ranking'
or who 'look down their nose at you'. Being an officer's wife myself, I felt slightly uncomfortable hearing those remarks and wondered privately what Sonia, whose husband belonged to
the 'other ranks', intended me to understand from them. Sonia subsequently explained that,
in fact, her small group of friends included a senior officer's wife who 'doesn't care about

that' and was 'just one of us'. Perhaps, then, Sonia was trying to establish how much *I* cared about the different hierarchical positions occupied by wives within the military community and whether or not I was open to listening to all respondents equally, irrespective of their partner's rank. In response, I was impressed by Sonia's apparent outspokenness. However, her narrative suggests that Sonia was less forthright elsewhere in the military community. For instance, far from challenging a situation that she described as 'not very fair', Sonia explained that she had 'to keep quiet otherwise I say something' – implying that, if she said anything, Sonia might say more than she ought. Maybe Sonia's circumspection, outside the relative safety of a research interview, was influenced by her husband's junior rank. My impression, however, was that while Sonia clearly found the military community hierarchy irksome, it carried less significance for her than her status as a *wife*.

Sonia repeated several times that servicemen's wives' problems are simply not considered. It certainly seemed that no one had really listened to Sonia's own concerns. Apparently even her husband had said, after hearing Sonia speak to me, 'I didn't realize that you felt like that' – an obliviousness possibly explaining his response several months earlier when finding Sonia crying, which she reported as: 'oh, I'm sending you 'ome' (i.e. back to Britain) rather than, more supportively, 'let's talk about this.'

When Sonia *did* speak out in the wider military community, her suffering was dismissed. Soon after arriving overseas she had approached the military housing authorities to seek accommodation more suited both to her family's needs and her own poor health. However, Sonia explained, 'everybody I turned to, didn't want to know.' She told an official how depressed she felt about her housing difficulties but said he replied, 'You've got a flat [i.e. apartment], what're you moaning about?' Sonia indicated that she felt as though she had been swatted away like a troublesome fly.

Sonia told me that after our first interview her husband accused her of saying 'too much'. She provided even more details about her feelings during our second interview but she increasingly referred to other people's emotional experiences, too, saying, for example, 'a lot of people are not happy' or 'a lot of people miss home.' Maybe Sonia's repeated inference that she was speaking for 'a lot of people' within the military community was a defensive response to her husband's accusation. It struck me, however, as being at odds with the limited social interaction that she described: 'I've got a couple of . . . basically stick to the same few friends . . . I don't really mix with a lot of people.'

There were other forms of repetition in Sonia's narrative that I overlooked initially. In our first interview Sonia mentioned her 'tears', 'homesickness' and 'depression' several times. Moreover, she often repeated phrases within the space of a single sentence, such as 'I was in tears every day, in tears every day.' When I noticed the frequency of Sonia's repetition, alongside her numerous references to how other people felt, I wondered whether both features were intended to add weight to Sonia's account of her experiences.

Whatever Sonia's conscious intention, however, I now believe that those unusual features of our interviews, together with her mixed messages about herself, were an unconscious attempt to communicate a dilemma that Sonia shared with other respondents about what can and cannot be said, or heard, within military communities. Although most respondents described painful experiences to me, they seldom discussed their feelings publicly for fear of being perceived as 'moaners' or 'weak'. Similar anxieties might explain why Sonia tended to be circumspect in public and to emphasize her strength (an attribute valued in military communities) while playing down her distress through inappropriate laughter and contradictions.

A possible explanation for why Sonia wasn't heard, when she talked about her feelings publicly, emerged as I revisited the transcriptions of our interviews, long after our meetings. This time, Sonia's many repetitions and allusions to others struck me forcibly, evoking a powerful reaction which I had not experienced previously. I suddenly felt unusually weary. My weariness might have been an unconscious resistance to re-engaging with Sonia's painful emotions. However, the extraordinarily soporific impact of reading her narrative again made me realize that, although I had empathized with Sonia during our interviews, her story now read to me as a litany of boring grievances. Rather than reinforcing Sonia's account of distressing aspects of military life, her repetition now had the opposite effect, diminishing my empathy with her suffering.

Of course, during my research project I *wanted* to learn about Sonia's feelings. Afterwards, however, how she had conveyed those feelings adversely influenced my interest.

I wonder, then, whether something similar occurred when Sonia described her emotions to people already uninterested in them, such as the housing official who dismissed her unhappiness as 'moaning'. I am suggesting that Sonia might have unwittingly sabotaged her attempts to be heard by, quite literally, saying 'too much' in a social context reliant on quiet stoicism and dutiful compliance.

Whether or not Sonia's narrative delivery is regarded as an unconscious communication, it highlighted for me a dynamic that typifies the dilemma faced by servicemen's wives. If they stay silent about the problems inherent in military lifestyles, their difficulties remain unrecognized and unresolved. Consequently, as Sonia said, 'wives get forgot about.' If, however, wives voice problematic issues, to seek solutions, they risk being perceived as disloyal 'moaners' and censured. This dilemma suggests that considerable anxiety arises in military communities when ordinarily concealed emotions are revealed. Moreover, it inevitably influences negotiations between the military and researchers engaged in exploring service families' underlying feelings.

Negotiating Research With the Military

Given the historic resistance in military communities generally, and military research specifically, to exploring service families' emotions, I was mildly apprehensive when as a courtesy I wrote to a senior military commander informing him about my research plans. To my relief, I received a supportive response. Later, I was grateful for similar cooperation from military staff, and from related data-gathering organizations, who provided useful statistical information. While the individuals with whom I communicated were helpful, few gave any impression of being especially interested in my research, so I suspect that their helpfulness derived solely from professional politeness. These contacts would have been aware, too, that I was entitled to obtain certain information under the Freedom of Information Act 2000 (UK Government, 2014).

I encountered some resistance, however, when I submitted an inquiry about the incidence of depression among relocated servicemen's wives to a British Forces Health Service. I was told that my question could not be answered, partly because no records were kept about spouses' mental health but also because my research required prior approval from the Ministry of Defence Research Ethics Committee (MoDREC).[1] Challenging the need for such approval, I explained that first, all respondents were civilians, who had provided informed consent; second, the MoDREC did not exist when my research began; and third, ethical approval had been granted by the university monitoring my studies. This explanation apparently satisfied the MoDREC.

Recently, however, some researchers have encountered more serious hindrances to conducting research with civilian spouses of service personnel. For instance, Blakely (2012) describes how, even though its own terms of reference suggested otherwise, the MoDREC insisted that her research required its approval. The MoDREC's extremely broad rationale was that Blakely's research 'could have an impact on MoD personnel' (2012: 90). Another researcher was denied access to spouses because a garrison commander feared that their participation in research might stir up 'discontent' (Jenkings et al., 2011: 45).

Clearly, then, some sections of the Ministry of Defence (MoD) and military perceive research into service spouses' lived experiences as potentially disruptive. Since spouses' feelings about military life arouse such anxiety, perhaps I should have anticipated that my research findings would not be universally welcomed. One response to my attempts to highlight wives' concerns and to suggest initiatives aimed at improving their satisfaction with service life left me feeling utterly dismissed, much as Sonia felt after describing her housing problems. I had sent a detailed executive summary of my research, along with an additional paper couched in less esoteric academic language, to the then deputy chief of defence staff (personnel). In his very short response, this senior officer included a remark to the effect that he recognized the necessity of looking after me and my 'chums' so that we wouldn't encourage our husbands to leave the military.[2]

I felt patronized and also disappointed that my research findings had not been taken seriously. On reflection, however, I wonder whether I might have made a mistake similar to Sonia's. Perhaps, in presenting my research findings as I did, I too unwittingly said too much. One might argue that my research was always going attract some disapproval in military circles because I was not only a researcher but also a military wife, and thus someone who was expected to be stoic and keep quiet about the anxiety-provoking emotions aroused by problems associated with service life, rather than write unsettling academic papers about them. However, I had not anticipated this potential problem (for a discussion about other pros and cons of my being an insider researcher, see Jervis, 2011: 119–121). Another possible explanation for why my research was so swiftly dismissed by this high-ranking officer is that such senior personnel tend to be so busy that they become accustomed to reading very short briefing documents, quite different from the research reports that I had submitted. Required to respond to life-and-death scenarios with practical and decisive action, senior officers simply do not have enough time to read lengthy papers that discuss detailed arguments. Instead, they rely on their staff to analyze the facts of a given situation and to present a range of options, including the likely outcomes and costs involved in each case. This established way of working might fit well with traditional quantitative military research methods that tend to produce snapshots of or statistical evidence about military life, but it presents a significant problem to qualitative military researchers who seek to communicate more nuanced accounts of the hitherto largely unexplored feelings experienced by service families.

While I was unsuccessful in my initial attempt to disseminate my research findings to a senior officer, whom I had hoped would use them to influence future policy, other military figures have engaged with my work. Subsequently, I have been asked about my research in several service settings, both in the UK and overseas. This interest has often come from personnel and associated welfare support staff whose responsibilities include the well-being of service families. Sometimes I have been asked to provide an electronic copy of my research, or an executive summary, which I have been happy to do. I found it striking, however, that one such request was accompanied by the disarming comment, 'I'm always on the look-out for free research.' On another occasion, I was asked 'which *single* research finding' I considered the most important to improve things for service wives. At face value, these remarks seem reasonable. However, for me they evidence a continuing attempt by the military to find inexpensive quick fixes for complex family issues that need to be much better understood before they can begin to be resolved. In order to achieve this understanding, I suggest that there needs to be greater willingness by the military to invest time and resources into first commissioning in-depth explorations of how today's service families really feel, and second, carefully studying all of the findings that emerge from such research.

The value of conducting detailed, qualitative studies is already recognized by some organizations that support military families. Certainly, each of the three armed forces Families Federations has expressed interest in my own research findings. In stark contrast to the dismissive response given by the senior military official mentioned earlier, the then chair of the RAF Families Federation found the same executive summary of my research 'a compelling read' and asked to see my entire thesis.[3] She subsequently endorsed the research, stating that all relocated partners of service personnel would recognize some of the feelings that it describes and that my recommendations could contribute significantly to improving service partners' welfare (McCafferty, 2011). Later, Recruit for Spouses, an organization working to improve service spouses' employment opportunities, used my research to evidence the detrimental impact of military relocation on the career prospects of trailing spouses and, consequently, on military family life (Jervis, 2012b). It is gratifying that these organizations, which represent so many service spouses, have affirmed my research findings. Perhaps the most important affirmation, however, has come from the numerous individual spouses who have told me that my research describes how they have felt.

Concluding Remarks

In addition to describing the process of conducting psychoanalytically informed reflexive research with service spouses, this chapter has suggested that research on the military is resistant to exploring the diverse feelings and emotional complexities of personnel and their families. One consequence of this resistance is that important welfare issues, which could impact on the military task, are being routinely overlooked. For example, the MoD conducts annual 'families' continuous attitude surveys' to elicit service spouses' opinions about certain aspects of military life, but the surveys completely disregard families' emotional needs (Fossey, 2012: 11). Those needs, which in common with other emotions are seldom discussed publicly in military communities, must become better understood if service families are to be properly supported. Since a commitment to provide such support is now enshrined in law by the armed forces covenant (MoD, 2011), future military research must find ways to enable the emotional responses of service spouses and families to military life to be fully explored and then heard. Given that wives often become silenced in military communities, psychoanalytically informed reflexive approaches that can both explore under the surface and address any unconscious influences operating there have the potential to achieve deep understandings of emotional experiences that would otherwise continue to remain unnoticed and unexpressed.

Acknowledgements

I would like to thank the editors for their comments on earlier drafts of this chapter, and Karnac Books for kindly giving permission for me to use previously published material.

Extracts from: Jervis, S. (2011). *Relocation, Gender and Emotion: A Psycho-social Perspective on the Experiences of Military Wives*. London: Karnac. Reprinted with permission.

Notes

1   Response to an enquiry about the incidence of depression among recently relocated military wives. Lt Col. N. Cooper, personal communication, 2 April 2007 (email).
2   Response to an executive summary of research. Vice Admiral P. Wilkinson, personal communication, 26 May 2009 (email).
3   Response to an executive summary of research. Gp Capt. D. McCafferty (Ret'd), personal communication, 1 June 2009 (email).

References

Alvesson, M. (1999) *Methodology for Close Up Studies: Struggling With Closeness and Closure*. Institute of Economic Research Working Paper Series. Lund, Sweden: Lund University.

Anon (2005) A Bird's Eye View. *British Community News*, September, 6.

Blakely, G. (2012) *The Impact of British Military Foreign Postings on Accompanying Spouses*. Unpublished doctoral thesis. University of Plymouth.

Boëne, B. (2008) Method and Substance in the Military Field. *European Journal of Sociology* 49(3), 367–398.

Clarke, S. (2000) On White Researchers and Black Research-Participants. *Journal for the Psychoanalysis of Culture and Society* 5(1), 145–150.

Clarke, S. (2002) Learning from Experience: Psycho-Social Research Methods in the Social Sciences. *Qualitative Research* 2(2), 173–194.

Clarke, S. (2006) Theory and Practice: Psychoanalytic Sociology as Psycho-Social Studies. *Sociology* 40(6), 1153–1169.

Clarke, S. (2008) Psycho-Social Research: Relating Self, Identity and Otherness. In Clarke, S., Hahn, H. and Hoggett, P. (Eds.), *Object Relations and Social Relations*. London: Karnac, pp. 113–135.

Finlay, L. (1998) Reflexivity: An Essential Component for All Research? *British Journal of Occupational Therapy* 61(10), 453–456.

Fossey, M. (2012) *Unsung Heroes: Developing a Better Understanding of the Emotional Support Needs of Service Families*. London: Centre for Mental Health.

Higate, P. and Cameron, A. (2006) Reflexivity and Researching the Military. *Armed Forces and Society* 32(2), 219–233.

Hollway, W. (2008) The Importance of Relational Thinking in Psycho-Social Research. In Clarke, S., Hahn, H. and Hoggett, P. (Eds.), *Object Relations and Social Relations*. London: Karnac, pp. 137–161.

Hollway, W. and Jefferson, T. (2000) *Doing Qualitative Research Differently: Free Association, Narrative and the Interview Method*. London: Sage.

Hollway, W. and Jefferson, T. (2013) *Doing Qualitative Research Differently: A Psychosocial Approach* (2nd edition). London: Sage.

Jenkings, K. N., Woodward, R., Williams, A. J., Rech, M. F., Murphy, A. L. and Bos, D. (2011) Military Occupations: Methodological Approaches and the Military-Academy Research Nexus. *Sociology Compass* 5(1), 37–51.

Jervis, S. (2011) *Relocation, Gender and Emotion: A Psycho-Social Perspective on the Experiences of Military Wives*. London: Karnac.

Jervis, S. (2012a) Parallel Process in Research Supervision: Turning the Psycho-Social Focus Towards Supervisory Relationships. *Psychoanalysis, Culture and Society* 17, 296–313; advance online publication, 1 September 2011. doi:10/1057/pcs.2010.36.

Jervis, S. (2012b) *The Effect Unemployment Has on Military Family Life*. Unpublished paper presented at the 'Recruit for Spouses' Launch of the Military Spouses Business and Employment Campaign, 7th November. London: House of Commons.

Jervis, S. (2014) Precious Gift or Poisoned Chalice: What Does Psychoanalysis Offer to Social Research? In Cullen, K., Bondi, L., Fewell, J., Francis, E. and Ludlam, M. (Eds.), *Making Spaces: Putting Psychoanalytic Thinking to Work*. London: Karnac, pp. 149–166.

Klein, M. (1946) Notes on Some Schizoid Mechanisms. In: Mitchell, J. (Ed.), *The Selected Melanie Klein*. Harmondsworth, Middlesex: Penguin, 1991, pp. 176–200.

McCafferty, D. (2011) Cover Endorsement. In Jervis, S., *Relocation, Gender and Emotion: A Psycho-Social Perspective on the Experiences of Military Wives*. London: Karnac, back cover.

MoD (2011) *The Armed Forces Covenant*. London: Ministry of Defence.

MoD (2013) *Statistical Series 6 – Other Bulletin 6.04 – Tri-Service Families Continuous Attitude Survey 2013*. Available at: https://www.gov.uk/government/uploads/system/uploads/attachment_data/file/285148/2013-main-report.pdf (Accessed 25 March 2014).

MoD (2014) *JSP 464 Tri-Service Accommodation Regulations (TSARs)*. Available at: https://www.gov.uk/government/publications/jsp-464-tri-service-accommodation-regulations-tsars (Accessed 25 March 2014).

UK Government (2014) Freedom of Information Act 2000. Available at: http://www.legislation.gov.uk/ukpga/2000/36/contents (Accessed 23 April 2014).

Winch, P. (1958) *The Idea of a Social Science; and Its Relation to Philosophy*. London: Routledge and Kegan Paul; New York: Humanities Press.

14

ETHNOMETHODOLOGY, CONVERSATION ANALYSIS AND THE STUDY OF ACTION-IN-INTERACTION IN MILITARY SETTINGS

Christopher Elsey, Michael Mair,
Paul V. Smith and Patrick G. Watson

Introduction and Chapter Summary

In this chapter we discuss what ethnomethodology and conversation analysis can contribute to studies of the military, specifically understandings of 'action-in-interaction' in military settings. The chapter is methodologically focused and explores how work in ethnomethodology and conversation analysis provides an alternative way of approaching the problems posed in studying the different forms of practice that constitute 'soldierly work'. Rather than approach these issues in the abstract, and in line with the central thrust of ethnomethodological (e.g. Garfinkel, 1967, 2002; Heritage, 1984; Lynch, 2007) and conversation analytic studies (e.g. Heritage, 1995; Sacks, 1995; Pomerantz and Fehr, 1997; Schegloff, 2007), we shall outline this approach through a discussion of the methods employed, and difficulties encountered, in the course of research we conducted into a specific case. This was a fatal 'blue-on-blue' or 'friendly fire'[1] attack on British infantry by American aircraft during the second Gulf War (see Mair et al., 2012; Mair et al., 2013). What initially drew us to the incident was the availability of a cockpit videotape that was leaked to the public during a controversial coroner's inquest in 2007, some four years after the attack took place. Crucially this videotape contained the audio communications between the two pilots involved in the attack and the ground forward air controller (GFAC) they were working with, providing unparalleled access to such an incident as it unfolded. Our interest in the footage was twofold. We wanted first to see what insights we could glean from data of this kind about combat as experienced first-hand, 'first-time-through'.[2] Second, we wanted to look at what the three official inquiries made of the incident (including two military boards of inquiry, alongside the coroner's inquest) and explore how they had used (and problematised) the video as a resource for analysing the actions of the pilots.

This methodological strategy reflects the 'duplex' forms of analysis that ethnomethodology and conversation analysis rest upon (Watson, 2009): in this case, an analysis of the pilot's communicative and sense-making practices coupled with an analysis of locally situated reconstructions of those practices by a number of authoritative auditors. This analysis of members' reconstructions of practices, rather than ours as researchers, involved us 'tacking' between the video and after-the-fact accounts of what the video could be said to show. In order to explain how we proceeded, we will initially discuss the problems we encountered in transcribing the video and what those difficulties themselves revealed about what the pilots were doing. After that, we turn to the ways in which we established links between the video and the reports published by the official inquiries, reports which offered competing and apparently conflicting interpretations of what happened and why. Through two examples we shall suggest, again, that this reveals something about what is involved in holding military operatives to account morally and legally, but also in opening up soldierly work to view, making it accountable (i.e. observable, reportable), and so available for inspection and evaluation in specific settings like military tribunals, courtrooms or even in the workspaces of journalists or academic researchers. Based on this, and having linked our research to wider work in the field as we go, we will conclude, finally, by returning to the question of what ethnomethodological and conversation analytic research adds to our understanding of action-in-interaction in military settings: namely, a focus on its specificities and the forms of organisation internal to it.

The chapter has a simple trajectory. The first section provides an account of how we came to embark on our work, providing an overview of previous military research informed by ethnomethodology (EM) and/or conversation analysis (CA) to demonstrate the types of data available and the scope of work in this field. We do so to outline the guiding methodological principles of these research traditions and the types of claims they seek to make, as well as to highlight lines of similarity and difference with other work in the field. Having provided some background, we then explore common research 'moves' and provide demonstrations using three empirical examples. Demonstration 1 looks at different transcription conventions and how presentational issues impact the reading and understanding of the materials documented in transcripts of action-in-interaction during combat (Mair et al., 2012). Demonstration 2 looks at the coupling of different data sources (the cockpit video and military reports) to explore how questions of 'evidence' and first-hand experience are resolved in military settings (Mair et al., 2013). Finally, demonstration 3 looks at how military practice is opened up and explained in practical terms via materials gathered under questioning.

Ethnomethodology, Conversation Analysis and Studies of Soldierly Work

In early 2007, like many others, our attention was caught by a controversy developing around a UK coroner's inquest into the death of a British infantryman, Lance-Corporal of Horse Matthew Hull, following an attack on his unit by American A-10 fighter planes just a few days into the second Iraq War. Andrew Walker, the Oxfordshire assistant deputy coroner who was overseeing the inquest, had requested that *all* of the information held by the UK and US forces relating to the case be released to him so he could arrive at a verdict. This request was not met. After a series of tussles that pitted judicial process against governmental interest, the US and UK eventually conceded the existence of additional evidence and turned over a cockpit video showing the attack from the perspective of one of the two pilots involved as well as the reports from two inquiries into the incident – the reports of the US Air Force's Friendly Fire Investigation Board and of the UK Ministry of Defence's Board of Inquiry (British Army, 2004). The video was subsequently leaked to the *Sun* newspaper in the UK, before it and the reports were finally and grudgingly released into the public domain later that year.

As the reports showed, the US and UK militaries had concluded that no one was ultimately to blame for the attack because the attack was an accident. By contrast the assistant deputy coroner Andrew Walker, based on the same evidence, concluded L. Cpl. Hull's death was an unlawful killing, an illegal act for which the pilots were criminally responsible. The UK press sided with the coroner and, in the period which followed Walker's verdict, the pilots' actions as well as the military response to them were widely denounced.

The controversy piqued our analytic curiosity. We were particularly interested in how one and the same piece of evidence – the video – could be marshalled in support of, not just divergent, but seemingly *incompatible* interpretations of what it could be said to be evidence of: the conclusions of the military boards versus the coroner's verdict and media commentary. An initial inspection of the video confirmed our suspicion that matters were not as clearcut or unequivocal as those promoting a given reading might suggest – events on the video were complex and far from easy to unpack – and we decided to pursue the issues further.

Just what the video could be said to show when viewed in particular ways for particular practical purposes became our focus, and we began to gather materials relating to the case from a variety of quarters. In addition to the video and the 'official' transcript of the exchanges between the military personnel involved (which we found to be frequently misleading and inaccurate – deficiencies we sought to remedy in producing our own transcript; see demonstration 1 as well as Mair et al., 2014), we obtained copies of both military board reports (after some digging on the UK Ministry of Defence website) and, from a different source, the minutes of the coroner's verdict. We also followed media coverage as well as emerging academic responses to the incident and its subsequent fallout. As we have no military background or training, we were particularly interested in instructional resources. Alongside useful media work, we thus read military training and field manuals and official literatures relating to the conduct of battlefield operations. But we also sought out and interviewed personnel either involved in the case or with experience of military air operations and/or the law to ask them to talk to us through the video to give us insights into how they made sense of what the pilots were doing. These interviews aided our understanding of military language and terminology in use, as well as insights into the various types of mission and their import, and thus provided background information from which our analysis could proceed.[3]

The distinctive character of our ways of working did not, however, lie in the data we drew upon. Despite bringing together the most comprehensive and diverse body of materials relating to this particular case that we know of, working with video, textual and interview data is by no means unusual. Nor were we the only academic researchers working on this particular case. The incident and the controversy which centred upon it have received a great deal of academic (let alone journalistic) coverage (see e.g. McHoul, 2007; Masys, 2008; Caddell, 2010; Howe et al., 2010; Kirke, 2012), including conversation analytic work (e.g. Nevile, 2009, 2013). Moreover, as there were no direct ethnomethodological precursors to our study, in developing our analyses we drew on work by those with a range of different approaches in order to get our analytical bearings: ethnographic studies (e.g. King, 2006; Hockey, 2009), organisational studies (e.g. Snook, 2002; Kirke, 2012), the 'normal' accidents literature (e.g. Hicks, 1993) and studies (including ethnomethodological and conversation analytic studies) of inquiries into the conduct of war and military operations (e.g. Benson and Drew, 1978; Lynch and Bogen, 1996; Boudeau, 2007, 2012; Rappert, 2012), as well as studies of representations of battle, the soldier's work and their public reception (e.g. Sacks, 1995 ['Navy Pilot' example pp. 205–222, 306–311]; Brown, 2008; Woodward and Jenkings, 2011; Mieszkowski, 2012). While we have taken a great deal from studies in these areas, we nonetheless depart from them in particular ways, and it is in those ways that whatever distinctiveness we can claim for our work lies.

Rather than produce an analysis of our own over and above the in situ accounts of the parties directly involved in the incident or involved in evaluating its consequences after the fact, we were interested in the methodic practices employed by the pilots, coroners and military investigators and how *they* (co)produced analyses and exhibited, demonstrated and displayed those analyses *as part of the work* they were engaged in, making those analyses *publicly available* to us as analysts in the process (Rizan et al., 2014). In a study of this kind, as Schegloff puts it (1991: 50–52),

> characterizations of [what] the participants [are doing has to] be grounded in aspects of what is going on that are demonstrably relevant *to* the participants, and at that moment – at the moment [that is] that whatever we are trying to provide an account of occurs . . . [the crucial question being] how to examine the data so as to be able to show that the parties were, with and for one another, demonstrably oriented to those aspects of who they are, and those aspects of their context, which are respectively implicated in the 'social structures' which we may wish to relate to [their actions and interactions].

This is no simple feat and, in the three demonstrations that follow, we want to show how such a study might be built up, focusing initially on the work of the pilots before showing how that work was taken up and analysed in the investigative context of the military boards.

Three Demonstrations

Demonstration 1: First Moves With Data

An obvious first move when working with audiovisual material – whether for research or other purposes – is to transcribe it. One of the great advantages of audio and video data, something long recognised within ethnomethodology and conversation analysis, is that it is possible to replay and scrutinise the interactions such materials capture, homing in on details that would be missed were we to solely rely on, for instance, our own impressions of what people were saying and doing. Transcripts are extremely useful in this, making it possible to identify particularly interesting aspects of the materials at hand and revisit them again and again as they play out in real time. It has been the 'abiding preference' of ethnomethodological and conversation analytic research to work with transcribed audiovisual materials precisely because of this – and such studies have generated deep insights into interaction as a social practice. However, as we shall go on to discuss, transcripts are never entirely analytically 'innocent' or neutral records: compiling a (written) representation of the material, entextualising it (Watson, 2009: 10), is always undertaken in the light of some particular set of practical aims. As such different ways of presenting interactional exchanges have consequences for how the exchanges themselves can subsequently be made sense of and understood, often in quite subtle but powerful ways.

As mentioned earlier, the data we were working with, the video of the incident, came pretranscribed[4] and, by and large, it was this transcript which was used by the coroner and invoked in the media debate in support of (caustic) judgements about the pilots' actions. When we came to watch and rewatch the video alongside that transcript, however, we realised it was deficient in several ways. Much of the dialogue was, for instance, mistranscribed and, in several places, who was actually speaking was misattributed, with the different parties confused for one another. As a consequence, we began to work on our own transcript, one that would correct these errors and enable us to arrive at a more accurate picture of what was going on. We did not appreciate just how difficult a task this would prove to be: despite the approximately fifteen minutes of footage having been substantially transcribed already, it took us almost a year to produce a transcript we were happy with.[5]

The difficulties we encountered were not technical distractions from our analytic task but proved to be instructive in and of themselves. As we came to see, confusions in the official transcript as to who was speaking at a given moment reflected confusions between the parties about exactly the same thing. Mistranscribed utterances were likewise tied to exchanges in which the sense of what was being said was also opaque to the pilots and the ground controller they were working with. In other words, the mistakes in the transcript helped us to see that this was a situation in which who was talking to whom, and for what purposes, was itself unclear throughout, something directly linked to the incident's tragic finale.

Our attempts to rectify these surface mistakes in the official transcript led us to something we came to see as a much more problematic feature of the transcript: the *linear* way it depicted the parties' exchanges and the inferences that were being made as a result of the deployment of its conventions as a particular kind of formatting device. Take for instance, the excerpt in Figure 14.1, which captures the opening set of exchanges in the video. In it

1	MANILA HOTEL	Eh POPOFF from MANILA HOTEL, can you confirm you engaged that eh
2		tube and those vehicles?
3		(1)
4	POPOFF 35	Affirm Sir. Looks like I have multiple vehicles in revets about {inhales} uh
5		800 metres to the north of your arty rounds. Can you eh switch fire, an uhm,
6		shift fire, try and get some arty rounds on those?
7		(1)
8	MANILA HOTEL	Roger, I understand those were the impacts that uh you observed earlier on my
9		timing?
10		(>1)
11	POPOFF 35	Affirmative
12		(>1)
13	MANILA HOTEL	Roger, standby ...
14		(1)
15	POPOFF 36	Hey, I got a four ship. Uh looks like we got orange panels on 'em though. Do
16		they have any uh, any eh, friendlies up in this area?
17		=
18	MANILA HOTEL	I understand that was north 800 metres

Figure 14.1 Linear transcript.

Source: Authors' own work.

we hear POPOFF flight (comprising POPOFF 35, the lead, and POPOFF 36, the wing) and MANILA HOTEL, the ground forward air controller they were coordinating with in supplying close air support to Coalition infantry in an area to the northwest of Basra (for more detail see Mair et al., 2012: 85–86).

Lacking any supporting instructions, the linear character of the transcript gives the reader the impression that the various parties are speaking to and can hear each other, meaning that they were all (potentially at least) apprised of what all the others were saying and doing. That is, it appears that all of the parties have and had equal access to – and subsequent awareness of – what was going on. However, as we came to see as we painstakingly worked through these materials, things were by no means as simple as this. This is brought out in the alternative version of the transcript we developed to represent the channels of communication as they were heard and understood by the military personnel in real time (Mair et al., 2012; Mair et al., 2013). Notice that there are two conversations occurring simultaneously in this transcript, marked by asterisks and arrows respectively.

What are the implications of these formatting decisions? When employed alongside the video, this transcript encourages a different reading of what, in the linear transcript, appears to be an anomalous exchange between lines 27–34, an 'exchange' explicitly discussed by the UK coroner in his verdict on the friendly fire incident. As he put it (Crown, 2007: 20–21):

> At 13.36 POPOFF36 said that he had seen a 'four ship'. MANILA HOTEL: 'I understand that was north 800 metres'. To assume that this meant that there were no friendlies in the area was a serious mistake. It would not and could not be reasonably taken as confirming that the area where POPOFF36 was flying was clear.

Based on the linear transcript which flattens out interactional differentiation, this exchange, as seen by assistant deputy coroner Walker, seemed to indicate that POPOFF 36 and MANILA HOTEL were talking past one another as the content of their talk did not match up: if the pilot was making a request for information in his turn, he should not have proceeded (to attack for instance) until his request had been clearly answered. MANILA HOTEL's 'reply' in no way constitutes such an answer. Read this way, the exchange does not furnish the pilots a warrant to attack the 'four ship' spotted.

This reading is clearly informed by the representation of the video in the transcript itself. Contrast this with the alternative 'reading' facilitated by Figure 14.2. The most obvious analytic lesson is that it becomes possible to separate the talk into two distinct conversations between different parties and with different practical intentions. Further, the fact that POPOFF 36 and MANILA HOTEL are not in direct contact is made visible to the reader. The rest of the excerpt shows that while POPOFF35 was interacting with MANILA HOTEL, POPOFF 36 was either technically or occupationally isolated from their interactions and so not involved in exchanges of any kind with the ground controller – that it appears they were in the linear transcript is an artefact of its formatting conventions.

		Air-to-Ground Communication	Air-to-Air Communication
* 1	MANILA HOTEL	Eh POPOV from MANILA HOTEL, can you confirm	
2		you engaged that eh tube	
3		and those vehicles?	
4		{Automated message}	
5		(1)	
* 7	POPOFF 35	Affirm Sir. Looks like I	
8		have multiple vehicles in	
9		revets about {inhales} uh	
10		800 metres to the north of	
11		your arty rounds. Can you	
12		eh switch fire, an uhm, shift	
13		fire, try and get some arty	
14		rounds on those?	
15		(1)	
* 16	MANILA HOTEL	Roger, I understand those	
17		were the impacts that uh	
18		you observed earlier on my	
19		timing?	
20		(>1)	
* 21	POPOFF 35	Affirmative	
22		(>1)	
* 23	MANILA HOTEL	Roger, standby. Let me	
24		make sure they're not on	
25		another mission	
26			(1)
→ 27		POPOFF 36		Hey, I got a four ship. Uh
28			looks like we got orange
29			panels on 'em though. Do
30			they have any uh, any eh,
31			friendlies up in this area?
32		=	
* 33	MANILA HOTEL	I understand that was north	
34		800 metres	
35		(3)	
* 36	MANILA HOTEL	POPOV, understand that	
37		was north 800 metres?	
38		(2)	
* 39	POPOFF 35	Confirm, north 800 metres.	
40		{Automated message}.	
→ 41		Confirm no friendlies this	
42		far north uh. On the ground	
43		(1)	
→ 44	MANILA HOTEL	That is an affirm.	
45		{Distortion, static} You are	
46		well clear of friendlies	
47		(.)	
* 48	POPOFF 35	Copy. I see multiple	
49		revetted vehicles. Some	
50		look like uh {inhales}	
51		flatbed trucks and others are	
52		uhm green vehicles. Can't	
53		quite make out the type.	
54		Look like maybe ZIL157s	
55		[	
* 56	MANILA HOTEL	Roger. That matches our	
57		intel up there	

Figure 14.2 Modified transcript.[6]

Source: Authors' own work.

The lesson we believe accompanies this demonstration is as follows: transcription procedures and decisions influence the presentation of the details of communicative exchanges and so shape how a given interaction is interpreted, understood and analysed (Gibson et al., 2014). Rather than impose a set of conventions on interactional data (as with the linear transcript), ethnomethodology and conversation analysis try to produce transcripts that bring out the transcribed parties' orientations to one another. As such the focus is to understand *their* methods for analysing and making sense of what is going on around them and what others are doing as part of that *while it is happening*. Our demonstration is, therefore, not about the pursuit of accuracy for accuracy's sake but about using the problems encountered in adequately transcribing complex battlefield interactions in order to arrive at a better understanding of the practical problems that accompany combat operations as variegated, real-world activities – something lost by linear formatting. In our case, in and through the work of producing a transcript for ourselves (instead of taking those produced by others at face value), we were able to develop a better grasp of real-time communication in this setting and the fragilities which accompanied the division of labour it was embedded in but also helped constitute.

Demonstration 2: Military Inquiries and Cockpit Videos

Our initial work on the transcript showed it was possible to glean a great deal from a direct analysis of the cockpit video (see also Nevile, 2009, 2013). However, as the earlier example of the coroner's verdict shows, we were not the only ones offering ways of reading and making sense of the video, and our interest was as much in how *those* analyses were arrived at as in producing a stand-alone analysis of the incident footage itself. Given the richness of the material it contained, the work of the US Air Force's Friendly Fire Investigation Board (hereinafter 'the Board') became a particular focus, offering an alternative way of approaching the cockpit data. 'Tab G' of the report contained partially redacted 'Witness Testimony and Statements' in which the two pilots separately responded to the Board's investigators' questioning about the incident. As part of their questioning, selected portions of the cockpit video were shown to the pilots and they were asked to explain what they were seeing and doing at particular moments of the mission.

The question of how the pilots and the investigators treated the video as 'evidence' and understood, interpreted and described the actions found within it is a fascinating one and central to the 'indigenous' or 'native' video analysis worked up in the course of the pilots' questioning. Contained within this material were explorations of how transparent or equivocal the video could be said to be, that is whether what was going on was clear to see or needed to be augmented by additional commentary and explanation. By linking the transcripts of the pilots' testimony, in response to the clips of the incident they were played, to our detailed transcript (discussed earlier), we were able to reconstruct and so consider the methodic work of the inquiry, particularly in terms of the problematic status of the video as evidence within it (for a comparative study see Goodwin, 1994: 753). By doing so, we came to see that the result of that work was to call into question the status of the video as an entirely objective or stand-alone document or 'imprint' of events.

Generally the video as evidence was treated in two contrasting ways by the various inquiry participants. First, on many occasions 'what the video shows' was treated as transparent or clear to anyone watching it (including the examiners who were not, of course, on the scene). However, there were moments in the video which caused confusion in that the footage was deemed unable to 'speak for itself' (Goodwin, 1994: 615–616) and therefore required the pilots to elaborate and expand by sharing their first-hand experiences of the moment-to-moment action as it had unfolded.

'What Happened' as Evidentially Transparent and Unproblematic

In the first kind of treatment, the emphasis was on 'what anyone could see' by 'looking at the tape' (i.e. what could be 'read off' the video footage unproblematically). Figure 14.3 provides an example.

1	Board Investigator	*Looking at the tape*, it looks like it would be maybe just some scattered clouds
2		off in the distance. *Clear vis ['visibility'],* probably at least I would say seven
3		miles and I don't know about the winds but would you say that that is an
4		accurate representation?
5	POPOFF 36	*Yes, definitely.*
6	Board Investigator	*Clear vis, and no cloud cover?*
7	POPOFF 36	*Right.*

Figure 14.3   Excerpt from POPOFF 36's testimony.

Source: USAF (2003: G24), emphasis added.

'What Happened' as Beyond the Videotape

However, at certain points during the pilots' testimonies it became apparent that their responses to the examiners' questions regarding the video went beyond the videotape and, in the process, highlighted limits to 'what anyone could see or legitimately deduce' from it. On these occasions, the pilots produced in-the-moment accounts that described and explained what they were seeing, doing and discussing at different points. As such, the pilots as witnesses came to operate as resources or conduits, supplying details missing or lost from the video (Lynch and Bogen, 1996: 155). Hence the pilot's-eye view trumped the seemingly objective record furnished by the cockpit video. This alternative orientation to the video as data is exhibited by the excerpts reproduced in Figures 14.4–14.6 (notice that, as these exchanges did not involve communication with MANILA HOTEL, a single column is used to display inter-pilot communication).

Space precludes a detailed analysis of these data excerpts (for a full-blown analysis see Mair et al., 2013). However, at present it is enough to point out that, taken together, these three excerpts demonstrate how the Board came to treat the video as problematic for certain purposes. That is, the video was not (and indeed could not be) treated as a stand-alone document of the incident. Instead the pilots' perspective was vital to understanding how the incident came to unfold as it did, with the pilots firing (unwittingly) on the British soldiers. As such we see an embodied familiarity with the plane, 'knowing my plane', on display (Figures 14.5–14.6), as well as a 'collaborative seeing' of the target and its presumed intended actions (Watson, 1999, Mair et al., 2013) that gradually emerges from the communication between the pilots. Here, then, the Board, when assessing the evidence, is heavily reliant

1	POPOFF 36	Uh, roll up your right wing, and look right underneath yah
2		(.)
3	POPOFF 35	I know what you're talking about
4		(.)
5	POPOFF 36	OK, well they have orange rockets on 'em
6		(1)
7	POPOFF 35	Orange rockets?
8		(.)
9	POPOFF 36	Yeah, I think so
10		(1)
11	POPOFF 35	Let me look
12		{Lines omitted}
13	POPOFF 36	I think killing these damn rocket launchers, it would be great. {Inaudible, heavy
14		distortion}
15		{Lines omitted}
16		(3)
17	POPOFF 35	Yeah, I see that eh. You see (?). I'm going to roll down, (see a break)
18		{Lines omitted}
19	POPOFF 36	OK, do you see the orange things on top of 'em?
20		{Lines omitted}
21	POPOFF 35	I'm coming off west. You roll in, and. *It looks like they are exactly what we're talking*
22		*about*
23		(2)
24	POPOFF 36	We got a visual (1) OK. I want to get that first one before he gets into that town then
25		(.)
26	POPOFF 35	Get him. Get him
27		(1)
28	POPOFF 36	All right, we got rocket launchers, it looks like, eh number 2 is rolling in from the south
29		to the north. And eh 2's in

Figure 14.4 Excerpt from video (emphasis added).

1	Board Investigator	OK, this time you're looking at?
2	POPOFF 35	I'm looking, you'll notice later on the tape there's periods where I'm not talking
3		a lot on the radio, cause I had both hands with the binos [binoculars] trying to get
4		a good look at what's going on...
5	Board Investigator	At 1341 and 31 seconds, he asks you if you see the orange things on top of them
6		... Did you actually see them?
7	POPOFF 35	Yeah, I did.
8	Board Investigator	OK, continuing on ... *1341 and 48 seconds, looks like exactly what we're talking*
9		*about. What is that exactly?*
10	POPOFF 35	Rockets. In my mind, what I'm looking to do here is roll in on the east side so I
11		can see if there's a development of shadows, because the sun, at this point, is
12		getting low on the horizon on the west side. And actually, what I do is I'm at a
13		low energy state to employ, so I pull off to the right on the east side, roll the
14		airplane up, have the throttles at max and throw in a couple of clicks of trim, and
15		let go of the stick, so I'm knowing my airplane, as airspeed increases, is going to
16		do this .... At that time, I roll the airplane up and I'm going down with the
17		aircraft, pulling up this way, and I'm looking at these things and I see vertical
18		developments with shadowing it looks like on the eastern side which is what I
19		would expect if in fact it was vertical development, so that's what I'm seeing.
20		{{Tape runs out}}

Figure 14.5 Excerpt from POPOFF 35's testimony.

Source: USAF (2003: G15–16).

1		{{Tape restarted}}
2	Board Investigator	*This time you were just describing that you were looking through your binoculars*
3		*and describing the type of target that you were seeing. You're saying what you saw*
4		...
5	POPOFF 35	Military vehicles looking like they have vertical developments on these things that
6		are consistent with what I thought were rocket tubes. The shadowing that I saw ... I
7		couldn't quite ... I wasn't low enough, you know I was kind of concerned about
8		having a manpad [manned portable air defence system] shoot at me. We were
9		briefed on threats to the southeast of there, and I didn't want to be the first guy shot
10		down during the war, so. The bottom line is, I didn't want to, while I was trying to
11		ID them, as best I could, I didn't want to put the aircraft in a position where it was
12		going to be an easy target.

Figure 14.6   Excerpt from POPOFF 35's testimony.

Source: USAF (2003: G16).

upon the access to 'war as work' that the video coupled with the pilots' testimonies furnishes when taken together.

A number of lessons can be drawn from this second demonstration. Methodologically speaking, we are advocating research that is rooted in the perspective(s) and understanding(s) of the participants themselves as evidenced by, in this case, *their* (not our) interrogation of 'the data'. By focusing on some of the ways the evidence was made to speak (or found to be unable to speak on its own), we come to see what hinged on the changeable epistemic status of the cockpit video in this context. That is, there is a practical lesson here for those who might think cameras could capture objective or complete versions of events, providing a record that recovers soldierly practice in all its complexity. In short, as the Board's investigations make clear, they do not – or rather cannot.

Demonstration 3: The Board's Recommendations; Making Soldierly Work Visible

Our final demonstration revolves around how the Board sought to analyse, explain and thereby arrive at ways of preventing such incidents in the future. The report clearly states that what happened was never under dispute. That is, the pilots themselves recognised they had mistaken their allies for enemies as soon as they saw the release of 'blue smoke', indicating friendly vehicles, from the targets on the ground (POPOFF 35 'got the smoke'). As their acknowledged starting point, the Board's task was to work out how and why this mistake occurred, an investigative task in which inspection of the cockpit video and the communicative activity contained within it played a central role.

Drawing together the evidence gathered in the course of their inquiries, the Board offered the following account of the cause of the incident:

> The Board found by clear and convincing evidence that the primary cause of the friendly fire incident on 28 March 2003 was *target misidentification* . . . From the pilot's perspective, they asked the GFAC on two occasions if there were any friendly forces in the area. On both occasions, the GFAC told the pilots that friendly forces

were 'well clear'. This *insufficient reference* to any potential or actual specific location of friendly forces *reasonably caused* the pilots to expect friendly forces would be a non-factor in the area they were targeting.

(USAF, 2003: 31, emphasis added)

In his closing statement, Brigadier General William F. Hodgkins, chair of the Board, elaborated on the implicit theme:

Communication discipline [i.e. indiscipline] is an issue that goes back to the advent of radio communications . . .Yet, *it still remains* an issue in this friendly fire incident, and is *embedded* as a factor in far too many aircraft mishaps and accidents – and resulting deaths. While it is difficult to imagine any more emphasis being placed on this issue, it is an area that should be re-evaluated.

(USAF, 2003: 33, emphasis added)

Here, then, we find an unequivocal conclusion placing the root cause in the (perennial) court of language use and communication. We see how the fatal misunderstanding or trouble is retrospectively discovered, rather than repairable in real time (Jordan and Fuller, 1975: 144). The implication is that conversational mechanisms of sense-making and understanding were not sufficient to highlight and repair the mistakes in this case. Of particular interest here is that while the recommendation is general in character (i.e. 'use coordinates when flying missions'), the Board is only able to arrive at it through an analysis of the pilots' and GFAC's actual location requesting and reporting practices. By focusing on their investigative and indeed analytical work, involving again recourse to the cockpit video *and* pilots' testimony, we are able to see how the interaction between the pilots and the ground air controllers became decoupled, and so the specific ways in which the sense of one set of parties' comments, 'the air', could be lost on the other, 'the ground' (i.e. via divergent understandings of the area the phrase 'well clear' was understood to refer to, see the excerpt in Figure 14.2). In short, where the inquiry highlighted a 'lack of effective and complete communication' linked to 'imprecise language' as well as 'undefined or non-standard terminology', we see here how this was traced back to the battlefield-specific interactional practices in which misunderstandings arose in this case. That none of the parties involved were fully on the same page (King, 2006), was made perspicuous through these inquiries. As a consequence, therefore, of the Board's attempts to explicate what happened and the role of language within it, we gain insights into location requesting and reporting as constitutive features of soldierly work, both in the normal course of things and where problems occur – practical features of combat we would not otherwise gain access to.

Conclusion: Prospects and Limitations

The purpose of this chapter has been to highlight what ethnomethodology and conversation analysis can contribute to studies of the military, particularly to understandings of action-in-interaction in combat settings. Through our three demonstrations, we have emphasised what we see as the particular strength of such studies: their focus on the specificities of action and interaction as well as the forms of organisation internal to them. We have also tried to show how different kinds of materials can be drawn upon for that purpose.

Given our nonmilitary backgrounds we were reliant on those materials. Unlike other researchers in the ethnomethodological tradition, as well as other social science disciplines, we have never encountered or trained within the worksite that we have analysed. Within ethnomethodology, this immersion process is referred to as the 'unique adequacy requirement' and finds examples in law, science, mathematics, software programming, professional musicianship and so forth, where the researchers learn to do the job of those they are observing (Burns, 1997: 43; Garfinkel and Wieder, 1992: 255; Sormani, 2014). This 'inside' knowledge assists in understanding the activities and practices characteristic of a given social setting (and see also here the work of Snook, King, Hockey cited earlier).

In the absence of first-hand knowledge of our own, our study had to rely on both primary and secondary materials, while remaining alive to the occasioned character of the commentaries the latter offered. The military boards' reports, like the coroner's verdict and media coverage, were not neutral descriptions but accounts offered to particular audiences for particular ends. This enabled us to explore a wider range of analytical problems than we would have been able to explore otherwise. As discussed in demonstration 1, we were, for instance, able to link the events documented by the video to the coroner's findings of an 'unlawful killing' and show how scrutiny of the evidence cited provided access to an alternative reading of the events. As discussed in demonstrations 2 and 3, we were also able to follow the ways in which the incident was taken up by the Board. On the one hand, the transcripts of the pilot interviews provided us with insights into the structure and organisation of both military inquiries in general and the methodic ways in which they produce accounts of combat incidents. On the other, the employment of the video during the pilots' testimony provided us with ethnographic commentaries on what was happening (both within the video itself and the pilots' first-hand experience of the incident) that further elaborated our understanding of the data under scrutiny. As a result, the methodological approach we have articulated is, therefore, a distinctive one, centred on the analysis of soldiering *and* its assessment as practical action or 'work' (Garfinkel, 1986).

As opportunities to engage first-hand with actual courses of soldierly work will remain rare (and where they arise, will often be circumscribed by the conditions of access the researcher will have to comply with), if we are to arrive at an understanding and offer 'convincing explanations of . . . what soldiers distinctively do' (King, 2006: 510), we will have to make as much use as possible of the materials we do have access to. Those materials include not only the increasing number of videos of combat operations now entering the public domain (via official channels and leaks), but also the increasing number of after-the-fact accounts relating to particular incidents that are also being made accessible. Ethnomethodology and conversation analysis are not the only approaches that can handle materials of this kind; however, in this chapter we hope to have demonstrated that they offer a distinctive approach with distinctive payoffs to their analysis.

Appendix: Transcription Conventions

{Beep, beep}: curled brackets contain background cockpit sounds and noises
((To MANILA HOTEL)): double parentheses contain transcriber's descriptions, and
    include such things as sighs, inhalations and so on

((inaudible)): indicates a stretch of inaudible talk

(Eh I see): words placed within single parentheses offer a possible but uncertain hearing of the talk

(1): numbers in brackets indicate time between turns at talk in seconds

(.): indicates a micro-pause, under half a second

(>1): less than one second, but more than half

Stress: emphasis in talk

=: 'latching'; one turn follows another immediately with no audible pause

[: single square bracket between lines indicates overlaps in talk

Notes

1   We do not intend to define friendly fire here – something which is itself the subject of controversy. For a standard account of what friendly fire (aka 'fratricide' or 'amicide') is, see Shrader, 1982.

2   The recommendation that we study practical activity 'first-time-through' stems from ethnomethodology's insistence on capturing the lived, moment-by-moment, still unfolding character of actual courses of action and interaction rather than on reconstructing already concluded activity via its outcomes after the fact (Garfinkel et al., 1981).

3   See footnote 5 for further discussion.

4   Both the video and the transcript released to the public can be found on the incident's Wikipedia page at https://en.wikipedia.org/wiki/190th_Fighter_Squadron,_Blues_and_Royals_friendly_fire_incident (Accessed 31 January 2016).

5   It was in the process of disentangling the technical aspects of the data that 'talk throughs' by experts and other instructional resources proved particularly useful; we had to seek such instruction out because much of what the pilots and ground controller were discussing was initially lost on us and had to be translated. For instance, the interviews along with media coverage and online resources enabled us to decipher talk of 'tubes' (i.e. missile launchers), 'goggles' (i.e. binoculars), 'revets' (i.e. fortified embankments) and so on.

6   The inspiration to use columns in the transcript came from Goodwin's (1993) work and the general transcription conventions or symbols are based on Jefferson's (2004) transcription system that is widely used in CA. See Mair et al., 2014 for a complete version of the modified transcript and the chapter appendix for the transcription conventions employed in the data excerpts in this chapter.

References

Benson, D. and Drew, P. (1978) 'Was there firing in Sandy Row that night?' Some Features of the Organisation of Disputes about Recorded Facts. *Sociological Inquiry* 48(2), 89–100.

Boudeau, C. (2007) Producing Threat Assessments: An Ethnomethodological Perspective on Intelligence on Iraq's Aluminium Tubes. In Rappert, B. (Ed.) *Technology and Security: Governing Threats in the New Millennium*. Houndmills: Palgrave Macmillan, pp. 66–87.

Boudeau, C. (2012) Missing the Logic of the Text: Lord Butler's Report on Intelligence on Iraqi Weapons of Mass Destruction. *Journal of Language and Politics* 11(4), 543–561.

British Army (2004) *Board of Inquiry Into the Death of the Late Lance Corporal of Horse Matthew Richard Hull, the Blues and Royals (Royal Horse Guards and 1st Dragoons) Household Cavalry Regiment*. London: Ministry of Defence.

Brown, K. (2008) 'All They Understand Is Force': Debating Culture in Operation Iraqi Freedom. *American Anthropologist* 110(4), 443–453.

Burns, S. (1997) Practicing Law: A Study of Pedagogic Interchange in a Law School Classroom. In Travers, M. and Manzo, J. (Eds.), *Law in Action: Ethnomethodological and Conversation Analytic Approaches to Law*. Aldershot: Ashgate, pp. 265–287.

Caddell, J.W., Jr. (2010) *Targeting-Error Fratricide in Modern Airpower: A Causal Examination*. Unpublished Master of Arts thesis, Georgetown University, Washington, DC, 15 November 2010.

Crown (2007) *In the Matter of an Inquest Touching the Death of Lance Corporal of Horse Matthew Hull*. Oxford: Oxfordshire Coroner's Court.

Garfinkel, H. (1967) *Studies in Ethnomethodology*. Englewood-Cliffs, NJ: Prentice Hall.

Garfinkel, H. (Ed.) (1986) *Ethnomethodological Studies of Work*. London: Routledge & Kegan Paul.

Garfinkel, H. (2002) *Ethnomethodology's Program: Working Out Durkheim's Aphorism*. Lanham, MD: Rowman & Littlefield.

Garfinkel, H., Lynch, M. and Livingston, E. (1981) The Work of a Discovering Science Construed With Materials from the Optically Discovered Pulsar. *Philosophy of the Social Sciences* 11(2), 131–158.

Garfinkel, H. and Wieder, L. (1992) Two Incommensurable, Asymmetrically Alternate Technologies of Social Analysis. In Watson, G. and Seiler, R. (Eds.), *Text in Context*. London: Sage, pp. 175–206.

Gibson, W., Webb, H. and von Lehm, D. (2014) Analytic Affordance: Transcripts as Conventionalised Systems in Discourse Studies. *Sociology* 48(4), 780–794.

Goodwin, C. (1993) Recording Human Interaction in Natural Settings. *Pragmatics* 3(2), 181–209.

Goodwin, C. (1994) Professional Vision. *American Anthropologist* 96(3), 606–633.

Heritage, J. (1984) *Garfinkel and Ethnomethodology*. Cambridge: Polity Press.

Heritage, J. (1995) Conversation Analysis: Methodological Aspects. In Quasthoff, U. (Ed.), *Aspects of Oral Communication*. New York: Walter de Gruyter, pp. 391–418.

Hicks, L. (1993) Normal Accidents in Military Operations Sociological Perspectives. *Sociological Perspectives* 36(4), 377–391.

Hockey, J. (2009) 'Switch On': Sensory Work in the Infantry. *Work, Employment and Society* 23(3), 477–493.

Howe, S., Poteet, S. Xue, P., Kao, A. and Giammanco, C. (2010) Shared Context-Awareness: Minimizing and Resolving Miscommunication during Coalition Operations. In *Proceedings of the 4th Annual Conference of the International Technology Alliance*, Imperial College London, September.

Jefferson, G. (2004) Glossary of Transcript Symbols With an Introduction. In Lerner, G. (Ed.), *Conversation Analysis: Studies from the First Generation*. Philadelphia, PA: John Benjamins, pp. 13–31.

Jordan, B. and Fuller, N. (1975) On the Non-Fatal Nature of Trouble: Sense-Making and Trouble-Managing in Lingua Franca Talk. *Semiotica* 13(1), 11–31.

King, A. (2006) The Word of Command: Communication and Cohesion in the Military. *Armed Forces & Society* 32(4), 493–512.

Kirke, C. (Ed.) (2012) *Fratricide in Battle: (Un)friendly Fire*. London: Continuum.

Lynch, M. (2007) The Origins of Ethnomethodology. In Turner, S. and Risjord, M. (Eds.), *Philosophy of Anthropology and Sociology*. Amsterdam: Elsevier, pp. 485–515.

Lynch, M. and Bogen, D. (1996) *The Spectacle of History: Speech, Text and Memory at the Iran-Contra Hearings*. Durham, NC: Duke University Press.

Mair, M., Elsey, C., Smith, P.V. and Watson, P.G. (2014) 190th Fighter Squadron/Blues and Royals Fratricide: Modified Incident Transcript. *Online Dataset*. doi:10.13140/2.1.4457.9206.

Mair, M., Elsey, C., Watson, P. G. and Smith, P.V. (2013) Interpretive Asymmetry, Retrospective Inquiry and the Explication of Action in an Incident of Friendly Fire. *Symbolic Interaction* 36(4), 398–416.

Mair, M., Watson, P. G., Elsey, C. and Smith, P.V. (2012) War-Making and Sense-Making: Some Technical Reflections on an Instance of 'Friendly Fire', *British Journal of Sociology* 63(1), 75–96.

Masys, A. (2008) Pilot Error: Dispelling the Hegemony of Blamism – A Case of De-Centered Causality and Hardwired Politics. *Disaster Prevention and Management* 17(2), 221–231.

McHoul, A. (2007) 'Killers' and 'Friendlies': Names Can Hurt Me. *Social Identities* 13(4), 459–469.

Mieszkowski, J. (2012) *Watching War*. Stanford: Stanford University Press.

Nevile, M. (2009) 'You Are Well Clear of Friendlies': Diagnostic Error and Cooperative Work in an Iraq War Friendly Fire Incident. *Computer Supported Cooperative Work (CSCW)* 18(2/3), 147–173.

Nevile, M. (2013) Seeing on the Move: Mobile Collaboration on the Battlefield. In Haddington, P., Mondada, L. and Nevile, M. (Eds.), *Interaction and Mobility: Language and the Body in Motion*. Berlin: De Gruyter, pp. 152–176.

Pomerantz, A. and Fehr, B. J. (1997) Conversation Analysis: An Approach to the Study of Social Action as Sense Making Practices. In van Dijk, T.A. (Ed.), *Discourse as Social Interaction*. London: Sage, pp. 64–91.

Rappert, B. (2012) States of Ignorance: The Unmaking and Remaking of Death Tolls. *Economy and Society* 41(1), 42–63.

Rizan, C., Elsey, C., Lemon, T., Grant, A. and Monrouxe, L. (2014) Feedback in Action Within Bedside Teaching Encounters: A Video Ethnographic Study. *Medical Education* 48(9), 902–920.

Sacks, H. (1995) *Lectures on Conversation*. Oxford: Blackwell.

Schegloff, E. (1991) Reflections on Talk and Social Structure. In Boden, D. and Zimmerman, D. H. (Eds.), *Talk and Social Structure: Studies in Ethnomethodology and Conversation Analysis*. Cambridge: Polity Press, pp. 44–70.

Schegloff, E. (2007) *Sequence Organisation in Interaction: A Primer in Conversation Analysis I*. Cambridge: Cambridge University Press.

Shrader, C. R. (1982) *Amicide: The Problem of Friendly Fire in Modern War*. Fort Leavenworth, KS: Combat Studies Institute, U.S. Army Command and General Staff College.

Snook, S.A. (2002) *Friendly Fire: The Accidental Shootdown of U.S. Black Hawks Over Northern Iraq*. Princeton, NJ: Princeton University Press.

Sormani, P. (2014) *Respecifying Lab Ethnography: An Ethnomethodological Study of Experimental Physics*. Aldershot: Ashgate.

United States Air Force (2003) *Investigation of Suspected Friendly Fire Incident Involving an A-10 and a United Kingdom (UK) Reconnaissance Patrol Near Ad-Dayr, Iraq, Operation Iraqi Freedom, 28 March, 2003*. Macdill Air Force Base, FL: United States Central Command, Office of the Commander in Chief.

Watson, R. (1999) Driving in Forests and Mountains: A Pure and Applied Ethnography. *Ethnographic Studies*. 4, 50–60.

Watson, R. (2009) *Analysing Practical and Professional Texts: A Naturalistic Approach*. Farnham: Ashgate.

Woodward, R. and Jenkings, K. N. (2011) Military Identities in the Situated Accounts of British Military Personnel. *Sociology* 45(2), 252–268.

15

RESEARCHING NORMATIVITY AND NONNORMATIVITY IN MILITARY ORGANIZATIONS

Aaron Belkin

Abjection, Camouflage and Mythology

When American civilians think about the armed forces, typically they may imagine cool fighter planes, masculine warriors, polished uniforms and well-spoken generals and admirals.[1] But there's of course another side to the US military that is neither shiny nor polite, despite appearances. My argument begins with the premise that American service members have engaged in a great deal of shameful conduct, that the public is willfully ignorant about such behavior, and that camouflage that conceals uncomfortable facts is implicated in almost every military policy, every military decision and every aspect of military culture, ranging from apparently trivial minutiae such as umbrellas and who gets to carry them, to large matters of state such as decisions to acquire weapons and wage war. Mechanisms that enable the public to remain in denial about shameful conduct are so deeply implicated in American culture that few aspects of the armed forces can be understood without an appreciation of the complicated interplay of abjection, camouflage and willful ignorance. In this chapter, I elaborate on these points and describe one prominent mechanism, normative alignment, that has played a role in hiding shameful aspects of service members' conduct in plain sight. While this brief meditation is about public delusion concerning the modern American military, I hope that my insights might be relevant for scholars analyzing myths about military organizations in other geographical settings and historical eras.

Without claiming to identify any new or underutilized methodology, a word about methods is appropriate given this contribution to the volume. A method, from my perspective, is a strategy for deepening understanding, for example a set of tools that may enable a scholar to verify the plausibility of a causal inference. What is interesting to me about this volume is that the editors have decided to assemble a collection on military methodology, not post office methodology or rail service methodology or Agriculture Ministry methodology. Why does the study of military organizations require a methodological toolbox that is any different than the assemblage of strategies that scholars have developed to understand other institutions, cultures and sites?

What is distinct about the military, I would suggest, is that military organizations are structured by contradictions, of which the contradiction mentioned earlier (civilized/barbaric) is just one of many. At the same time that military organizations may be (and may be expected to be) civilized and barbaric, they can also be (and be expected to be) protective/destructive; filthy/clean; violent/peaceful and the list goes on. Furthermore, military organizations depend on structuring contradictions to carry out tasks that other types of organizations, in general, do not have to master. In particular, structuring contradictions enable military organizations to train personnel to be willing to sacrifice their lives and to take the lives of others (Belkin, 2012). Finally, while structuring contradictions are central for military operations, they only work properly when hidden in plain sight. Military organizations, in other words, disavow structuring contradictions at the same time that they depend on them. Myths play central roles in the process of disavowal, but interrogating those myths is not always easy.

Other organizations certainly may be structured by contradictions. In the United States, for example, the Department of Agriculture is at least partly responsible for the public's nutritional well-being, but also for the profitability of the processed-food industry. Military organizations are different because:

1   Typically, military forces are structured by many contradictions, not just a few.
2   The contradictions are extreme. For example, American service members have been exceptionally noble, but also sadistically barbaric.
3   Because of the number and extremity of structuring contradictions, and the need to disavow their existence, it takes a lot to hide contradictions in plain sight. Camouflage that sweeps structuring contradictions under the rug is complex and interesting.
4   The production, operation and concealment of structuring contradictions is a high-stakes business. For those militaries engaged in population control at home and/or in overseas ventures, the viability of domestic order as well as empire itself may hinge on contradictions that structure military organizations and how well they are hidden.

What this means, methodologically, is that to the extent that the foregoing observations are valid, scholars may not be able to understand much about the relationship between power, social control and military organizations, either at home or abroad, without opening ourselves up to the possibility not only that military forces are structured by contradictions, but that such contradictions are organizationally functional yet hidden in plain sight. From my point of view, it doesn't really matter if scholars seeking to understand military organizations avail themselves of literary analysis, ethnography, statistics and/or case study comparisons. What matters is whether we are open to an appreciation of contradictions that structure military organizations and how and why those contradictions and their concealment operate.

Civilized/Barbaric

The 2013 documentary film *The Kill Team* is difficult to watch. Its focus is an American unit in Afghanistan that murdered civilians on purpose and, apparently, for the sport of it. The unit carefully staged executions after collecting weapons that they neglected to report to commanders. In small villages, they selected targets, shot them and then approached the corpses so that they could plant unreported weapons they had obtained elsewhere. In their reports to superiors, unit members claimed that the civilians in question had fired at them and that they had responded with deadly force as a protective measure. While the motives behind the murders remain murky and seem to have differed among various members of the unit, perpetrators alluded or pointed explicitly to a number of factors including boredom, rage, social pressure and a desire to demonstrate masculinity. When one member of the unit threatened to report the murders, his unit mates threatened to kill him. Near the conclusion of the film, one of the perpetrators acknowledges that his unit's behavior in Afghanistan was not exceptional:

> Everything that we did was put into the media as we're horrible, we're the kill team, we're [a] rogue platoon, we're all these things. I don't care what the military wants to say, but this goes on more than just us. We're just the ones that got caught.
>
> *(Krauss, 2013)*

American service members have engaged in shameful behavior on a patterned basis, during both peacetime and wartime. They have murdered, tortured and raped enemy soldiers and civilians, as well as each other, sometimes in sadistic, gratuitous and policy-driven ways. They have, in their domestic lives, manifested high rates of shaken baby syndrome, meaning that they have become so angry with their babies that they have shaken them to death (Gessner and Runyan, 1995). They have beaten their spouses at rates that exceed the incidence of domestic violence among civilians by a considerable degree (Enloe, 2000). They have engaged in disproportionately high rates of binge drinking and unprotected, promiscuous sex (DoD, 2011). They have defecated in rivers that they knew civilian populations depended on for drinking water (De Bevoise, 1995). They have lied when speaking to members of the media and while testifying in Congress (Parrott, 1982). They have insulted citizens who do not serve in uniform, whom they may refer to as 'civilian scum' (Mazur, 2011). They have demonized women, people of color and anyone whom they construed as weak (Traynor, 2010). And I am not talking about exceptions or 'bad apples'. I am talking about patterned practice in the US Armed Forces during the modern era.

Despite extensive evidence of shameful conduct that has been part of the public record since the Winter Soldier investigation of 1971, very few Americans think about service members as perpetrators of violent crimes. That said, readers who doubt my claims might consult Nick Turse's recent book, *Kill Anything That Moves* (2013). Based on a decade of archival and ethnographic research, Turse shows that the My Lai massacre, in which American service members murdered several hundred civilians in Vietnam, including women and children, was the rule, not the exception. To the extent that they think about it at all, many Americans believe that the massacre was an unfortunate, atypical and even singular event. Turse shows that this is simply not true, and he demonstrates that "the indiscriminate killing of South Vietnamese noncombatants . . . was neither accidental nor unforeseeable" (2013: 12). While estimates vary, Turse concludes that approximately two million civilians were killed during the war. And he shows that murder, torture and rape often were carried out in gratuitous, sadistic ways. Senior officials were well aware of the scandalous conduct of the troops, and Turse uncovered extensive evidence that the army engaged in carefully choreographed and typically successful cover-up efforts to hide the conduct of combat forces. Torture, murder and rape were not exceptions and were not the effects of a few bad apples. The number of victims was not in the dozens, hundreds, thousands or even tens of thousands.

The behavior of military personnel must be understood in context, as shameful conduct does not occur in a vacuum. In Vietnam, killing civilians was policy driven, not accidental. The Pentagon's designation of the number of Vietnamese killed as the metric for assessing progress in the war created a powerful incentive to murder civilians, and military officials all the way to the highest level were well aware of this. In practice, the policy-driven incentive to kill civilians played out in terms of competitions that rewarded units that amassed the most kills. More broadly, the US military has tolerated and encouraged disgraceful behavior, sometimes implicitly and sometimes by policy, across a wide range of historical moments and geographic settings. In some cases, the military has enacted and enforced policies that led to the murder, torture and rape of foreign citizens (Kramer, 2006; Mayer, 2008; Turse, 2013). It has fostered desperate and abusive organizational cultures that have driven service members mad (Belkin, 2012). It has polluted bases and lands with toxic waste and mightily resisted pressures to clean up its messes (Lutz, 2009). It has ignored court orders (Shilts, 1999). It has resisted civilian oversight and said that it should be held to lower constitutional standards than other governmental agencies (Mazur, 2011). It has discriminated against minorities and lied about its reasons for doing so (Belkin, 2011). It has lied about its pattern of lying, even in the context of the cancellation of a program designed explicitly to lie to civilians (Center for the Study of Sexual Minorities in the Military, 2002).

And yet, despite layer upon layer of barbarity, members of the American public see the US military as a noble institution. The American public is in denial about these patterns, and the US military would have difficulty functioning if the public's denial was shattered. Public opinion polls show consistently that soldiering is the most respected profession and that the military is the most respected institution in the United States (Pew Research, 2013). The military depends on popular support and adulation to reward politicians who vote to increase the military's budget, generate enough recruits to fill vacant slots, and discourage officials from snooping too closely into how the Pentagon spends its money. Military officials rightly fear that if public opinion were to change, they would be held more accountable and provided with fewer resources, and it would be more difficult for the organization to fulfill its mission. So, popular support is important. And willful ignorance about shameful military conduct is important for maintaining popular support. This is the reason why military officials invest significant efforts in the camouflaging of disgraceful behavior. But how, exactly, does the camouflage work?

Willful Ignorance and Normative Alignment

To address this question, I would like to posit that camouflaging disgraceful behavior is not just about hiding shameful facts, but also about the resilience of favorable impressions of service members and the organization more broadly. Shortly after the abuses of Abu Ghraib were revealed in 2003, opinion polls revealed that the public continued to believe in the nobility of soldiering and of the armed forces (Pew Research, 2004). Perhaps one way to think about the public's denial about military atrocities, then, is not just in terms of the hiding of uncomfortable outcomes, but also the ways in which favorable impressions persist even when highly embarrassing matters are revealed. Favorable attitudes about service members and the armed forces are not static, and the reputation of the military has dipped precipitously in certain eras, such as in the aftermath of the Vietnam War. That said, for most of the last century, favorable attitudes have been quite buoyant, and the reputation of the armed forces and the profession of soldiering has not suffered serious damage despite eras that featured never-ending strings of scandals, such as the 1995 rape of a 12-year-old girl in Okinawa and the abuse of over 100 women during the 1991 Tailhook scandal. How is the military's reputation preserved? How and why does the public remain in denial?

Here I want to turn to a discussion of normative alignment, a mechanism that has played a prominent role in concealing shameful aspects of the US armed forces, and sustaining the resilience of favorable public impressions. Normative alignment is the lining up of distinct yet related categories to make it seem as if the categories go together, that their alignment is natural, and that positive portrayals of each category are interconnected. By repeatedly conflating the distinct categories as well as the idea that positive aspects of each individual category are interconnected, the favorable portrayal of each individual category can come to conjure favorable understandings of the others. To the extent that the conflation is repeated frequently enough to saturate public consciousness, this illusion of continuity can become conventional wisdom, such that when any individual category comes to seem dirty in public opinion, it can be cleaned up via references to positive characterizations of the others. Normative alignment, then, is a mechanism that involves a great deal of 'smoothing over' and 'lining up', of constructing the interconnectedness of categories that might not otherwise go together.

To develop my discussion of normative alignment in greater depth, I underscore the related and well-established concept of heteronormativity: "the institutions, structures of understanding, and practical orientations that make heterosexuality seem not only coherent . . . but also privileged" (Berlant and Warner, 1998: 547–566). Fundamental to heteronormativity is the normative alignment of three categories – sex (man or woman), gender (male or female), and sexual orientation (gay or straight) – at certain cultural sites (Warner, 1991). Heteronormativity is premised on the belief that men are supposed to act male and desire romantic relations with women, that women are supposed to act female and desire romantic relations with men, and that anything else is unnatural or nonnormative. Mocking the sexual orientation of effeminate men or butch women reflects the heteronormative assumption that when an individual's display of gender is seen as inconsistent with their biologically defined sex, this must have implications for their sexual orientation as well. Judith Butler narrates an example of this alignment when she asks readers to

> consider the way in which heterosexuality naturalizes itself through setting up certain illusions of continuity between sex, gender and desire. When Aretha Franklin sings, "you make me feel like a natural woman," she seems at first to suggest that some natural potential of her biological sex is actualized by her participation in the

cultural position of "woman" as object of heterosexual recognition. Something in her "sex" is thus expressed by her "gender" which is then fully known and consecrated within the heterosexual scene. There is no breakage, no discontinuity, between "sex" as biological facticity and essence, or between gender and sexuality.

(Butler cited in Fuss, 1991: 27)

The critical point is that these alignments are, as Butler observes, illusions. There is of course no natural or innate reason to believe that men act or should act male, or that they desire or should desire sexual relations with women. Once one realizes that these alignments are illusions, it becomes possible to identify cultural sites as mundane as Aretha Franklin's song, "A Natural Woman," at which sex, gender and sexual orientation are lined up as if their alignment were natural. The dense accumulation of practices, representations and institutions that construct their alignment as a natural phenomenon shapes a collective sense about what is normal and beyond question. As Serlin notes, "the grandiose illusions of normalcy that are so deeply cherished by our culture" depend heavily on this idea of lining up, or "making identical" (Serlin cited in Smith and Morra, 2006: 159–160). Of interest is not just the perception of alignment but also the practices that make perceptions seem natural in the first place. Why and how do some people come to believe that being a woman naturally implies acting feminine and desiring sexual relations with men? How do such beliefs get established and reinforced? How do they come to seem natural?

Heteronormativity is a variant of normative alignment, and it has, of course, unraveled somewhat in recent years. If, for much of the twentieth century, many people assumed that men are supposed to act male and desire women, women are supposed to act female and desire men, and anything else is unnatural, then it is also the case that in recent years the alignment of sex, gender and sexual orientation has loosened. For many members of the public, exceptions (such as women who act male or men who desire men) may now seem normative. Normative alignment, then, can be dynamic over time, and those who fear its unraveling have worked hard to preserve it. (Consider, for example, the hundreds of millions of dollars and decades of effort invested into campaigns to prevent gays and lesbians from obtaining the right to get married.) In a military context, normative alignment can be fragile as well, which is why military officials have to work hard to reinforce it. What does normative alignment look like and how does it operate in a military context?

My argument is that negative aspects of service members, the military-as-organization, the American state and American foreign policy more broadly get swept under the rug via discursive and other practices which make the alignment of positive characterizations of all four seem natural and preordained in the first place. Two specific maneuvers deserve mention. First, scholars can be on the lookout for practices through which "the warrior becomes the emblem of a nation's identity" (Hedges cited in Godfrey, 2009: 206). Normative alignment, then, depends on the conflation of the soldier, the military, the state and the imperial project, as if they were the same thing. Second, scholars can be aware of instances in which favorable and unfavorable depictions of the troops have been connected with positive and negative representations of the military, state and empire. To the extent that service members, the military-as-organization, state and empire stand for each other as if they were the same thing, and to the extent that positive characterizations of them seem interconnected, it becomes that much more difficult to disable the public's uncritical adulation of the military and to notice unfavorable military conduct.

Consider several brief examples of how positive images of the troops can line up with virtuous representations of the US military, state and empire. In 1945 and 1946, a photographic exhibit titled *Power in the Pacific* toured the United States. Amid photographs of navy warships and convoys was a large photograph of smiling sailors with the accompanying text: "Yesterday these men were boys; today they are seasoned warriors" (Huebner, 2008: 70). Former President Ronald Reagan attended the burial of the Unknown Soldier from Vietnam on Memorial Day, 1984, and declared: "We may not know his name, but we know his courage. He is the heart, the spirit, and the soul of America" (Jeffords, 1989: 125). For Reagan, the anonymous soldier's body was glorified as a part of the national body, almost as if there were no difference between the two. From 1964 until 1993, the words "Bring Me Men" were emblazoned in two-foot-tall metal letters on an archway under which Air Force Academy cadets marched during their first day on campus (Emery, 2003). Echoing their origins in an 1894 poem titled *The Coming American*, the words suggested that a strong military and nation needed strong men to deliver the hopeful and civilizing promise of empire to a grateful world: "Bring me men to match my mountains / Bring me men to match my plains / Men with empires in their purpose / And new eras in their brains" (Foss, 1894).

At these and many other sites, the soldier, military, state and imperial project are conflated, as are their nobility, normativity and masculinity. Each example, then, is an illustration of normative alignment, of how and why the lining up of the related but distinct categories of service member, military, state and empire has come to seem natural. Through the endless repetition of such conflations, the reputation and resilience of each category have been enhanced. Perhaps an appropriate metaphor is the interlinking of four bank accounts such that when one is overdrawn, reserve funds from the others are automatically transferred and used to erase the deficit.

Conclusion

The American public, and perhaps other publics as well, is in deep denial about its own armed forces, and the denial is organizationally functional and highly constructed. An appreciation for the dynamics implicated in public denial may open up new ways of studying how the armed forces work. For scholars studying the American military and its relationship with the public, an appreciation of normative alignment and its role in camouflaging impermissible facts may open up opportunities for rethinking the meaning of various sites. For example, why were there only a half dozen people in the movie theater when I saw *The Kill Team*? Why, when I was walking through a terminal at Chicago's O'Hare Airport in 2013, trailing a group of enlisted service members, did civilian onlookers erupt in rounds of spontaneous and sustained applause as the troops made their way through the terminal? Why, during the early stages of the Iraq war, did Democrats and liberals spend so much energy expressing their support for the troops? Why, during preboarding announcements, does United Airlines invite active duty military personnel to board the plane early, even before customers sitting in first class? Why have so few commentators on the high rates of post-traumatic stress disorder

(PTSD) among veterans of recent wars failed to ask more explicit questions about the possible causal role of perpetrating violence against civilians? The discussion of the PTSD almost always includes vague references to exposure to violence in combat. But why do observers fail to ask more specific questions about the possible perpetration of illegal violence? And why are constant reports of rape among American service members narrated in terms of institutional failures to address the problem rather than an institutional success in creating a desperate culture in which socialization and training depend on the establishment of a master/slave dynamic among drill sergeants and recruits?

More broadly, my aim has been to illustrate one example of a contradiction that structures military organizations, and to interrogate how that contradiction came to be hidden in plain sight, or swept under the rug. My colleague David Serlin has argued that military organizations are genius at making contradictions disappear, as if there were no contradictions at all. This chapter has been a meditation on one such site.[2]

In my own research, I have benefited from close readings of fictional and nonfictional texts and a willingness to seek evidence from a range of sources and media. More important than the types of data that scholars seek and the methods they use to arrive at and verify the plausibility of inferences, however, my argument is that military analysis requires a critical eye, in particular an openness to the creation, maintenance and concealment of structuring contradictions and how they sustain military organizations. Military forces are structured by many contradictions whose existence and operation may be shrouded by many mechanisms and many normativities. My aim is not in any way to restrict a sensibility for structuring contradictions to the US Armed Forces or to the observations or themes addressed in this chapter. Rather, my hope is to persuade scholars working in the field of military studies that the choice of method is much less important than an appreciation for the complicated interplay of contradiction and concealment.

Notes

1 Portions of this chapter have been excerpted from Belkin, 2012.
2 Due to limitations of space, I have not elaborated on how this contradiction and its concealment have been central of military operations. For elaborations on that point, please see Belkin, 2012.

References

Belkin, A. (2011) *How We Won: Progressive Lessons from the Repeal of "Don't Ask, Don't Tell."* New York: Huffington Post Media Group.

Belkin, A. (2012) *Bring Me Men: Military Masculinity and the Benign Façade of American Empire*. New York: Columbia University Press.

Berlant, L. and Warner, M. (1998) Sex In Public. *Critical Inquiry* 24(2), 547–566.

Center for the Study of Sexual Minorities in the Military (2002) Rumsfeld's Claim That Pentagon Would Never Lie Seen as Untrue. *Center for the Study of Sexual Minorities in the Military*, 27 February. Santa Barbara: University of California Press.

De Bevoise, K. (1995) *Agents of Apocalypse: Epidemic Disease in the Colonial Philippines*. Princeton, NJ: Princeton University Press.

Department of Defense (2011) *Health Related Behaviors Survey of Active Duty Military Personnel*. Washington, DC: Department of Defense. Es-4; 3.

Emery, E. (2003) AFA Ends 'Bring Me Men' Era; Longtime Sign Falls in Shake-Up at Academy. *Denver Post*, 29 March.

Enloe, C. (2000) *Maneuvers: The International Politics of Militarizing Women's Lives*. Berkeley: University of California Press.

Foss, S.W. (1894) *The Coming American*.

Fuss, D. (1991) *Inside / Out: Lesbian Theories, Gay Theories*. New York: Routledge.

Gessner, R. and Runyan, D. (1995) The Shaken Infant: A Military Connection?' *Archives of Pediatrics and Adolescent Medicine* 149, 467–469.

Godfrey, R. (2009) Military, Masculinity and Mediated Representations: (Con)fusing the Real and the Reel. *Culture and Organization* 15(2), 203–220.

Huebner, A. J. (2008) *The Warrior Image: Soldiers in American Culture from the Second World War to the Vietnam Era*. Chapel Hill: University of North Carolina Press.

Jeffords, S. (1989) *The Remasculinization of America: Gender and the Vietnam War*. Bloomington: Indiana University Press.

Kramer, P.A. (2006) *The Blood of Government: Race, Empire, the United States & the Philippines*. Chapel Hill: University of North Carolina Press.

Krauss, D. (2013) *The Kill Team*. Dvd F/8 Filmworks.

Lutz, C. (Ed.) (2009) *The Bases of Empire: The Global Struggle against US Military Posts*. London: Pluto Press.

Mayer, J. (2008) *The Dark Side: The Inside Story of How The War on Terror Turned into a War on American Ideals*. New York: Doubleday.

Mazur, D. (2011) *A More Perfect Military: How the Constitution Can Make Our Military Stronger*. Oxford: Oxford University Press.

Parrott, E.H., Jr. (1982) CBS News, General Westmoreland, and the Pathology of Information. *Air University Review*, September.

Pew Research Center for the People and Press (2004) Pew Research Center for the People and Press. Washington, DC, June 17.

Pew Research (2013) Public Esteem for Military Still High. *Religion & Public Life Project*, 11 June.

Shilts, R. (1999) *Conduct Unbecoming: Gays and Lesbians in the US Military*. New York: Fawcett Columbine.

Smith, M. and Morra, J. (2006) The Prosthetic Impulse: *From a Posthuman Present to a Biocultural Future*. Cambridge, MA: MIT Press.

Traynor, I. (2010) US General: Gay Dutch Soldiers Caused Srebrenica Massacre. *Guardian*, 19 March.

Turse, N. (2013) *Kill Anything That Moves: The Real American War in Vietnam*. New York: Picador.

Warner, M. (1991) Fear of a Queer Planet. *Social Text* 29(4), 3–17.

SECTION 3

Experiences

16

THE AESTHETIC OF BEING IN THE FIELD

Participant Observation With Infantry

John Hockey

A useful and concise definition of participant observation is that offered by Emerson et al. (2002: 352): 'Participant observation – establishing a place in some natural setting on a relatively long-term basis in order to investigate, experience and represent the social life and social processes that occur in that setting.' The literature on using this method with infantry (e.g. Pipping, 1947/2008; Little, 1964; Ben-Ari, 1998; Segal, 2001; King, 2006; Tortorello, 2010; Irwin, 2012; MacLeish, 2012) and the similar occupation of private military contractor (e.g. Higate, 2012) is not extensive. One of the reasons for this lack of research is the still relatively closed nature of military institutions and the problem of gaining access to troops (Jenkings et al., 2011), but another reason may well be that researchers are put off from engaging with this occupational group due to their perceptions of the difficulties of engaging in fieldwork with them. This chapter will explore the reality of undertaking such research.

During 1979–1980 I carried out three months of participant observation with British Army infantry, fieldwork which covered initial organizational socialisation at a basic training depot, everyday life within an operational battalion in the contexts of barracks, field exercises in UK/Canada and operational deployment in South Armagh (Northern Ireland) while the conflict with the Provisional Irish Republican Army (PIRA) was ongoing. The great majority of this fieldwork was of the 24-hour variety, an immersion that provided the material for an ethnography (Hockey, 1986). My vocabulary of motives (Mills, 1940) for doing the research was a complicated one. As with most fieldworkers I did not come to the field empty-handed. I wanted to produce an ethnography of the subculture of the infantry and what I had in mind was a work which fitted with the Chicago School of occupational ethnographers who had pioneered urban participant observation. I chose that method, as I evaluated it as the only means which would allow me to chart the complexity of the subculture, in particular the relationships within it between conduct, context and change, with change being measured not in years but in days and hours. Change assessment is motivated, as Vidich (1971: 171) notes, by the 'desire of, and necessity for, individuals . . . to act in terms of what is possible in specific immediate situations'. That was the general research approach I aspired to and was inspired by. As a result the main theoretical lens I brought to participant observation was essentially a Meadian sociological social-psychology (Mead, 1934) which underpinned much of the Chicago School's output and which has a strong focus upon the interactional foundations of social processes. In addition, prior to entering university as a mature student I had been an army corporal and encountered sociology by chance, an epiphany which changed the course of my life. So I brought to the research a powerful desire to investigate sociologically my former life and to find out who I had been. The biographical impulse also meant that what I produced analytically, I now realise is not so much an ethnography of UK infantry but one essentially about infantry privates and noncommissioned officers (NCOs). I am struck now by what little material there is within my ethnography on officers, relatively speaking – a state of affairs propelled by my desire to tell a research story about 'other ranks' which at the time had not previously been told in a sociological fashion. After all, I had been one of them. What follows is based on my ethnographic fieldnotes, particularly those on my experience of doing participant observation, in other words my craft practices in the field.

On Context, Self and Trust

Doing participant observation effectively demands a combination of analytic and social skills. Crucially the former cannot be practiced unless a sound social relationship with one's participants has been established via the use of the latter, otherwise social closure is highly likely with the outcome being no insightful data and the possible demise of one's research. Thus, I knew the establishment of *trust* between myself and troops was to be the highest priority. The research context in which I found myself had a number of features which are unlikely to be found in civilian occupational environments in such an encompassing combination. First, the infantry is overwhelmingly a physically demanding environment. Second, the infantry was (and perhaps still is) an environment saturated with traditional masculinity (there are no women in the UK infantry) and heterosexuality. Third, there are episodic features of risk to participants within it.[1] All these features influenced my research process and I will point out their impact as this narrative progresses. Entering this particular social world as a researcher and 'stranger', I was met with a collective curiosity and evaluated not by my own standard but those of my research participants. This was aptly put to me by one private: 'We wanted to know what kind of lad you were and were interested to see what this civvie [civilian] had about him. We thought you might be some kind of short haired hippy given you were from the University!'

Prior to the research I had suspected that this kind of evaluation might take place and that there were likely to be tests of me before I could be defined as a 'good bloke' (or otherwise, and thus detrimental to my research), and thus someone who could be trusted and in turn engaged with openly. I therefore needed to pass the tests. Fortunately I had a number of resources available to deal with these challenges; these were a lifetime knowledge of working-class culture, military experience prior to entering university and a serious engagement with competitive distance running. Using these resources in Goffman's (1974) terms I presented a particular kind of self when in the field. This self was a thoroughly embodied one, for as Crossley (1995: 47) stresses, the mind is inseparable from the body as they remain 'reversible aspects of a single fabric'. The most habitual medium of testing was how I talked with troops. That communication had certain narrative and linguistic features rooted in wider UK male working-class culture but which were accentuated within the infantry subculture. For example, swearing was pervasive. Language was direct, immediate and often loud, and for the most part bereft of the more mediated, arguably stilted, linguistic forms characteristic of UK middle-class culture. Language in this way was used as a challenge to me to see how I responded, particularly to aggressive humour in the form of banter (see Collinson, 1988), in troops' argot: 'having the piss taken' or 'getting 'em going'. These rapid bantering forms demand a particular response in kind, so one takes it and gives it back and one then may or may not 'pass the test'. Showing embarrassment – or even more so, anger – constitutes failure, and provokes more aggressive banter. Some of these occasions were quite elaborate and on occasion constituted ambushes of a particular ilk for the researcher. For example, there was great interest in my sexuality based on a general perception of university life being something akin to 'a posh Play Boy Club' (as one private commented):

> I got wound-up earlier by the lads. Talking to Corporal G, who slowly brought up women and sex, and 'positions' (that was the trap!), this imperceptibly lead onto masturbation and when asked I admitted to it. Whereupon, the remainder of the Section who had been sitting ostensibly preoccupied with the intricacies of cleaning SLR's [weapons] suddenly collectively chanted 'Dirty Wanker!' Accompanied by vigorous, collective hand simulations . . . To which I replied with my best V sign and 'at least I am not having to go buy IT (sex) in Edmonton when we go on R&R!' [rest and recuperation]. The latter strategy being a stated objective of the Section. All then grinning and I got told by M that I was not a bad lad for someone who was Welsh.
>
> *(Fieldnotes, Training Area, Alberta, Canada)*

The last line of this example, of course, produced more banter about supposed Welsh characteristics; on we went, as these exchanges were pervasive within the subculture, and as an outsider one had to be always ready for them, perennially – on one's toes, so to speak – so as to be able to play this particular language game. To do this effectively I drew upon the biographical resources I have previously mentioned. The only comparable experiences my participants had of civilian outsiders had been a visit by various journalists, a group who in a lance corporal's words had 'hung about for a few hours, had drinks with the officers, and fucked off after being nosey'. In contrast I was staying for an extended period of time and, because I was also participating and not just observing, my physical performance was also viewed with intense interest. Simply put with few exceptions I did what 'the lads' did physically. So I marched for long distances carrying heavy loads, clunked along in heavy boots on morning runs, clambered over assault course obstacles, sweated, got cut, bruised, blistered, wet, frozen and fried on ranges and training areas. I stank with them after no showers for long periods. I defecated in the open air with them (sometimes collectively). I postured on early morning seaside promenades with them after nightclubs had thrown us out and 'friendly fighting' was 'just the job to end a good night out John!' (Private). In essence all this constituted a corporeal performance to my participants, illustrating not so much that I was fit enough to do the activities but rather more importantly that I was prepared to and would *endure with them.* What was interesting to discover about this process was that while I was fit (due to distance running) that did not mean that I was fit to TAB (march) with a large bergen (rucksack). Fitness comes in particular forms, so I had to rapidly relearn that particular form. In addition, lots of field exercise activities were completed on little sleep, again a somewhat uncomfortable relearning for the researcher. I knew from my previous military service that within the infantry world a fundamental theme of stoicism was prevalent. Simply put, one soldiered on in the face of adversity and, one did not 'jack one's hand in'. This stoicism also encompassed verbal interaction, as I was already aware that frequently complaining in the face of physical adversity was not a characteristic that was admired by troops. So while I may have complained lots internally, I did not habitually externalise those thoughts. The result of all this verbal and corporeal work was to realise a certain presentation of self, resulting in me receiving over time, indicators which pointed to me being categorised as a good bloke. The following examples I regarded as particularly salient:

> The Company is dug-in and has dropped off bergens, sleeping bags, all the heavy gear. At dawn the attack phase of the exercise commences and that means everyone will be having to move fast. Tonight we are freezing. Those who are not 'on stag' [sentry] get out of their trenches, and huddle together below the ridge-line in hollows. Jack a Lance Corporal, Jim and Dave – both privates, and me share a single poncho . . . With three it's just about effective and the mutual body heat is comforting. With a 4th body, the poncho is stretched too far, and the heat escapes, so you just freeze more slowly. Yet there is no suggestion I be excluded, rather I am told 'get in here dipstick [stupid]!'
>
> *(Fieldnotes, Training Area, Alberta, Canada)*

> Warrant Officer, Corporal and me standing having a brew waiting to watch 'Hot Gossip' [a dance group] on Top of the Pops [TV]. No seats left as this is the big highlight of the week and all who can watch pile in, some securing rickety chairs half an hour before kick-off. W.O to me: 'When you are out (on patrol) with the boys John, if there is a contact (with PIRA) do what they say, OK?' Before I can reply Cpl B responds: 'He'll be alright he's a good lad, not daft like'. HG starts, talk stops, the lads wolf whistle and all have 'eyes on' for the girls.
>
> *(Fieldnotes, Crossmaglen Base, South Armagh, Northern Ireland)*

Ultimately when doing participant observation of this type, if one trusts one's participants one has to commit to them keeping one safe, a commitment about which there seems to be relatively little fieldwork literature (e.g. Kovats-Bernat, 2002). Trust involves one making 'the "leap of faith" that brackets ignorance and doubt' (Mollering, 2006: 372); the following instance ended a long session of what was described as 'rough and tumble' on one morning parade:

> Lots of intermittent running, jumping, climbing, carrying kit and each other this morning. Ending in various I suppose what you can call confidence and team building exercises. The last was where the lads lined up in two lines and linked arms so

there was a sort of long cradle. Then everyone took turns in sprinting and diving headfirst into the cradle. On my turn the Sergeant shouts 'right lets drop John on his big university head!' – just before I sprinted. Lots of laughter. One, two three and into their arms! Very symbolic.

(Fieldnotes, Barracks, North West)

On Knowing

How I coped with being immersed in the infantry world was by using the aforementioned resources to talk to troops and engage in their activities. I also used these resources to look after myself mundanely. So for example, I knew how to look after my feet which often took a hammering given that I had not worn boots for years. I also knew to keep hydration and slow-burn carbohydrate levels topped up all the time, and also that possessing curry powder made any combination of composite rations palatable – even an all-in-one of rice pudding and Irish stew which I partook in collectively one freezing night on a Northern training area. This knowledge about care of the self was not only about physical comfort, for it also helped foster a relative psychological ease with aspects of the surrounding militarised environment:

> I have been struck by how I have been reacting to the lack of privacy when doing this research. I wonder how another fieldworker with a different biography would have reacted to that feature? My whole living space for me and my bit of kit is the middle tier of a 3 tier bunk. Top bunk bloke can touch the ceiling and we can all touch the side walls with our heads if we roll over. Two long lines of bunks with a small isle in the middle. No windows. Constant noise, light, patrols going out and coming in on a 24 hour cycle. Every spare inch festooned with weapons, bergens, etc. The bloke above me keeps his loaded mags [rifle magazines], spare ammo and smoke flares stuffed between his mattress and the springs. If I lift up slightly I can almost touch them with my nose! Of course I know what that kit can and cannot do, so I am ok with it being there. I realise I have slotted back into a way of living I intimately knew once.

(Fieldnotes, in the 'Submarines', Crossmaglen Base)

This knowledge and the commonplace practices which it facilitated meant that *crucially* I was not a nuisance to troops, and they did not have to look after me in situations in which there were often huge demands on their own resources. This was an additional factor which I believe contributed to their positive perception of me. Doing the research meant there was a certain degree of risk attached to it. On an everyday level, being around people who have loaded weapons constitutes said risk. Interestingly, when initially with armed troops I found myself watching how they moved around with weapons, as my past knowledge seemed automatically to be invoked in evaluating weapon handling procedures for safety. I did this even more so when in the same kind of situation with recruits, who after all are learners. I also used my prior knowledge to make decisions about my activity when risky situations manifested themselves. This knowledge and the decisions that flowed from it did not always contribute positively to the research. The prime example was when CS (riot gas) training was underway, which involved troops being exposed to the gas. I knew from my military experience that such exposure was highly unpleasant. Previously I had told troops that 'I would do things with you,' but in this instance I did not do so, motivated by a concern for my distance runner's lungs. I then experienced a period of social closure from them putting my research in jeopardy. Subsequently I had to work hard to recover my position of being accepted. In the main, however, my military knowledge worked productively for me when evaluating how much exposure to risk I was prepared to tolerate.

> Off with the Mortar boys! Who were looking forward to a Boss [excellent] time as lots of ammo was available. On mortar line with five teams firing, everyone happy. I am with one team and after about 10 minutes firing a bomb dropped in the tube does not fire. Misfire! Serious as the live bomb has to be extracted with a special canvas sling, taken away and exploded by the team NCO. He says to me: 'Time for you to fuck off John'. Do I stay or do I go? J circles the tube arms out like a plane shouting 'flap time, flap time', laughing and winding it all up. They were all looking at me so maybe I had a runaway stare on my face. I make the decision to stay because I know I will get cred with the boys but also because I know that the chances are overwhelmingly that the thing will not detonate when it's extracted.
>
> *(Fieldnotes, Alberta Badlands Training Area)*

This kind of evaluation was even more imperative in Crossmaglen, because when out with a patrol I made a mistake. The mistake was to concentrate sociologically on making sense of what was happening with the patrol. What that meant was that I lost any tactical awareness and that manifested itself in becoming too close to the next man in the patrol. This mistake meant that I increased jeopardy for both myself and the other members, much to their considerable ire. This occurrence caused much soul-searching about my responsibilities to myself and to my research participants. I was forced to consider these issues and do a form of cost-benefit analysis about the kind of information being out on patrol gave me sociologically. The result was that I made the decision to curtail my exposure to that level of risk unless the information I perceived myself to need was absolutely vital. The most significant features influencing the latter decision was first that my attention had become divided between two tasks which was folly in an operational environment. Second, but perhaps even more fundamentally, was the salutatory realisation that I was no longer skilful in corporeal military practices (Hockey, 2009), as civilian life had radically transformed my previous military embodiment.

On Trying to Be Analytic

There is a large literature on doing fieldwork as an insider researcher, and there is no space to trawl through these voluminous debates at this juncture. That said, I feel it might be useful to say a few words about my particular position in relation to the specific subcultural field of infantry. A salient point in this literature is to ask: how is one an insider? (Hockey, 1993). On one level I was an outsider due to my being a civilian, but on another level I had once been an insider and therefore was able to deploy insider knowledge in my research role. So for example, I understood a lot of the technical terms and argot troops used, but not all of the latter, some of which was unit-specific. Thus the term FUBS initially made no sense (fat, useless bastards) but Gimpy (general purpose machine gun) did. The latter had attached to it numerous other sedimentations of understanding. Thus I knew something of its range, what it weighed and how that felt over prolonged periods, how it was loaded with ammunition and so forth. This grasp gave me a bedrock of understanding so that my intellectual feet had something relatively stable to stand on initially. When one is doing participant observation there is no 'slo-mo' button: one cannot put the slow motion on, although when starting in the field I fervently wished for one. Initially, with infantry life ceaselessly unfolding around me, I was sometimes bewildered by its pace of change, and getting analytic purchase was difficult even with the bedrock of former membership and experience. What I held was something akin to a member's understanding of military reality, an understanding which admittedly as the years had passed had become somewhat blurred and inconsistent. Yet this understanding could still map out the main parameters of soldierly existence and it allowed me to relate very quickly to the meaning and import of troops' conduct.

Some members of the unit I was with knew I had done previous military service, but the great majority seemed not to, and I did not volunteer this information unless asked. What this allowed me to do was to ask naive questions about their life, and it allowed them to volunteer simple statements about organisational practices, and thus explain their world to me without expectations that I already had that knowledge. One of course can only fully take advantage of this approach if one is not naive. The advantage was that information flowed more freely between myself and the lads. Moreover, using this approach the possibilities that I was being, in unit argot, 'rubber dicked' (i.e. deceived) were minimised as my undisclosed previous knowledge constantly monitored what was presented to me. Additionally, by using this approach I hoped to minimise the degree to which I imposed my definition of events upon things. Volunteered statements I saw as much less open to my influence and manipulation by the military knowledge I already held; as Becker (1977: 60–61) notes, there are good sociological advantages in playing dumb. How much all this actually worked was at the time hard to judge, but I did try to maintain a reflexive awareness which kept me alert to the dangers of using my previous cookbook knowledge in a sociologically abusive fashion (see Ben-Ari, 1998: 135–136, for a similar anthropological concern). An event would occur and I would make an evaluation of what was happening, and then try and situate it against other similar occurrences in the research process so as to try and establish a pattern. At the same time I would contrast the happening with instances which I recalled from my own military biography (or not), and ask myself, 'is this *really* what is going on?' – a phrase which became something of a mantra for me, as I knew what the danger was. I had after all had been a soldier, and that was quite a seductive trope running through my consciousness. It was one which threatened to superimpose my past onto the troops' present, with the danger of making a distorted and lazy analysis of the field.

The combination of using participant observation and my own partially militarised biography then produced, arguably, an ethnographic narrative of some analytic depth. I was able to construct this narrative because I was saturated with data which came at me on occasion literally twenty-four hours a day. Subsequently I have asked myself the question, 'could I have done the research with a different method?' For example, I could have produced an account based on interviews. However, what this would not have given me was the thickness of the data I accrued. What I mean by that is the everyday minutiae of soldierly existence occurring in their contexts of use repeatedly:

> The section are lying up, watching a track for the exercise 'enemy'. Afternoon winter rain has been sweeping over us, everything brown, green, grey, in fast falling light. Waterproofs are not really waterproof in this. D__ is lying next to me and after a while motivated I suspect by my chattering teeth he thrusts one of his water bottles at me, whispering to me with a grin 'here this will sort you out!' It's full of barley wine, brilliant!
>
> *(Fieldnotes, Training Area North)*

This instance constitutes a 'document of' (Wilson, 1974: 68) an underlying pattern of troops' behaviour, one which was aimed at making their lives easier in all contexts, one which was summed up by their use of their phrase, 'Stupid soldiers can always be uncomfortable.' Yet that pattern of behaviour was essentially context dependent for its operationalization. Thus, later when on operations in South Armagh I asked D__ if he was still carrying barley wine, he replied: 'No not me, I would never do that, here it's the sharp end, you can't fuck around like that it's too risky.' It is this kind of thick data that participant observation can give the researcher and which would be very difficult to access even by other qualitative methods of data collection. One does participant observation by degree, by which I mean that the level of involvement with research participants can vary (Hockey, 1993). I chose to 'do things with the lads', and apart from the odd drill parade and the CS gas incident previously depicted, my participation in their activities was high. I could have done much more observation than participation, but I am again of the view that my understanding of the complexity of the meanings specific to their social world would have been less. Immersion in activities produces not only an understanding of their meanings, but it also produces exposure to sets of feelings and emotions which in turn feed back into the construction of meaning:

> At the end of a 12 mile speed march with bergens and weapons, it was flat out 'balls to the wall' as the lads put it. We drop the bergens and head over to the truck on the back of which there is a brew [tea] urn, where the Colour Boy [Colour Sergeant] is also dishing out choci bars. You can tell everyone is absolutely knackered because no one is talking, and everyone has that kind of drained look. Brews and choci do the business though and soon smiles and 'that was Mega rough' comments emerge.
>
> *(Fieldnotes, Alberta, Badlands)*

What this episode produced in the researcher was gross fatigue, and I therefore had not just a cognitive but also a sensory understanding of that as a normal feature of the infantry life. Also in that instance I understood why the Colour Boy was defined by troops as a good NCO, because he was, to cite a private, 'always looking after the lads'. In addition, the particularly high level of meaning attached to mundane treats became revealed as the combination of chocolate and hot sweet tea had its visceral effect upon bodies which had just been pushed to their physical limits. The point here is that all this this analysis came from one particular instance full of thick data.

On an Aesthetic Engagement

In the cult film *Repo Man* (Cox, 1984), about the occupation of repossessing automobiles in Los Angeles, a novice learns from a veteran (brilliantly played by Harry Dean Stanton) that 'the life of a repo man is always intense!' In a similar fashion one can claim that doing partic- ipant observation constitutes the most intense research method available to those wishing to do ethnographic work. The researcher's self is the main research instrument via which data is obtained and one is liable to spend long periods of time up-close and personal with one's research participants. Having opted for that method, if one then goes and makes the decision to do it with infantry soldiers and commits to engaging in their activities, one had better be prepared for conditions of heightened intensity, particularly in relation to the features of phys- icality and episodic risk, in addition to a heterosexual rumbustiousness which characterises their general behaviour (Hockey, 1986: 112–122). I constructed and deployed a way of being with infantry to which there was a particular aesthetic. Traditionally, an aesthetic way of being has largely been equated with activity described as expressive, evocative, beautiful, sacred, sublime and artistic (Haapala, 2005: 39). However, this position neglects other important dimensions of experience, namely struggle and the mundane. As Leddy (2005: 8) states when calling for an aesthetics of the mundane, such an analytic lens should include not just the evocative but also displeasure. Dewey's (1980: 2) work is of help here, as he places aesthetics firmly in the realm of everyday life so that any kind of experience can be aesthetic as long as it constitutes an intensification of ordinary experience. He also notes that people are often struggling to maintain an equilibrium with their surrounding environment (Dewey, 1980: 12). That striving, that intensification, that constant adaptation and re-adaptation constitutes a process out of which a particular aesthetic consciousness can be formed. For Dewey, essential to the latter are experiences of 'heightened vitality' (1980: 18). In the infantry world, mun- dane work requires considerable vitality compared to most civilian occupations. Training for operational deployment and that deployment itself demands a concentrated focus of energy, individual and collective, in which absolute alertness to and immersion in the task at hand is needed. This is what Dewey terms 'wholeness', which for him constitutes the core of aes- thetic being as individuals are totally immersed in their activity.

My fieldwork experience was an aesthetic one in Dewey's terms, for it embraced a com- bination of physically and intellectually demanding activity, together with a need for stoicism when grappling with that combination often in difficult conditions both environmentally and psychologically. It demanded unremitting attention to both my physical self and to my socio- logical eye. This fieldwork alertness was evident, for example, in the mundane instance of me adjusting the position of my bergen to alleviate the debilitating load and simultaneously trying to figure out why a private opposite me is habitually pronouncing himself to be in a state of NFI (not fucking interested), while simultaneously insisting on carrying the section's heaviest weapon (GPMG) for nearly all of our twenty-two-mile march until he was stopped by cramp at seventeen miles. One is immersed in the moment and its meaning when a drunken private waves a beer bottle at one while proclaiming one to be 'a spy for the officers'; or when one has an overwhelmingly poignant conversation with someone who had recently accidentally killed a friend at the end of a South Armagh operational patrol; or when one squats impotently looking on in Crossmaglen as medics work feverishly on a casualty of a PIRA attack and one knows that this for him is touch-and-go. Throw in periods where I was often short of sleep and I would periodically realise that I had done no sociology for a few hours at a time, then becoming immersed in intense frustration at failing with the job at hand. My fieldwork then consumed me with its intensity, and I now realise I was probably at the age limit for doing it at least physically, at the time being thirty-four years old, while the privates who surrounded me averaged out at eighteen to twenty years. That intensity of the experience resulted in after-effects which I had not foreseen. I recreated and deployed parts of a dormant self during the fieldwork, but upon returning to the overwhelmingly middle-class, mediated milieu of university, life was not without problems. Initially I was louder, brasher, more spontaneous and I cursed habitually, all forms of conduct which, in small ways, are akin to what Garfinkel (1967: 58) has called 'breaching practices'. As my female partner of the time noted, 'what was upsetting was it wasn't the you I knew.' There is an intensity to the experience of doing participant observation generally given the amount of time and energy needed to arrange being in the field with a group of people. Once in the field, that intensity will fluctuate depending on the degree of 'commitment' (Becker, 1977: 261–273) the researcher displays and thus involvement in the activities of the group being studied. The more committed, arguably the more encompassing the fieldwork experience will be, and thus in Dewey's terms the more *aesthetic*. Participant observation as a method for gathering data harbours the potentiality for the researcher to be consumed or saturated by the fieldwork experience. This is an experience which can bring both costs and benefits, intellectually, physically and emotionally. There is then a need for a greater portrayal of the aesthetic dimension of this particular ethnographic method. This is particularly so when engaging in research with the military, as the costs under certain conditions have the potential to be extensive.

Given these kinds of intensity, can one's ethnographic research questions be answered using methods other than participant observation? If one decides to do the latter, *how* is one going to be involved, and what are the limits? What resources does one have to have to be able to deal with the kinds of activities and stressors I have outlined? Are the research benefits liable to be worth the physical and psychological exposure? These are issues which fieldworkers need to give serious thought to before making the decision to put their booted feet into the infantry world.

Acknowledgements

Thanks are due to Rachel Woodward for her sharp editorial eye, also once again to 'A' Company and of course the Mortar Boys!

Note

1   A fourth feature for the research context was ordered by military law, and I have written elsewhere (Hockey, 1996) on how I as a researcher responded to the treatment of troops by superiors, which by civilian standards was authoritarian and on occasion draconian.

References

Becker, H. S. (1977) *Sociological Work: Method and Substance*. New Brunswick, NJ: Transaction Books.

Ben-Ari, E. (1998) *Mastering Soldiers: Conflict, Emotions and the Enemy in an Israeli Military Unit*. Oxford: Berghahn.

Collinson, D. L. (1988) Engineering Humour: Masculinity, Joking and Conflict in Shop-Floor Relations. *Organization Studies* 9(2), 181–199.

Cox, A. (Dir) (1984) *Repo Man*. Venice, CA: Edge City Productions.

Crossley, N. (1995) Merleau-Ponty, the Elusive Body and Carnal Sociology. *Body & Society* 1(1), 43–63.

Dewey, J. (1980) *Art as Experience*. London: Perigee.

Emerson, R. M., Fretz, R. I. and Shaw, L. L. (2002) Participant Observation and Fieldnotes. In Atkinson, P., Coffey, A., Delamont, S., Lofland, J. and Lofland, L. (Eds.), *Handbook of Ethnography*. London: Sage, pp. 352–368.

Garfinkel, H. (1967) *Studies in Ethnomethodology*. Englewood Cliffs, NJ: Prentice Hall.

Goffman, E. (1974) *The Presentation of Self in Everyday Life*. Harmondsworth: Penguin.

Haapala, A. (2005) On the Aesthetics of the Everyday: Familiarity, Strangeness and the Meaning of Place. In Light, A. and Smith, J. M. (Eds.), *The Aesthetics of Everyday Life*. New York: Columbia University Press, pp. 39–55.

Higate, P. (2012) Cowboys and Professionals: The Politics of Identity Work in the Private Military Security Company. *Millennium* 40(2), 321–341.

Hockey, J. (1986) *Squaddies: Portrait of a Subculture*. Exeter: Exeter University Press.

Hockey, J. (1993) Research Methods: Researching Peers and Familiar Settings. *Research Papers in Education* 8(2), 199–225.

Hockey, J. (1996) Putting Down Smoke: Emotion and Engagement in Participant Observation. In Carter, K. and Delamont, S. (Eds.), *Qualitative Research: The Emotional Dimension*. Aldershot: Avebury, pp. 12–27.

Hockey, J. (2009) Switch On: Sensory Work in the Infantry. *Work Employment and Society* 23(3), 477–493.

Irwin, A. (2012) There Will Be a Lot of Old Young Men Going Home: Combat and Becoming a Man in Afghanistan. In Amit, V. and Dyck, N. (Eds.), *Young Men in Uncertain Times*. Oxford: Berghahn, pp. 59–78.

Jenkings, N., Woodward, R., Williams, A. J., Rech, M. F., Murphy, A. L. and Bos, D. (2011) Military Occupations: Methodological Approaches and the Military-Academy Research Nexus. *Sociology Compass* 5(1), 37–51.

King, A. (2006) The Word of Command: Communication and Cohesion in the Military. *Armed Forces & Society* 32(4), 493–512.

Kovats-Bernat, J. C. (2002) Negotiating Dangerous Fields: Pragmatic Strategies for Fieldwork Amid Violence and Terror. *American Anthropologist* 104(1), 208–222.

Leddy, T. (2005) The Nature of Everyday Aesthetics. In Light, A. and Smith, J. M. (Eds.), *The Aesthetics of Everyday Life*. New York: Columbia University Press, pp. 3–22.

Little, R. W. (1964) Buddy Relations and Combat Performance. In Janowitz, M. (Ed.), *The New Military*. New York: Russell Sage, pp. 195–223.

MacLeish, K. T. (2012) Armor and Anesthesia: Exposure, Feeling, and the Soldier's Body. *Medical Anthropology Quarterly* 26(1), 49–68.

Mead, G. H. (1934) *Mind, Self and Society*. Chicago: University of Chicago Press.

Mills, C. W. (1940) Situated Actions and Vocabularies of Motive. *American Sociological Review* 5(6), 904–913.

Mollering, G. (2006) Trust, Institutions, Agency: Towards a Neo-Institutional Theory of Trust. In Bachmann, R. and Zaheer, A. (Eds.), *Handbook of Trust Research*. Cheltenham: Edward Elgar, pp. 355–376.

Pipping, K. (1947/2008) *Infantry Company as a Society*, trans. P. Kekale. Helsinki: Finnish National Defence University.

Segal, D. R. (2001) Is a Peacekeeping Culture Emerging Among American Infantry in the Sinai MFO? *Journal of Contemporary Ethnography* 30(5), 607–636.

Tortorello, F. J. (2010) *An Ethnography of 'Courage' Among U.S. Marines*. PhD dissertation, University of Illinois.

Vidich, A. J. (1971) Participant Observation and the Collection and Interpretation of Data. In Filstead, W. J. (Ed.), *Qualitative Methodology: First Hand Involvement With the Social World*. Chicago: Markham, pp. 164–173.

Wilson, T. P. (1974) Normative and Interpretive Paradigms in Sociology. In Douglas, J. D. (Ed.), *Understanding Everyday Life*. London: Routledge & Kegan Paul, pp. 57–79.

17

ETHNOGRAPHY AND THE EMBODIED LIFE OF WAR-MAKING

Kenneth MacLeish

Stewart had a lot of problems, each tied haphazardly to the next, with many of the ostensible solutions constituting new problems in their own right. It wasn't just that he had been to war – several times – and come back hurt and anxious. It was that every salient aspect of that hurt and the life that surrounded it seemed to be wrapped up somehow with the Army, this institution that had put him in harm's way in the first place, that he was now dependent on for his care and healing and livelihood, and that couldn't seem to make up its mind between hanging onto him and sending him packing. The Army was Stewart's problem, and he was its problem.

This mutual predicament of survival, endurance and the management of life exposed to death is, for me, a central aspect of the critical study of military institutions. War is a tool of policy, a product of macro-level social forces, and an object of aestheticization, narrative and mediated representation. But it is at the level of institutional life that war shapes and takes shape in the lives, feelings and bodies most directly bound up in its production and effects. The details of how war is made may at first seem to have little to say directly about war's ostensive roots, aims and meanings. But critical, ethnographically grounded accounts of war as embodied experience provide both a destabilizing challenge to the things we think we already know about war and a generative, concrete anchor to the questions we want to answer about it. I'm not claiming here that 'experience' is an antidote to obscuring or misbegotten 'representation', but rather that taking embodiment seriously as both a product of and foundation for social life (Scheper-Hughes and Lock, 1987) is crucial to understanding the overdetermined conditions of military life and the dynamic, unruly, traumatized and tedious ways that people live with those conditions. Because bodily experience is, like culture itself, always already 'out in public' (Geertz, 1973), the embodied experience of war and military life – the capable, vulnerable, maimed, wondering, desiring, grieving and killed bodies that are war's tools and materials – represents a crucial intersection of the phenomenology of experience with systems of disciplined violence production and the public culture representations and discourses that make war legible.

Stewart's problems were just one such intersection.

Mutual Provocation

As Stewart described his situation to me, we sat in warm afternoon sun out in front of the building housing a civilian soldier support organization where I worked as a volunteer – a low, slightly decrepit structure finished with cheap linoleum and panelling, down at the west end of the US Army's Fort Hood in central Texas. It was 2008 and I was in the middle of twelve months of fieldwork looking at the everyday life of war-making there. Many of the 50,000 or so soldiers who were at Fort Hood then had seen multiple war-zone deployments. Stewart himself had been to Iraq three times. He was a truck driver, or rather a motor transport operator, what soldiers call an 88M, after its official military occupational specialty code – per the NATO phonetic alphabet, the M is pronounced 'mike'. First with the Texas National Guard and then with an active-duty Army unit, Stewart worked convoy security for fuel and supply trucks traveling between Kuwait and southern Iraq and drove Heavy Equipment Transporters (HETs), the monster-sized flatbed tractor trailer trucks that are used to move Abrams tanks and Bradley Fighting Vehicles. They ran days-long convoy missions, a misery of tedium, anxiety and regular ambushes and roadside bomb attacks.

Stewart was tall and rangy, with jutting jaw, sharp nose, shaved head, and a dark narrow bristle of regulation moustache. Like many of the soldiers who patronized the organization and availed themselves of the TVs, couches, Internet access and free snacks on offer in its homey, chaotic lounge, Stewart had been assigned to Fort Hood's Warrior Transition Unit (WTU). The WTUs were created in the wake of the 2007 Walter Reed Army Medical Center neglect scandal (Hull and Priest, 2007) to provide a uniform, force-wide system of care for soldiers the Army deemed too sick or injured to do their regular jobs. They were cycled into WTU to be treated and evaluated under the close supervision of caseworkers and commanders while the able-bodied took their places in deploying line units. This holding pattern was meant to last no more than ninety days, but it rarely took less than six months, and frequently ran to over a year (cf. Dao and Frosch, 2010). Stewart's most immediate impairment was a partially torn right anterior cruciate ligament that left his knee swollen to the size and proportions of a cantaloupe – as he showed me by removing his knee brace and pulling the digital-print camo fabric of his uniform trousers tight around it. It was painful and immobilizing, but according to the doctor he had just seen, the surgery to correct it was 'elective' and wouldn't be happening any time soon, if at all. A ligament in his other knee was damaged too, and his rotator cuff was torn – prosaic injuries born of the sitting, standing, jumping and lifting that his job demanded. He had also been diagnosed with posttraumatic stress disorder (PTSD) and a traumatic brain injury (TBI) from a roadside bomb blast.

His body had been doubly vulnerable and doubly hurt: taxed, worn out and damaged not only by the enemies trying to kill him but also by the apparatus that had put him there in the first place. And now this third round: doctors confirming the source of his pain and impairment but offering no solution, an Army that regarded him as useless but would not yet allow him to leave. The less Stewart's body conformed with the Army's needs, the more tightly it was wrapped up in mechanisms of monitoring and management, drawn further in even as it was being denied and cast out. Every new moment of their interaction affirmed both the Army's control over Stewart and Stewart's alienation from the Army.

I knew that Stewart had been sent home early from his most recent deployment, and I asked if it was because of his injuries. It was not. He explained to me that while he was in Iraq the last time, his recently ex-wife back home in Texas was systematically looting his house and his bank account, and there was nothing he could do to stop it. He appealed repeatedly for help to his immediate superior, his squad leader, with whom he already had a fractious relationship. But he was told, in the Army vernacular's typical combination of macho dismissal, tough-love solidarity and metaphorics of work and duty, to simply 'suck it up and drive on.' The squad leader's indifference continued until one day Stewart sarcastically demanded whether he had to slit his wrists to get someone to pay attention to his problem. Whether from obtuseness, spite, a sense of bureaucratic obligation or an opportunistic desire to be rid of a soldier whose problems he could not solve, the squad leader elected to ignore the obvious theatre and exasperation of Stewart's remark and interpret it literally. At the squad leader's instigation, Stewart was medically evacuated from Iraq, and he was back in Texas within seventy-two hours.

Before I had a chance to remark on the paradoxical tension between the Army's urgent response to Stewart's false suicidality and its ponderous indifference to his physical injuries, he went back to complaining about his doctors, his WTU sergeant losing his paperwork, and the motorcycle licensing course he was going to miss because he had been forbade from doing anything that might further endanger his knee. The Army's apparent ownership of his body was so thorough that, as he somewhat hyperbolically (and apocryphally) exclaimed, "I could get an Article 15 [a nonjudicial disciplinary citation] for damaging Army property!" At the same time, he also downplayed the severity of his own experience: he had been in combat, but only a bit; he had been hurt, but not nearly as badly as others he knew; he'd had it hard, but there was far worse out there.

Each of Stewart's problems testifies to a different burden or predicament of contemporary modes of organized war-making. Each is a 'signature wound' marking what Jennifer Terry calls the 'mutual provocation' of technology, geopolitics and the biopolitical management of life (Terry, 2009), the latest generation in the modern interrelationship of tools for killing, techniques for lifesaving, and arguments about who should be the objects of such things and why. The convoys Stewart was driving were a vital part of a US counterinsurgency doctrine that thoroughly interweaves combat, security and reconstruction operations with one another while, paradoxically, providing an abundance of targets for insurgents as it spread heavily armed and armoured American personnel all over the country. His injuries came from long days of physically demanding work often performed while wearing thirty pounds of unforgiving, movement-restricting Kevlar and ceramic plate armour, and from the particular combination of that armour with insurgent bomb technologies that left his flesh relatively intact but his brain concussed and damaged in poorly understood ways. His body, a valuable if malfunctioning piece of 'Army property', was being worn out in ways both acute and gradual. Stewart's marital breakup and his absence during its aftermath were precipitated in part by the Army's gruelling operational tempo of continuous multiple 12- to 15-month deployments with little time at home in between. Even the continuing presence of the bullying, indifferent squad leader, if he was truly as unfit for his position as Stewart's descriptions suggest, can be read as a sign of personnel demands pushed by a relentless operational tempo and the imperative to 'make numbers' and 'fill slots'. Stewart's apparently spurious medevac and subsequent languishing in the WTU testify to a military medical system struggling to manage an unprecedented volume of challenging medical conditions while maintaining the maximum utility of soldier bodies and trying to minimize both cost and public outrage.

Whether any of these problems is defined as originating within or outside the Army, inside or outside the concern of Stewart's commanders and leaders, inside or outside of Stewart's own strained capacity to continue his work is itself arguably a product of the institution's capacity to determine its objects of intervention and the limits of its responsibility amid the halting messiness of everyday life. Stewart's embodied experience was the ground for this exercise in distinction. He waited, stuck, he and the Army not done with one another.

Agent, Instrument, Object

One approach to stories like Stewart's is to imagine that the facts can simply speak for themselves, giving evidence of war's brutality and unreason, its insult to human life and dignity. But such a framing is only possible with other frames of reference already firmly in place: political narratives asserting what war is *about* (security, liberation, geostrategic opportunism, cronyistic plunder), doctrines that affirm war's necessity, inevitability or rightness, and lore that exalts, pities or pathologizes those who fight it. All these things present their own unshakeable doxa, making certain statements about war obligatory (it is hell, it is a tragedy, it is unavoidable, *support the troops!*) and others unsayable (that it might seem sexy or fun, that it might be a good way to get rich or at least avoid being poor). There is no end to the things we think we already know about war. Social science itself is not devoid of such assumptions,[1] reinforced in its prejudices by the fact that the normalized violence production of military institutions remains a relatively novel and underexplored topic in the broader anthropology of violence (Lutz and Millar, 2012).

Stories like Stewart's, products of ethnographic presence and engagement, force a reckoning with these reigning articles of faith. His unruly trajectory begins in the morass of medical bureaucracy and prosaic but disabling pain; flashes back to the terror and boredom of the war zone; gains a different momentum in the rancour and vulnerability of ex-marital conflict; takes an abrupt left turn through the nonemergency of nonsuicide; and reaches its murky, inconclusive end back in a present in which no clear narrative has yet been crafted and yet much is clearly at stake. The various aspects of Stewart's embodied and intimate suffering shift in and out of significance: suicide is called into existence with great urgency and then vanishes; his injuries prompt diagnosis and management but not treatment; he is both a problem for the Army and the privileged object of its care and interest. Each kind of story that might be told about Stewart would fit him into a different role – malingerer, dupe, burnout, schemer, unsung hero, victim of neglect – and each of these would suggest a different version of what his experience was *really about* and which other details were therefore secondary: 'collateral damage' that is in fact the whole point.

In few places is life as explicitly instrumentalized and rapturously valorized as it is in military institutions, in which service members are subjected to a disciplined unfreedom and empowered to forms of violence often completely contrary to the values of the civilian society they come from and defend, but as a result of which, paradoxically, they come to embody the ideal of citizenship and the epitome of individual virtue (Lutz, 2001). Marcel

Mauss characterized the body as "man's first and most natural instrument" (Mauss, 1973: 75), and the soldier's body arguably the most necessary and most carefully managed component of the war apparatus. And while civilians are the primary casualties of most modern wars, the soldier *is* that apparatus in its most indivisible, cellular form. The soldier is at once the agent, instrument and object of state violence. He or she is coerced and empowered by discipline, made productive by being subject to countless minute and technical compulsions (Foucault, 1979). The soldier is permitted by the sovereign power to kill in the name of upholding the law, but is also allowed to be killed, placed outside the law's protection as he is sent into harm's way (Schmitt, 1985; Agamben, 1998). And he or she is the subject of extensive measures to protect and maintain life, to keep him or her alive and able to continue working, fighting and killing effectively, a biopolitical subject not merely kept from dying but also made to live (Foucault, 2003). The same mechanism is responsible for both sustaining and endangering soldiers' lives. Throughout all this the soldier also contends with the mundane, chronic bodily burdens of doing military work that often do not rise to the level of straightforward injury – the condition of being slowly worn out that Lauren Berlant calls 'slow death' (Berlant, 2007). And his or her general condition is one of profound *vulnerability* to violence and institutional intervention (MacLeish, 2013).

Stewart's account clearly illustrates the interconnection of the violence of the foreign battlefield, the ambivalent care and bureaucratic indifference of the institution, and the interpersonal attachments of the home, with all their attendant frictions. In the tense and uncertain present of ongoing war-making, the present in which Stewart found himself, the differences between rule and exception, function and dysfunction, or goal and side effect were not terribly clear. They had to be asserted with great force by doctors, sergeants, commanders, and even Stewart himself as a backward-looking projection. In such a setting, the significance of the body is not merely that it is a symbol that is "good to think with" (Scheper-Hughes and Lock, 1987; Douglas, 1996), but that it is a "subject that it is 'necessary to be'" (Csordas, 1993: 135). Stewart's problems don't merely *stand for* something else. Rather, their phenomenology constitutes him as a product of the military's world of necessity, his sensations and experiences and bodily self arising from where the Army has told him to be and what it has told him to do.

Taking military embodiment as an object to be explored in its own right rather than a mere substrate to or reflection of seemingly more fundamental factors makes it possible to tell very different stories about war and soldiers' experiences. Stories about the careful management and deliberate exposure of human life force a recognition of the centrality of the destroyed bodies through which war's "force and status of material 'fact'" arise (Scarry, 1987: 63). This concrete embodiment of state power, including its prosaic and unspectacular forms described earlier, suggests a much broader politics of war and its analysis, one in which ideology, myth and structural prerogatives are always already wrapped up in the organized harms, visceral contradictions and mutual provocations of normal military life.

Nervous Systems and Persistent Tensions

Actually getting methodological purchase on those mutual provocations in the context of everyday life can be challenging. In my own work, navigating my way aimlessly around the sprawling and seemingly anonymous landscape that was a common feature of Fort Hood and its host city, Killeen, it was at first incredibly daunting to try to imagine where I would actually find the object of my analysis, the place where war showed up and was made real. I felt this anxiety again when I actually began getting to know people and discovered that the glaringly obvious presence of the military was an object of soldiers' constant commentary and critique, even as its role in their lives was utterly routinized, normalized and hidden from an outsider by a veritable fortress of shorthand, slang, technical terminology and an encyclopaedic mastery of rules. It was hard to tell whether the institution was the basis for normalcy or an egregious intrusion upon it. Even the stereotypes of military folk as debased, dangerous and out of control (cf. Lutz, 2001) – stereotypes I had conscientiously tried to gird myself against – turned out to be a basic feature of how people talked about one another and themselves. Not only was there no way for me, a civilian, to be 'inside' of this thing, but it began to seem as though the military itself, predicated in so many ways on distinctions and sharp boundaries, might not have an inside, at least as I had imagined it.

I eventually came to the notion that this apparent contradiction might actually be the best way to characterize much of Army life. Fort Hood and places like it can be framed as what Michael Taussig (1992) calls a 'nervous system' – shaped by the continuous, generative tension between the imposed rationality of an institution like the military, its excessive and contradictory rules and regulations, and the irreducible exigencies of daily life within the institution. It is a *system* because it presents itself as a comprehensive, unfeeling and monolithic order; it is *nervous* because life within it is dynamic, agitated and full of unruly feeling. It is the space marked out by the rule, the exception to the rule, and the practice of living with the rule and its exception. Everything is structured, but there is much that can't be counted on. There is a rule for everything, but not always a reason. What is 'normal' is not necessarily tolerable, yet one lives with it anyway.

Perhaps this ambiguous relationship to being in the military should not have come as a revelation. Goffman writes of the total institution, that such entities always exist in relation their boundaries and their others, and may not need in practice to be as 'total' as they seem. Rather, "The full meaning of being 'in' or 'on the inside' does not exist apart from the special meaning . . . of 'getting out' or 'getting on the outside.'" Total institutions do not demand "cultural victory," but instead "sustain a particular kind of tension between the home world and the institutional world and use this persistent tension as strategic leverage on the management of men" (Goffman, 1968: 13). Indeed, even in a setting where the 'home world' *is* the institution, those who inhabit it remain constantly attentive to the ways that the two sides unduly infiltrate on another (cf. Hawkins, 2005; MacLeish, 2013: Ch. 4). The total institution is, in this sense, always already 'nervous': inescapable but deliberately incomplete. Its intrusiveness, excess, irrational hyperrationality and frequent self-contradiction shouldn't be shocking or scandalous then, even if they are hard to live with. I would argue that a different kind of critical perspective is possible when we don't see these things as resolvable problems and treat them instead as constitutive features: these things are not side effects; they are just part of the nervous tension by which the system bends life to its purposes and life pushes back.

What the ethnography of military institutions can offer here is a view not of the imagined inside, but of the constant policing, performing and imagining of the boundaries between in and out. The presence of an outsider – one who embodies an *elsewhere* to the ethnographic mise-en-scène (Marcus, 1998) – can even serve as a salutary provocation for the definition of these boundaries. The ethnographer's awkward presence and clueless questioning can help call these common-sense definitions into articulation.

But this is not just a methodological issue. Soldiers and those close to them know how capable the Army is of defining and redefining the boundary between inside and outside, between its institutional obligation and the personal accountability of the individuals who labour on its behalf. And they themselves constantly and variously assert, with words and actions, their own boundaries, their own notions of what the Army is or ought to be responsible for, what they do or do not owe it, what it can and cannot claim of their lives, and what civilians get right and wrong about them. They may claim, again like Stewart, that they are proud *and* in need of help, cynical about *and* satisfied with their work, disillusioned *and* deeply patriotic. These kinds of ambivalence, irony and self-consciousness about outsider perceptions are not merely important in their own right, but are themselves a kind of expression of 'cultural intimacy' (Herzfeld, 1997) – an expression testifying to the shifting and uncertain quality of life in this realm of shifting and uncertain boundaries. This agitated and sprawling nervousness is central to the blatant visibility and hyperrational excess of the system, proliferating boundaries, thresholds, divisions between inside and outside, and productive effects at every scale and level, drawing lines that bind and divide between a thing and its observer, between an institution and its servant.

These same dynamics were a signal feature of the Foundation, the volunteer, on-post organization where I met Stewart and many of my other interlocutors. It was started by Debbie, a local civilian woman in her mid-forties who began handing out cookies to soldiers at departure manifests at Hood in 2003. She eventually carved out a hybrid position as charitable organizer, surrogate family, and institutional advocate for a broad and heterogeneous mix of soldiers who had, to use a phrase frequently invoked at the Foundation, 'fallen through the cracks'. Her many roles testified to the basic unmet needs of many soldiers and military families – needs that were not supposed to exist, in a sense, because the Army was meant to tend to them.

The Foundation's exceptionalism – a civilian entity in the midst of a military post, outside the chain of command – made it a sort of collecting place for things that had overrun standard categories and procedures and troubled the distinction between inside and outside – people, problems and relationships that the institution seized on for leverage and others that it chose to leave alone. This included the priorities and burdens of 'real' life, like family and livelihood. The disciplinary power exercised over soldiers' bodies itself possessed a kind of nervous incompleteness as it encountered flesh that was resistant simply by virtue of being hurt and tired (Foucault, 1979). Even as bodies, lives and relationships fell through the cracks, they somehow fell further *in* to new categories that they had to navigate, disavow or work themselves out of. As with Stewart, the categories of the sick soldier, the hurt soldier, the simply unlucky soldier and the 'bad' soldier all bled into and mutually produced one another. If the exception proves the rule, the Foundation was a place to see the rules along with their effects and excesses, a place where rules, procedures and explanations were experienced as having full, formative force but also as alien to any sense of normal life, and where, as a result, they were talked about constantly. Was it better to let the Army docs fuse your damaged vertebrae, or should you wait to get out and do it at the Veteran's Health Administration Hospital? Could the Army say your sleep apnoea made you nondeployable and then deploy you anyway? Why hadn't you gotten a paycheck in twelve weeks? Why did your furniture get sent to Alaska when you moved to Texas? Could a first sergeant you didn't know really force you to cut your lawn? Could your husband's command make him take his meds? Why did your head still hurt six months after a bomb blast? Each question like this was inevitably surrounded by a pocket constellation of conflicting answers, opinions, commentary, jokes and more questions. It was an intensified version of what seemed to be a general military condition: people who were avid critics of the contradictions of their condition, expert navigators in the impersonal but intrusive world of closely governed life, often as a matter of survival.

This was very much a world that, from an ethnographic perspective, was not so much exotic as it was "fully understood by no one [outsider or native] and that all [were] in search of puzzling out" (Marcus and Fischer, 1999: xvii). The proverbial text of cultural knowledge and practice that the ethnographer aims to read over the native's shoulder, to borrow Clifford Geertz's famous metaphor (Geertz, 1973), was on the one hand, glaringly accessible and obvious – there was an incredible abundance of regulations, symbols, protocols and folklore that people I met were eager to tell me about when they learned that I studied 'culture'. But this text was also perilously opaque in its bureaucratic hyperrationality, and these same rules, signs and traditions were frequently objects of frustration, critique and vexed wonderment in everyday military talk. Getting access to this text did not confer any great insight or translate into being straightforwardly 'in' anything, in the classical immersive ethnographic sense. But what it did allow for was a view to the many thresholds and distinctions that seemed to me to define life at Fort Hood, and that so often hinged on differential exposure to harm, violence and the vicissitudes of power – distinctions between soldier and civilian, enlisted and officer, those who had deployed and those who hadn't, the injured and the healthy, the green and the experienced, the ignorant and the wise, the dedicated and the lazy, those who saw combat and those who stayed inside the wire, soldiers and spouses, and men and women. These categories didn't line up or nest neatly. Rather, they were multiple – combining into a restless whole that was "more than one, but less than many" (Mol, 2002: 82; Deleuze and Guattari, 1987). They sat uneasily with one another from moment to moment and circumstance to circumstance, with claims of often high-stakes difference carved from shared possibilities.

This ethnography of exception and breakdown departs from a systematic critique of political economy and ideology in which a monolithic notion of 'militarization' serves as the ultimate explanation for all aspects of military life. It also avoids the hazards of an overly narrow commitment to the closed world of the military that leaves both the corporate culture and political role of the military unquestioned. Instead, it favours a partial and situated approach to the exigencies and bodily conditions of military life itself (Haraway, 1991; Strathern, 2004), 'in the gap' (Stewart, 1996) of dynamics like persistent tension, perpetually unfinished discipline, the endurance of slow death, and the contradictions of US global geopolitics. The perpetually constructed and reconstructed distinctions between being in and out described earlier work as tools of the institution, features of life lived within it, and terms for how it is figured and represented.

Involvement and Imagination

Americans, military and civilian alike, insist that only those who have been to war 'really know' what it is like, and we often take this presumption to mean that only those who have been to war are qualified to comment on it. This is an assumption from which even scholarly inquiry may not be immune, and it limits that inquiry in at least two significant ways.

The first limitation is structural. The division of inside and outside – those who have been and seen and those who have not – configures soldiers as heroes, monsters or victims, and civilians as cheerleaders, concerned onlookers or blithe free-riders in relation to the violence of war. As Catherine Lutz has elaborated, such stereotyped oppositions do tremendously powerful work distancing civilians from the violence done in their name and confining it to the bodies and lives touched by it first-hand (Lutz, 2001). At a certain point, some scholars have argued, such disavowal of knowledge and authority arguably becomes an abdication of responsibility – if we cannot know, then we need not be bothered to know, and indeed our ignorance becomes a form of solemn respect (MacLeish, 2013: Ch. 5; Wool, 2015). Just as the military institution draws selective boundaries around its care for and ability to dispose of soldierly life, so this sort of benign-seeming, common-sense refusal to know draws selective boundaries around the deployment of and responsibility for lethal violence. Both cases elide the fundamental complicity between protecting life and instrumentalizing it. A story like Stewart's, we might then say, is not only about how he and the Army are yoked together in the entangling work of his wearing out and survival, but about how militaries and liberal politics (and the kinds of common-sense but limited critiques it affords) are similarly entwined, producing law, right and reason that are inevitably underwritten by violence, the deaths of soldiers, enemies, and, overwhelmingly, innocent civilians. Valuing and dignifying Stewart's experience means recognizing the deliberate and routine conditions that produced it, and acknowledging those conditions can prompt us to interrogate our own standards of value and dignity, of the worth that is attached to using some bodies to kill other bodies.

The second, related problem is epistemological. The ethnography of embodied war experience offers a way to explore and even to bridge the perpetual construction of war's mutually exclusive inside and outside, and to trouble the categories by which it is known in the first place. When so much work is being done simply to make and reproduce categories, it is better not to take them for granted, and if as scholars we believe that it is not possible to imagine experiences different from our own, we may as well stay home. Because platitudes about the limits of understanding war are so engrained, it may seem presumptuous or subversive to produce scholarly work that invites readers to identify with the experience of others' exposure to or participation in it. But as philosopher and medical ethnographer Annemarie Mol suggests, it can also serve as a provocation to disidentify with the solidity of our own experience and expectations. Her book *The Logic of Care*, an extended essay on the management of chronic illness, is peppered with second-person descriptions of patient experience, a stylistic choice that she prefaces with this disclaimer:

> In this book you will not find sentences such as: 'We cannot imagine what it must be like to have a chronic disease.' Sentences like that are nasty. They do not state explicitly that the author and reader are in good health, but they imply it all the same. That is not what I am after. On the contrary, I want to avoid unmarked normality. [. . .] 'I' am not un-mortal or immune to disease. And your normality, dear reader, is not

presupposed here either. Instead, I will use all my rhetorical skills to seduce you – whatever your current diagnosis – to take up the patient's position while you read. The unspecified 'you' in this book tends to be someone with diabetes. Whether or not you happen to have that disease, I kindly invite you to imagine yourself involved in the situations described.

(Mol, 2008: 13)

Critics and scholars of war and the military might ask how 'sentences like that' imply that both author and reader and their world are untouched by war. In place of such presupposition, ethnography can seduce and persuade and invite in the interest of avoiding the 'unmarked normality' that typically keeps war at arm's length. War and its organized institutional forms are a fundamental part of the order and logic of contemporary political life, and we are all potentially 'involved in the situations described' in scholarship about it. That is to say, sufficiently seductive ethnographic writing can cause the inside-outside distinction to disappear as we consider the ways that we are all of us already inside the systems that produce war, and it can allow our imagination and identification – and care – to extend to those whose relationship to war is different from or more immediate than our own. In the process, we can begin to make sense of war's predicaments of survival in ways that would otherwise seem extreme, startling or unfathomable.

Acknowledgements

This chapter is a revised version of an article titled 'The Ethnography of Good Machines' (MacLeish, 2015), which appears in *Critical Military Studies* 1(1). Portions of this work originally appeared in *Making War at Fort Hood: Life and Uncertainty in a Military Community* (MacLeish, 2013) and in papers and remarks presented at the meetings of the American Anthropological Association, the American Ethnological Society, and the Society for the Social Studies of Science. Particular thanks to Can Aciksoz, Omar Dewachi, Alex Edmonds, Erin Finley, Derek Gregory, Sarah Hautzinger, Bea Jauregui, Cathy Lutz, Marcel LaFlamme, Seth Messinger, Vinh-Kim Nguyen, Rachael Pomerantz, Jean Scandlyn, Tyson Smith, Emily Sogn, Nomi Stone, Lucy Suchman, Catherine Trundle and Zoë Wool for their thoughts. The fieldwork for and write-up of this research were supported by a National Science Foundation Graduate Research Fellowship, a University of Texas Continuing Fellowship and a National Institute of Mental Health postdoctoral fellowship at the Institute for Health, Health Care Policy and Aging Research at Rutgers University.

Note

1   The lively debate – which in US anthropology is still ongoing – about the engagement between scholars and military institutions is beyond the scope of this piece, but interested readers should see Network of Concerned Anthropologists, 2009; González, 2010; Kelly et al., 2010; Hoffman, 2011; McNamara and Rubinstein, 2011.

References

Agamben, Giorgio (1998) *Homo Sacer: Sovereign Power and Bare Life*, trans. Daniel Heller-Roazen. Stanford: Stanford University Press.

Berlant, Lauren (2007) Slow Death (Sovereignty, Obesity, Lateral Agency). *Critical Inquiry* 33(4), 754–780.

Csordas, Thomas (1993) Somatic Modes of Attention. *Cultural Anthropology* 8(2), 135–156.

Dao, James and Frosch, Don (2010) Feeling Warehoused in Army Trauma Care Units. *New York Times*, 24 April, sec. Health. Available at: http://www.nytimes.com/2010/04/25/health/25warrior. html?_r=1&th&emc=th

Deleuze, Gilles and Guattari, Félix (1987) *A Thousand Plateaus: Capitalism and Schizophrenia*, trans. Brian Massumi. Minneapolis: University of Minnesota Press.

Douglas, Mary (1996) *Natural Symbols: Explorations in Cosmology*. London: Routledge.

Foucault, Michel (1979) *Discipline and Punish: The Birth of the Prison*. New York: Vintage Books.

Foucault, Michel (2003) *Society Must Be Defended: Lectures at the Collège De France, 1975–76*. New York: Picador.

Geertz, Clifford (1973) *The Interpretation of Cultures; Selected Essays*. New York: Basic Books.

Goffman, Erving (1968) *Asylums*. Chicago: Aldine.

González, Roberto J. (2010) *Militarizing Culture: Essays on the Warfare State*. Walnut Creek, CA: Left Coast Press.

Haraway, Donna Jeanne (1991) *Simians, Cyborgs, and Women: The Reinvention of Nature*. New York: Routledge.

Hawkins, John P. (2005) *Army of Hope, Army of Alienation: Culture and Contradiction in the American Army Communities of Cold War Germany* (2nd edition). Tuscaloosa: University of Alabama Press.

Herzfeld, Michael (1997) *Cultural Intimacy: Social Poetics in the Nation-State*. New York: Routledge.

Hoffman, Danny (2011) The Subcontractor: Counterinsurgency, Militias, and the New Common Ground in Social and Military Science. In *Dangerous Liaisons: Anthropologists and the National Security State*. McNamara, Laura, and Robert Rubinstein, (Eds). Santa Fe, NM: School for Advanced Research Press, pp. 3–24.

Hull, Anne and Priest, Danna (2007) The Hotel Aftermath. *Washington Post*, 19 February. Available at: http://www.washingtonpost.com/wp-dyn/content/article/2007/02/18/AR2007021801335. html

Kelly, John D., Jauregui, Beatrice and Mitchell, Sean T. (Eds.) (2010) *Anthropology and Global Counterinsurgency*. Chicago: University of Chicago Press.

Lutz, Catherine (2001) *Homefront: A Military City and the American Twentieth Century*. Boston: Beacon Press.

Lutz, Catherine and Millar, Kathleen (2012) War. In Fassin, Didier (Ed.), *A Companion to Moral Anthropology*. New York: Wiley-Blackwell.

MacLeish, Kenneth (2013) *Making War at Fort Hood: Life and Uncertainty in a Military Community*. Princeton, NJ: Princeton University Press.

MacLeish, Kenneth (2015) The Ethnography of Good Machines, *Critical Military Studies* 1(1), 11–22.

Marcus, George E. (1998) On the Uses of Complicity in the Changing Mise-En-Scène of Anthropological Fieldwork. In *Ethnography Through Thick and Thin*. Princeton, NJ: Princeton University Press, pp. 105–131.

Marcus, George E. and Fischer, Michael M. J. (1999) *Anthropology as Cultural Critique: An Experimental Moment in the Human Sciences* (2nd edition). Chicago: University of Chicago Press.

Mauss, Marcel (1973) Techniques of the Body, *Economy and Society* 2(1), 70–88.

McNamara, Laura and Rubinstein, Robert E. (Eds.) (2011) *Dangerous Liaisons: Anthropologists and the National Security State*. Santa Fe, NM: School for Advanced Research Press.

Mol, Annemarie (2002) *The Body Multiple : Ontology in Medical Practice*. Durham, NC: Duke University Press.

Mol, Annemarie (2008) *The Logic of Care: Health and the Problem of Patient Choice*. London: Routledge.

Network of Concerned Anthropologists (2009) *The Counter-Counterinsurgency Manual: Or, Notes on Demilitarizing American Society*. Chicago: Prickly Paradigm Press.

Scarry, Elaine (1987) *The Body in Pain: The Making and Unmaking of the World*. New York: Oxford University Press.

Scheper-Hughes, Nancy and Lock, Margaret M. (1987) The Mindful Body: A Prolegomenon to Future Work in Medical Anthropology. *Medical Anthropology Quarterly*, New Series 1(1), 6–41.

Schmitt, Carl (1985) *Political Theology: Four Chapters on the Concept of Sovereignty*. Cambridge, MA: MIT Press.

Stewart, Kathleen (1996) *A Space on the Side of the Road: Cultural Poetics in an "Other" America*. Princeton, NJ: Princeton University Press.

Strathern, Marilyn (2004) *Partial Connections*. Walnut Creek, CA: Altamira Press.

Taussig, Michael T. (1992) *The Nervous System*. New York: Routledge.

Terry, Jennifer (2009) Significant Injury: War, Medicine, and Empire in Claudia's Case. *WSQ: Women's Studies Quarterly* 37(1), 200–225.

Wool, Zoë (2015) *Emergent Ordinaries at Walter Reed*. Durham, NC: Duke University Press.

18

BITING THE BULLET

My Time With the British Army

Vron Ware

So many power structures – inside households, within institutions, in societies, in international affairs, are dependent on our continuing lack of curiosity.

(Enloe, 2004: 3)

The sociological imagination enables its possessor to understand the larger historical scene in terms of its meaning for the inner life and external career of a variety of individuals. The sociological imagination requires us to grasp history and biography and the relations between the two in society.

(Mills, 1959: 5–6)

In the late autumn of 2008 I was travelling back to London after a day's fieldwork at a military training centre in Surrey. When I got off the train at Waterloo I noticed two uniformed soldiers in front of me and recognised one of them as the officer in charge of a diversity recruitment programme whom I had recently interviewed. I caught up with him at the ticket barriers and we chatted for a few minutes in the crowded rush-hour concourse before going our separate ways. As the two men vanished into the throng, I had the immediate sensation of seeing myself standing there as though I were naked. Ashamed to be thought of as complicit with the British government's war machine, I felt acutely self-conscious that I had just been talking to soldiers in public. Three or even two months earlier this would not have been fathomable. Apart from the fact that men in camouflage suits were seldom seen on public transport, I did not think of myself as someone who was able to cross that extraordinary divide between the familiar social world and the hostile apparatus of military power. At that moment I felt undeniably uncomfortable, but I resolved to put this new awareness to good use and to maintain that visceral sense of estrangement during the rest of my research.

Earlier that year I had experienced another shock when I arranged to meet my contact officer at a university building in Camden Town, north London. I planned to interview him about his experience of diversity management since he was working on employment policy,

231

and we were also going to discuss a schedule for my fieldwork. I had only met him once before when he was wearing battledress, the patterned khaki uniform used for everyday wear, and of course when he arrived in a suit he looked completely different, as though he was in disguise. His ability to 'pass' as a civilian left me feeling slightly wrong-footed and unsure how to relate to him. Since there were no refreshments in the building, he suggested going over to the Pret A Manger round the corner to get a cup of coffee, and this threw me as well. A voice in my head asked in astonishment: but this is our world, how does he know his way about in it?

These experiences of traversing the psychological line between what was civilian and what was military – both of which seemed to involve 'them' encroaching on 'my' home ground – were the result of an auspicious encounter at the start of that same year. In January I had requested a meeting with the adjutant general of the British Army after he had invited me to come and speak to him about the increasing numbers of Commonwealth migrants working in his organisation. The context of our original contact was important. We had met, in February 2007, at a weekend conference on 'Britishness' organised by the British Council and the Ditchley Foundation. I was finding it a rather dispiriting affair, with many self-important individuals keen to sound off on their pet themes and few opportunities for dissident voices to be heard. The country was embroiled in two wars in Iraq and Afghanistan, and on learning that there was a senior military figure present I was even more alienated. However, during the course of the afternoon, I was struck by the heartfelt tones in which the general spoke of the young men and women under his charge. While this may have made him easier to approach, I had my own reasons for listening to him. After years of protesting against war in all its manifestations I had come to question my ignorance about Britain's military institutions, not least how they could be organised and equipped to attack another sovereign country with no democratic mandate.

As Cynthia Enloe has famously pointed out, the moment one becomes newly curious about something is also a good time to think about what created one's previous *absence* of curiosity (Enloe, 2004: 3). An opportunity to speak directly to an army officer was too good to miss, and I found a chance to talk to him on his own. I was officially attending the event in my role as a writer, commissioned to produce a book about Britishness which would somehow encompass all these debates (Ware, 2007). To my surprise it turned out that the general was keen to talk to me too since, in his words, the army wanted to be part of that conversation as well. It was not until some time later that I understood that he was referring to the controversy caused by the rising proportion of non-UK citizens in the army, a situation that was later resolved by capping their numbers in certain sections rather than across the board. However, in the meantime I put his card safely away until I was in a position to take up his invitation.

Our subsequent meeting the following year took place in a different climate. The profile of the armed forces had altered considerably in that period due to several factors. There had been direct interventions such as the Military Covenant Campaign, the launch of Help for Heroes, and the then Prime Minister Gordon Brown had just announced the governmental inquiry into the 'National Recognition of our Armed Forces' (Davies et al., 2008). British fatalities in Helmand had continued to escalate, and the crowds gathering to pay respect to the coffins being repatriated through Wootton Bassett were beginning to attract media attention. In the intervening time I had started to think about what it might mean to investigate the army as a social and cultural institution. I was wondering where the military belonged in relation to the rest of the public sector, for example, and if it was so important as a supremely national body, why was it so secretive and opaque?

An Otherwise Crowded Map

As I was soon to discover, the modernisation of the British Army from 1998 onwards is not a story that is widely known. The task of piecing it together entails detective work on many levels: from researching the enactment of old and new laws governing equality and diversity to tracking procedures for reporting bullying or harassment; from scrutinising employment tribunals to making use of ever more detailed collections of statistics and monitoring reports. It means learning about the particularities of military culture with all its hierarchical structures and bonding rituals, which is not an easy task for an outsider. It then involves asking how racism, homophobia and sexism might be factors that prevent cohesion between soldiers, as well as contributing to violent crimes committed against detainees and civilians in combat zones. To compile this story also demands an analysis of the impact of secularisation, not least the provision of multifaith chaplains, since military service is steeped in Christianity through links with Crown, church and state. And finally, exploring this recent history also requires an alertness to the ways in which the impetus towards diversity, far from being an imposition, can actually acquire its own momentum in a military setting.

This last point can be illustrated in several ways: first, a visible degree of diversity allows the army to promote its multicultural and gender-neutral workforce as an index of its professionalism and proficiency. Second, the significant presence of minority ethnic soldiers means that the institution can claim to be inclusive and reflective of society, as long as few questions are asked about the demographic make-up of that diverse cohort. Third, minorities, and this includes women, are often utilised for cultural skills that can be promoted as assets in communicating with civilians in the combat zone.

Grasping the complexity of these different strands and agendas might have been overwhelming had I been better informed at the start. Fortunately I was a novice when it came to studying the social situation of the armed forces and the complex terrain known as the civil-military relationship. Later I would suffer occasional flashes of dread when I realised what I had taken on, but by then it was too late to turn back. As I began to educate myself on how to study military organisations, I realised that there was relatively little material available on social relations within the contemporary army and virtually nothing on the history of institutional racism. Looked at from a sociological perspective, the UK Armed Forces were certainly absent from academic discussions about social cohesion, institutional racism, national identity, gender differences, equality and diversity. In short, as far as I could see, the British military sector represented a blank space on an otherwise crowded map, much like the areas that Google camera cars are forbidden to enter. And as I became better acquainted with a rich literature on the politics of military service in different national contexts and in different historical periods, the ingredients of a particularly British discourse became more discernible.

This introduction intentionally underlines the fact that my research began as a leap in the dark. At that time the world I entered was so unfamiliar that it took many months before I was capable of piecing together what I found. Following the initial jump, the investigation took the form of a journey – guided by interlocutors as well as by intuition – to discover how the institution worked and how the different parts fitted together. It was also a foray into the heart of Britain. I visited many places I had never been to before and learned more about the UK's internal geography. I also saw recognisable landscapes with utterly new eyes. The real sensation of exploring happened when I went through military checkpoints into otherwise inaccessible training centres and regimental headquarters. I did find a different world, but at the same time I knew I was deeper inside the same country. As a result of my travels I came to understand the colossal imprint of military history, language, memories and ways of thinking on mainstream British culture.

Half the Battle

Following the publication of *Military Migrants* (Ware, 2012) the most common question I was asked was: how on earth did you persuade them to let you in? There are several layers of explanation here, but the simple answer is that I met the adjutant general (AG) at a conference where he invited me to investigate the situation of migrant soldiers. That same book on Britishness, published soon after we met, had ended with an account of a peace vigil organised by Women in Black and a call for global revolution (Ware, 2007). The AG had a copy on his desk when I went to see him. However, it was not just a decision for a high-ranking individual with carte blanche to invite curious writers to nose around the establishment. I was obliged to wait several months before a commercial publishing contract with the Ministry of Defence (MoD) was negotiated and the necessary ethical research protocols carried out.[1] Securing permission to carry out interviews among serving personnel on army property is really half the battle. Without the endorsement of the right office, let alone the right officer, it is not possible to get past the security checkpoint, or even to find somebody willing to talk.

Obtaining the contract was my first indication of how hard it was to get access to the institution, and it was made clear that my manuscript would have to be submitted to the MoD for reasons of security and accuracy before publishing. Eighteen months into the fieldwork there was an attempt to challenge my credentials on the basis that I had not gone through the right channels, and this made me realise that I had just been lucky.[2] By the time I finished I felt as though I had slipped through a crack in a wall that magically opened up for a few seconds and then closed behind me. Not only did I have the endorsement of the head of the army's human resource department, I was also placed in the care of a senior employment officer whose posting fortunately lasted as long as my research. This meant I had some continuity in an organisation in which people seemed to move on every three years and then vanish without a trace.

Once the question of access and permissions was all sorted, my formal interviews began in July 2008. An account of my early days of fieldwork conveys something of the learning process that I undertook at the time. My contact officer knew that I had no previous experience of military institutions, and set up the first round of interviews with officers with overall responsibility for recruiting and training 'foreign and Commonwealth' soldiers (FCs). As I painstakingly wrote up my notes in response to their PowerPoint presentations and briefings, I tried to familiarise myself with army acronyms and institutional habits, acutely sensitive to all that was strange and different in this new environment. I will never forget my first day, when I saw a tall uniformed man striding through the corridors with a tiny dachshund at his side. I subsequently realised that it was normal for staff – including civilian secretaries – to bring their dogs to work, and so I came to accept the presence of a bed under the desk as a common sight. The absence of cats spoke for itself.

There were plenty of other idiosyncrasies that struck me, not least the use of language. Of course I wasn't party to the more demotic versions of army slang more commonly used by soldiers, but it was fascinating to hear certain words and phrases being picked up by those for whom UK English was not their first language. Learning to banter was a particular cultural challenge, and several people complained about the difficulties of remembering not to use swear words when they went home to their families. As well as acquiring a new language, recruits undergoing the first phase of training often spoke of significant physical changes, such as increased fitness, improved deportment and a new-found ability to get up early. But perhaps the most striking feature of military life that I noticed was the way in which individuals related to each other in accordance with the chain of command.

For a start, I observed that when two people in uniform pass each other they perform a sequence of actions that acknowledge not only difference in rank, but also the degree to which they might know each other or how often they encounter each other. Most of the time this looks like a version of saluting, accompanied by a verbal greeting which can vary between a perfunctory grunt to an informal exchange, although it is always reciprocal. On a training base, therefore, new recruits have to be inducted into this holistic system of deference which requires acting as a soldier at all times. When moving from one block to another, I was frequently accompanied by young trainees, self-consciously swinging their arms in an exaggerated fashion, just as they had been instructed. On one occasion I observed a new boy – he appeared to be quite a fresh recruit – admonished for his less-than-upright deportment as he crossed paths with a senior officer. And this training in physical discipline begins even before the recruits step into their uniform. I spent a few days visiting the army selection centre, where recruits undergo final tests before signing up. I watched as the candidates, still wearing their civilian clothes, assembled to walk over to the army canteen. Eager to impress, the young men followed instructions to form ranks and then set off in step, arms moving stiffly in rhythm. The fact that we were all civilians made no difference at that point, and I was the one left feeling a bit awkward as I had to break into a trot to keep up.

Here I must also add that my observations were not all fixated on what was different, as I began to look with new eyes at the university workplace and institutional practices too. After an absence of more than seven years from a British university, I was becoming acquainted with the ever-changing procedures for annual career appraisals and the internal complaints system. When it came to new buildings there were some striking similarities such as the demarcation of staff refectories as 'The Hub' or the use of faceless multinationals like Sodexo for catering. As more and more reconstruction was being carried out on military bases I also learned that the blocks for single soldiers were built along the same lines as much of the new student accommodation proliferating across London and on university campuses. These continuities and comparisons were just as important as the jarring disjunctions in learning to situate military institutions within a larger landscape of national, educational and professional bodies undergoing neoliberal forms of transformation.

I Thought I Knew About Britain

My interviews with Commonwealth soldiers began at the Army Training Centre at Pirbright, Surrey, where recruits destined for the 'trades' sections[3] spend fourteen weeks undergoing the first phase of their education. In early August 2008 I arrived at the security gates at the preordained time and, although I was expected, I distinctly remember the adjutant responsible for facilitating my interviews expressing frustration that nothing had been organised as he had instructed. It was there that I understood that the army was basically like any other workplace, with all manner of inefficiencies and miscommunications – the underlying joke being, of course, that this was the army. Needless to say, a visit from an academic researcher interested in Commonwealth recruits was hardly going to be a top priority in a training establishment processing a hectic turnover of students at a time when the military machine was stretched to the limit.

However, on this first occasion some candidates were quickly rounded up and I was taken to a room where five men were waiting: four from Fiji and one from St Vincent. Since they were sitting in a row facing the front of the room, I was obliged to go and sit opposite them. This meant I had my back to a chalkboard so that I was positioned as a teacher, or at least as someone in authority. I was so concerned to put us all at ease that I didn't tape the conversation – it somehow seemed impolite – and had to make notes as we spoke. Perhaps due to the novelty of the situation, I felt a certain bond with these particular men and sought them out several times again before I attended their graduation ceremony, or 'passing out', a few weeks later.

After the allotted time for this interview ran out, I was shown into a different room where I met the next group. I would later become familiar with this generic setting: the portraits on the walls, photos of winning teams, gleaming sporting trophies in glass cabinets and in the centre, a highly polished wooden table. Usually known as the history room, this was a repository for regimental record-keeping which usefully functioned rather like a front parlour for receiving guests. On this occasion, the young men ranged around the table were evidently more eye-rubbingly disoriented than I could claim to be. They had arrived from Grenada and St Vincent the previous week and this was only their second day – they had only just received their uniforms and undergone the obligatory haircut. Their palpable disorientation was hardly a surprise since a military training regime was likely to be a shock to any civilian, but perhaps I was more open to their awkwardness as I too was a complete outsider. However, I quickly learned that some of them also felt alienated from their younger and less well house-trained British peers, as my notes on that day recorded:

> I thought I knew about Britain but it is different. The other recruits are untidy, don't like to shower. They are used to a level of things in life. In the Caribbean we tend to adapt. We are surprised at the level of drugs and smoking. We have discipline at home, corporal punishment.

Thus although I was able to sympathise with their process of adjustment – something I could barely imagine – their responses to what they saw as British cultural norms also highlighted things that might have been unremarkable to me had they not pointed them out.

This realisation prompted a further degree of reflexivity that would underpin my subsequent research.

The third set of interviews, with another group of new entrants, this time from St Vincent, Grenada, Gambia and Nigeria, passed in a similar vein. After this I felt confident enough to ask if there were any women (or females as they were called) available, and a group of three from Malawi, Fiji and Zimbabwe was quickly assembled. Since they were further into the course they were a good deal more relaxed and talkative, and they also appreciated a reprieve from drill practice. My notes from that meeting corroborate my vivid recollection that, rather than keeping to the earlier format where I asked all the questions and individuals answered in turn, a dynamic quickly developed between the four of us so that issues and perspectives emerged as a result of the interaction. This was to be the first example of the most rewarding and stimulating group sessions that I would experience during fieldwork. At times I would feel that I was chairing a seminar rather than holding interrogations.

I have outlined my first day of interviews not simply to provide a frank account of beginning my research project but also to underline that this was not a simple ethnography. I cheerfully left the premises with plans to come back the following week, but when I reached home that evening, famished and mentally exhausted, it took some time to reacclimatise. It was as though I had travelled to a very distant place over which the shadow of war hung low and heavy. At the same time, there was something unutterably mundane and yet undeniably 'other'. On every occasion that I set off to garner more information and testimonies, regardless of whether I was visiting somewhere I had been before, I would have to steel myself for the day's ordeal. I quickly learned to pick up a sandwich on the way out so that I could devour it as soon as I sat down in the train on my return journey. In spite of there being no shortage of brews (hot tea) in a gay assortment of mugs there was rarely time to stop for lunch, and even if I went to the canteen with my interlocutors, there was very little edible vegetarian food to be had in any case. But the point is that my research was not strictly ethnographic. That is to say, it did not entail living on an army base or require immersion in a military community. On two occasions I stayed in the officers' mess because of the distance from alternative accommodation, but apart from a couple of nights in bed and breakfasts in Yorkshire, it was possible to digest the materials in the sanctuary of my own home before planning my next venture. In this way I was able to maintain the equilibrium between estrangement and familiarity that I had early on identified as a crucial component of my research ethic.

My fieldwork lasted from the summer of 2008 until February 2011, although 2009 was the most intensive year. During this time, I completed my study of the training centres and then focused on particular trades, such as the Royal Logistics Corps and Royal Artillery before moving on to some of the infantry regiments where there was a high proportion of FCs. I made a trip to Germany, spent a day in Sandhurst and visited the Gurkha regiment in Folkestone. I stayed overnight in the Infantry Battle School in Brecon and attended the annual conference of the Armed Forces Buddhist Society in Hampshire. After weeks of begging to meet spouses I attended a 'wives' meeting' in a sergeant's mess and visited a number of military families in their homes. And between all the prearranged meetings I would talk to the drivers dispatched to ferry me to and fro, occasions that bore their own ethnographic fruit but which also helped to ease my transition between life inside and outside the military ecosphere.

As time went on, I began to realise that I was documenting a period of social and cultural history that might otherwise remain unexplored. I was aware of my responsibility to locate and contextualise this chapter of military recruitment and institutional attention to 'diversity' within the longer sweep of Britain's colonial and postcolonial past. In other words, it was essential to impart a sense of temporality, or at least construct a matrix of overlapping and intersecting timelines, in order to make sense of the disparate forms of evidence I was accumulating. This applied as much to the policies and practices of the army, the MoD and government as to the shaping of the wider political and cultural narratives. It was also true when it came to tracing the story of Britain's newest Commonwealth soldiers as well. With this in mind, I was able to know exactly when my fieldwork was completed.

Forensic Fieldwork

The decision to extend military recruitment to citizens of Commonwealth countries was announced in February 1998 when the then Home Secretary John Reid told the House of Commons that, after a period of review, the existing five-year residency requirement for military recruits was to be suspended. But the impetus for turning directly to countries like Fiji and Jamaica began after a specific episode which took place at the Edinburgh Tattoo in August that same year. Impressed by their performance at the Tattoo, recruiters from the Royal Scots approached members of a Fijian marching band with an invitation to join their regiment. The anecdote was mentioned time and time again, but I could never corroborate the details with first-hand testimony. In 2009 the Fiji Support Network was formed to support the Fijian military community, and this made it easier to contact particular individuals. By 2010 I was finally able to locate some of the people who were involved in the early days. This was important in terms of establishing an authenticated oral history of Commonwealth recruitment in this phase, but there were important continuities as well. It was no coincidence that some of these first recruits from Fiji had family connections to the UK military, and through them I could trace relatives of the contingent of 200 Fijians who were recruited in 1960, many of whom had stayed in the UK.

Listening for oral memories of the same events, incidents or practices from different perspectives is an inevitable part of forensic fieldwork. During the course of my conversations with senior officers I had sometimes heard anecdotes which were intended to illustrate how seriously they took the issue of cultural diversity and the need for mutual respect. After about eighteen months, I began to hear other versions of these incidents from individuals who had been involved. One such occasion concerned a funeral for a Fijian woman who died in service. Her husband, also a soldier, decided that she should be buried in the UK rather than her body returned to Fiji, which was the more common practice. Since he was of royal lineage, the funeral arrangements acquired a diplomatic element and also provided an opportunity for the regiment, sanctioned by the MoD, to show their respect for the culture of Fiji. It was initially described to me in passing by the commanding officer on an early visit to the base, but I was to hear the husband's full account when I met him to talk about his experience of joining the army in 1999. Thus by establishing the beginning of this phase I was able to access an oral memory of the longer process. But this was not the only starting point for my book.

By the end of 2009 I had interviewed many men and women at different stages of their careers, but I had never been present at a signing-in ceremony. I then discovered that the candidates who had been preselected in their own countries, whether in the Caribbean or in Fiji, were summoned to undergo a week of final tests in an establishment located in the same premises as the Army Training Centre in Pirbright. In other words, this was the missing link between candidates arriving in the UK and the start of their Phase 1 training next door, where I had begun my apprehensive interviews with the dazed recruits described earlier. It so happened that one of the last contingents of recruits preselected in Belize was due to arrive the following week, and I was able to accompany a driver to meet one straight off the plane. The sight of this young man slumped on a bench at the terminal, waiting to be collected by his military host, remains indelibly imprinted on my brain. Once I had met these individuals and followed them through their first week of tests and contracts, I knew I had the beginning of my book. The ending was similarly clear.

By late 2010 the Coalition government had announced that the army would be reduced in size following the publication of the Strategic Defence and Security Review (SDSR, 2010). The first round of redundancies, announced in the spring of 2011, were due just as I was finishing the final chapters. Although the decision to terminate the recruitment of Commonwealth soldiers was not made until 2013, it was already evident that foreign-born troops were regarded as vulnerable (Harding and Kirkup, 2012).

Throughout this period of fieldwork I noted any number of changes, both in the material environments of bases I visited as well as in institutional practices. Just as in higher education, the military sector is constantly subject to various forms of neoliberal restructuring, and it was not immune to the impact of the financial crash in 2008. One of my last conversations with my contact officer took place in the newly built Land Forces HQ in the shade of potted palms, and the fact that I was finally able to order herb tea in the cafeteria I took as my signal that it was time to wrap things up. But the fieldwork that I carried out within the confines of army premises was not the only dimension of research necessary to write this story.

The point of this chapter has been to reflect on the 'craft practices' of researching the military as an institution and examining its wider relationship to society. But the writing process did not begin until many months of fieldwork had passed, and in the meantime, the achingly difficult job of formulating a theoretical framework for the project grew out of discussions, readings and archival research that took place miles away from the military's strange environments. The status and profile of military work in the broader society had been changing on many levels. It was notable, for example, that when I started in 2008 there was very little media representation of military work, but the following year it seemed that you could not turn on the TV without encountering a reality documentary programme about life in the army – whether in Helmand or in the training centres. Men and occasionally women in military uniform became regular features of big sporting events, and occasions such as Armistice Day acquired an increasingly affective – and some would say, a coercive – force. I began monitoring these developments more diligently after I started research for my first article about the social aspects of soldiering in 2009 (Ware, 2010a, 2010b). It was then that I came to understand the myriad ways in which the population at home were being fully incorporated into the wars ostensibly being fought in distant lands.

Playing It Straight

One of the hardest aspects of holding together the disparate types of research was the endemic problem of weighing the agency of the individuals who gave interviews against the deeper structure of the organisation, as well as the historical, cultural and political contexts beyond that. Guided by the wisdom of veterans like Cynthia Enloe, who simply urged me to 'feel your way', I resisted the pressure from senior colleagues to describe my 'methodology' in advance and to elaborate on the inevitable 'research questions'. Instead I sought advice from seasoned ethnographers whom I knew and studied a range of books that I admired, and ones I disliked, in order to develop the approach that felt most appropriate (Gill, 2004; Hewitt, 2005; Back, 2007; Frühstück, 2007; Trimbur, 2013).

The key to finding the right tone, I discovered, was to keep an open mind but remain true to myself: to convey in the simplest terms what I saw and what I heard, organised in a structure that provided a historically and sociologically grounded analysis (using material that is all in the public domain), but which also supplied a critical context that challenged the injustices brought by the abuses of power – racism and war in particular. That sounds all very well, but I also knew that the proof of the pudding would lie in my ability to throw light onto the dark recesses of the military interior in a way that did not ridicule or deride the motives of those who made their living from the profession of legally sanctioned violence. I also knew I had to satisfy the scrutiny of the MoD at the end of the day, although I banished this thought until the time came to submit the manuscript. It made sense, therefore, to anticipate the prospective reader as one who might be following their own journey into this organisation, deliberately starting from a familiar place before venturing into the unknown, the unheard and the normally invisible.

My decision to begin *Military Migrants* by describing an event in Trafalgar Square was an intentional device to locate the subject at the heart of public life rather than parachute straight into the confines of an army base. The first chapter does indeed open with a scene taking place behind military lines, positioning the reader as a witness to a group of young men observed in the act of swearing loyalty to the British Crown in the absence of their families and friends, many miles from home and in a country they barely knew. Glimpsed on the cusp of their military careers, these individuals serve as guides to the process of becoming soldiers, while their preparedness to take that life-changing step asserts the agency of all those in their situation within the broader morass of forces beyond their control. As in any documentary account, the spoken words of the interviewees, reproduced in the context of a particular scene or setting, are often able to bring other types of material to life by reinforcing a vivid sense of first-hand experience. However, I was concerned not to present dialogue or reported speech in instrumental ways that simply illustrated or reinforced my own arguments; instead I tried to allow the conversations and observations to drive the argument forward, or at least to suggest an angle or perspective that needed to be explored. It was disheartening, then, to receive this comment from the committee of readers assembled by the MoD:

> The book does cause some frustration in that it does appear to take at face value what is being said by the various interviewees and this subsequently appears to be reported as fact.

This critique, which required me to travel to the MoD building in Whitehall to explain the concept of standpoint theory, was a salient reminder that the book would be addressing some very different publics and that it was difficult to predict how the contents would be read. This point was further underlined for me when I was invited to a theatre workshop to advise in the development of an updated production of the 1980s play *Black Poppies*, which was based on verbatim interviews with serving soldiers and veterans.[4] Excited that the book had been taken up in a way I had not anticipated, I watched as Ben, one of the actors, took his place at the front of the room and began to address us as though we were in some kind of presentation. I immediately recognised the scene as a rendition of the training session I had described in a section on equality and diversity law.

Ben had read the passage with clinical precision, although he was now adding his own interpretation to animate the character of the trainer. The PowerPoint presentation evidently doesn't work, so he is forced to speak from his own notes. His students are not helping either. The atmosphere is pregnant with a sense of obligation mixed with cynicism and a certain weariness. As the instructor struggles to deliver his interactive presentation, his invisible and tight-lipped students will not be drawn. Inevitably his own ambivalence starts to show and he makes concessions to the men. In an exercise designed to discuss a real-life situation, which in this case concerns the reaction to an openly gay colleague, he pleads with them: 'It's well known that gay men don't fancy straight men.'

By this time everyone was laughing. It had become more ridiculous as it went on and the director called it to a halt. I have to admit that I was feeling a little uncomfortable. Ben was following my account, pretty much word for word. And it was true that the instructor was getting minimal response so that he had to work against a thick silence, something that I emphasised in the text. My motives for describing this scene were deliberate. I had intended to convey the scepticism with which older, experienced military men treated the latest developments in equality and diversity management, but I was also keen to underline the sincerity of the instructor who radiated his new-found commitment to the subject – except that I had not meant it to be quite so funny. If anything, the subtext was that nobody likes to be lectured about such things, and old soldiers were no different.

As it happened, I had also described the use of theatre workshops to illustrate more innovative aspects of equality and diversity training within the army, and it was no coincidence that it was the acting out of the 'old soldiers' that got the most laughs among audiences of young corporals and lance corporals there as well. But in this reflexive account of my own experience, perhaps the final point to emphasise is that the sociological practice of researching the armed forces must steer a path between two undesirable outcomes: the first is to make anything to do with soldiering utterly remote and yet exotic; the second minimises the distinctions between what is military and what is not. If we can grasp the way that military values, practices, perspectives, priorities and policies are increasingly becoming camouflaged within our everyday social lives, perhaps then we can imagine more effective ways to hold powerful institutions to account. Only then can we resist the corrosive effects of allowing our politicians to fight wars in our name.

Notes

1   I arranged this through the Open University – then my employer – rather than the more usual MoD Research Ethics Committee (MoDREC).

2   An academic researcher working within the defence establishment was invited to a meeting which I also attended, and queried my credentials since she could find no trace of me through MoDREC. She asked me to stop all my interviews until my situation had been clarified. I was later told that if my book caused a stir, and the generals started asking who let me in, it was no good saying that I was 'a nice person'. This reminded me that when I was negotiating my contract, the (civilian, ex-army) administrator told me on the phone that we had to iron out the details beforehand as he didn't want 'a smacked bottom' when the book came out.

3   The infantry regiments train recruits at the Infantry Training Centre in Catterick, Yorkshire.

4   *Black Poppies* was a dramatisation of the experiences of black servicemen from World War II to the late 1980s. Originally produced by the Royal National Theatre Studio, this special recreation for television was filmed on the Broadwater Farm estate in North London, http://explore.bfi.org.uk/4ce2b7c17ea45 (Accessed 11 June 2014).

References

Back, Les (2007) *The Art of Listening*. Oxford: Berg.

Davies, Q., Clark, B. and Sharp, M. (2008) *Report of Inquiry Into National Recognition of Our Armed Forces*. London: Crown Copyright MoD.

Enloe, Cynthia (2004) *The Curious Feminist: Searching for Women in a New Age of Empire*. Berkeley: University of California Press.

Frühstück, Sabine (2007) *Uneasy Warriors: Gender, Memory and Popular Culture in the Japanese Army*. Berkeley: University of California Press.

Gill, Lesley (2004) *The School of the Americas: Military Training and Political Violence in the Americas*. Durham, NC: Duke University Press.

Harding, Thomas and Kirkup, James (2012) Battalions With Foreign Bias Face Axe in Army Cuts. *Telegraph*, 29 June. Available at: http://www.telegraph.co.uk/news/uknews/defence/9366306/Battalions-with-foreign-bias-face-axe-in-Army-cuts.html (Accessed 6 November 2014).

Hewitt, Roger (2005) *White Backlash and the Politics of Multiculturalism*. Cambridge: Cambridge University Press.

Mills, C. Wright (1959) *The Sociological Imagination*. London: Oxford University Press.

Strategic Defence and Security Review (2010) *Securing Britain in an Age of Uncertainty*. London: HM Government.

Trimbur, Lucia (2013) *Come Out Swinging: The Changing World of Boxing in Gleason's Gym*. Princeton, NJ: Princeton University Press.

Ware, Vron (2007) *Who Cares About Britishness? A Global View of the National Identity Debate*. London: Arcadia.

Ware, Vron (2010a) Lives on the Line. *Soundings*, 45. Available at: http://www.eurozine.com/articles/2010–09–28-ware-en.html (Accessed 6 November 2014).

Ware, Vron (2010b) Whiteness in the Glare of War: Soldiers, Migrants & Citizenship. *Ethnicities* 10(3), 313–330.

Ware, Vron (2012) *Military Migrants: Fighting for Your Country*. Basingstoke: Palgrave Macmillan.

19

RESEARCHING MILITARY MEN

Stephen Atherton

This chapter draws upon research that sought to identify the complexities of military masculinities and to problematise the notion of a hegemonic military masculinity (see Atherton, 2009). The research identified how the military inculcates embodied practices such as drill, parade and training, and addressed how these could also be used as a form of discipline and to 'break down' new troops. In the research, I aimed to challenge more traditional theorisations of military identities through an examination of domesticity as both a conceptual term and as a practical set of domestic skills inculcated during military service. This examination counterpoints the hegemonic masculinities that one would expect to encounter in the highly masculinised institution of the military and seeks to identify how domesticity was used as a means of breaking in new soldiers to the military hierarchy. The research also explored how domestic skills are subsequently utilised outside of military domestic spaces. Once demobilised many of the soldiers in this study returned to a 'family home', yet they had to readjust to civilian life and to acclimatise to new domestic routines and spaces. The various performances and narratives of domesticity uncovered through this research give an insight into the sometimes problematic transitions from soldier to civilian.

A common theme among academics writing on the military is that a unique set of hypermasculinised subjectivities are produced and performed through the day-to-day activities of the soldier (see e.g. Collinson and Hearn, 1994; Barrett, 1996, 2001; Woodward and Winter, 2007). This maintains a 'closed-off' military sphere for its members which, nevertheless, proceeds to seep into the popular imagination in the form of macho icons. And yet, Connell (2006) identifies recent studies in masculinities (see Collier, 1998; Wetherell and Edley, 1999) that have drawn upon psychological, postmodern and poststructuralist theories to suggest that men perform a cultural repertoire of masculinities that are specific to particular situations. What such studies hold in common is an emphasis upon the fluidity of masculinity as an embodied, practiced concept; that is, while articulating a broad-based notion of masculinism as an ideology that justifies and naturalises unequal power relations between men and women, they analyse how in practice particular rhetorics and behaviours not only undermine a simplistic, binary formulation of masculine/feminine identities, but also allow for power relations to emerge among men. It is important therefore to consider the methods in which we can capture the proliferating masculine subjectivities that are developing within the military to better understand these power relations.

In order to gain insight into the manner in which military men negotiate the issue of masculinity, and particularly domesticity, I analysed the formal policies and practices in place around various 'domestic' practices, the manner in which these were animated by the men involved, and how these men thought and felt about such practices. The bulk of my methodology, then, addressed talking with and observing these men as they narrated their experiences of recruitment, mobilisation, serving in the military and demobilisation. In order to do this, I chose former military personnel who had all undergone the sometimes turbulent process of returning to civilian life. In all, I interviewed thirty-two former army men over a period of a year and a half, in addition to off-the-record discussions with various other organisations and army officials. I also drew on various military archives and consulted official reports and statistics released by the Ministry of Defence. In the next sections, I discuss in more detail the pragmatics of this research process, as well as some of the pressing issues that emerged. In this chapter I will specifically examine the importance of interview location, the mechanics of conducting interviews with ex-military men, and issues regarding the positionality of researcher and researched. I will then go on to examine how the research process was imbued with emotion, and the challenges that were faced with particularly challenging interviews. The final aspect of this chapter focuses on the ethical frameworks in which the research was conducted.

Talking With Men

All of the men interviewed had left the army in the 25-year period 1985–2010, with many having a service history beginning before those dates. Within this time period many had experienced deployment in Northern Ireland, the Falklands, the 1991 and 2003 Iraq conflicts, Afghanistan 2001 onwards and the former Yugoslavia from 1992; many also noted involvement in antiterror operations post-9/11. This is a particularly interesting period insofar as, first, there has been a significant change over this time period in the productions of masculinities, particularly in domestic spaces in Britain. Jackson (1991) notes that the socially constructed identity of the 'new man', for example, is a particularly important change in the cultural repertoire of masculinities among British men. As much of this research was conducted around themes of the home and domestic spaces, this time frame provided for some interesting themes to emerge concerning the manner in which such new forms of masculine identity intersected (or not) with military masculinities. Second, Coker (2001) and Kronsell (2005) suggest that in Europe, post–Cold War defence budgets were cut as a consequence of changing security situations globally, and the budget cuts in turn led to changes which they have argued has 'feminised' the military. Kronsell (2005) suggests that the 'manly warrior' image of the military has diminished as a result of these changes in conjunction with new forms of warfare. However, she also notes that post-9/11 politics and the 'war on terror' have given power back to defence and security agencies and that the decline in the traditional ideas of men and military may be short-lived. My chosen time covers some interesting transitions in the role and perceived function of the military, and key events such as the break-up of the Soviet Union, the emergence of 'democracy' in former Eastern European countries, civil war in various parts of the globe, the first and second Gulf wars and the current conflict in Afghanistan and Iraq. All of these have repercussions as to how military men perceive their own role and identity, in light of the conflation of masculinity, military power and nationalism, but also some very vocal criticism from various publics regarding the post-9/11 conflicts in Iraq and Afghanistan. The men interviewed and referred to in this chapter had left the army at the time of the research, and although they had left at different periods and after different conflict and policy changes, collectively their experiences provided an entry point into how military masculinities are constructed in place, and change with place.

Interviewing was my primary method of data collection. Across the social sciences, it is generally accepted that interviews can provide useful insight into the experiences of people, which is particularly pertinent as this research was conducted within the disciplinary traditions of human geography. Gaskell (2002), Baxter and Eyles (1997), and Hoggart et al. (2002), for example, all suggest that interviews can gain an in-depth understanding and enable exploration of complex relationships. Flick (2002) suggests that interviews can tap into the participant's implicit knowledge and beliefs and, via the process of talking them through, make them more explicit. For some (Gibson-Graham, 1997; McDowell, 1997; Stacey, 1997), this enables the researcher to bring to light (that is, to an academic or policy audience) 'hidden' voices, or to explore complex relations within the workplace that would not otherwise be apparent. Therefore interviews were chosen as the most appropriate research method, as this allowed a more detailed examination of the complex negotiations of space and place experienced by this group of former soldiers.

In interview contexts, the gender of interviewer and respondent are always issues to be accounted for, given the social construction of gender roles which frame the interview as an encounter between two people (see e.g. Winchester, 1996; Horn, 1997; Pini, 2005). In my own role as researcher I also encountered complex power-laden positionalities, as I was required to talk to this sample of former military men in order to garner the information required for the research, but was also placed in the position of listener. This distinction was crucial in analysing my own identities within the research process. For example, during the research process I often found that I took on a supporting role, where many men would talk about the problems they faced upon leaving the army in particularly emotional displays. As I explore later in this chapter, this became particularly problematic and created a situation where I unwittingly came to be seen as a counsellor figure by some of the men in the cohort. Therefore, to an extent, I felt that a 'feminised' positionality had been projected upon me, but this was further complicated by other aspects of the interview where there seemed to be a shared understanding and a simultaneous production of masculinities. It is clear that the simultaneous productions of masculinities were due, in part, to the very nature of the relationship involving two males interacting, and the associated homosociability (see Monaghan, 2002a, 2002b). However, the shared understandings were equally related to my own positionality in that I was interested in exploring the military pasts of the interviewees. For many the interviews were the first opportunity they had to talk about their military service outside of their established networks of friends and ex-service personnel.

These feelings of trust and empathy were facilitated by the pragmatics of the research process; that is, how I located and recruited my interviewees. My initial efforts to recruit discharged soldiers under the aegis of various organisations and charities that offer support to former military personnel were unsuccessful because of legal issues of privacy. Instead, I was able to use personal contacts to snowball a series of meetings with individuals and small groups. Specifically, my initial interviewees were recruited in three ways. First, through previous research on masculinity on construction sites in the north of England I had found that a number of the men I engaged with had been in the army. Second, a family friend based in the Midlands in England was able to provide me with contacts among his former cohort. And third, an acquaintance working within the Territorial Army in Wales was able to introduce me to a series of his former colleagues. Hence, each new contact was enrolled with the explicit approval of the former one, and was often heralded by one contact phoning another in order to vouch for myself as a researcher interested in their military experiences. Recruitment or participants through this more informal snowballing approach made recruitment of interviewees perhaps much more straightforward than had I gone through more official channels to recruit. I was particularly aware that, as a civilian, it would be problematic gaining access to these men through official channels. Furthermore, through informal discussions with veterans' organisations it became apparent that the army did not necessarily maintain records of where veterans were located.

I asked all interviewees where they would like to meet, and subsequent interviews took place in interviewees' own homes, pubs, cafes and restaurants. Interviews conducted in the home tended to occur when partners and children were not present; on those occasions when family were within the house the interviews took place in a smaller room away from the main living area for the duration of the meeting.

The Importance of Interview Location

Within the literature on interviewing, little attention has been paid to the location of interviews and there is little guidance as to how different interview locations have implications for the positionality of researcher and participant (Elwood and Martin, 2000; Herzog, 2005). I felt, however, that the location of the interviews was tremendously important not only in understanding the particular contexts in which answers were given, such as the division between public and private spaces in which respondents might feel more at ease answering particular questions, but also in regards to how particular settings, such as the home, provide a symbolic 'reservoir' of objects that can be enrolled within a person's narrative. Interview sites can, therefore, be thought of as producing 'micro-geographies' of spatial relations and meaning (Elwood and Martin, 2000: 649). In a participant's home, they go on to suggest, the interviewer may observe artefacts, posters and other belongings that may give further insight into beliefs and commitments held by the participant and that location can enable the researcher to obtain richer and more detailed information than from the content of the interview alone. They also suggest that participants hold multiple identities in relation to the places in which they are being interviewed, and communicate these identities differently within varied interview locations.

In addition to recording and transcribing interviews, I took detailed notes in a research diary to provide further contextual information. The following example is taken from my research diary of notes during Robert's interview:

Sitting in his living room had the feeling of being in a legion pub with rosettes, pictures and military reading materials on display. It was clear from the positioning and volume of this material that it was part and parcel of Robert's life and not placed there for my benefit. The magazines were all related to various aspects of military life and weaponry and I note that they were all organized chronologically. The photographs on the walls were of Robert in various uniforms throughout his service period, some smart dress uniforms and others of him in battle gear.

Moreover, interview location had a direct influence on the topics that this sample of military men were comfortable discussing. There were distinct differences in the information disclosed between those men interviewed in more private settings, such as meeting rooms or domestic spaces, than those in more public spaces such as cafes and restaurants. One such example was the interview with Steven, which took place in a cafe. The following extract from my research diary identifies his uncomfortable body language and awkward narratives when discussing deployment:

> Although the cafe was quiet with only a handful of customers I felt Steven did not feel at all comfortable with the situation. When I asked about his experiences of deployment he seemed to become tense and looked around before answering some questions. He became flustered in places and the answers were inaudible at times. Had the cafe been empty, or the interview taken place in a different location, then this part of the interview might have run more smoothly.

The possibility of conducting the interview in another location might have yielded different results, yet Steven did not want to attend an interview in my own home and did not invite me to his home. This in itself raises questions about the separation of military and a domestic space, a factor that has often been typical of existing research on military subjectivities (see e.g. Hockey, 1986; Dandeker et al., 2006; Soeters et al., 2006; Colville, 2009; Gagen, 2009). Through analysing the location of interviews, I was therefore able to gain an initial insight into how the participants viewed their own military and civilian identities as separate and distinct. This separation of the military from civilian life also occurred in interviews which took place in participants' own homes.

The following two excerpts from my research diary were written during and after the interview with Andy:

> The interview took place in the kitchen, which wasn't ideal as there were was a steady stream of people entering and leaving the room. Andy explained that he didn't want his parents to hear the interview as his mum was disabled and he didn't want her to get upset when recalling the accident he suffered while in service.

Also:

> It was useful actually conducting the interview in Andy's house as he was able to point out objects or the direction of rooms that provided further context to the

narratives. When discussing memory blanks he was able to physically demonstrate where he was in the kitchen when he had taken milk out of the fridge but then forgotten why. By demonstrating this Andy moved away from the table where we were sat, which may have helped him to talk about this difficult issue in more detail as he could distance himself from the recorder.

The first excerpt here demonstrates that by conducting the interview in the kitchen, rather than in the main living area, Andy sought to distance his parents from his encounters in the army, particularly those in relation to his injuries. Andy felt extremely guilty that he was not able to adequately care for his disabled mother because of these injuries. This factor became apparent through Andy's narrative and provided further evidence of his desire to maintain the distinctly separate spaces.

The second excerpt here provides a more practical element when noting the importance of interview location. Andy was able to demonstrate, in situ, particular behaviours that he had found troubling. It is therefore clear that adding interview location as a further factor in the analysis adds further depth to the narratives, and allows the spatial context of the interview to be examined. Therefore by exploring military masculinities as spatialised, space and place were a significant focus for analysis as well as being a key focus as part of the data collection process. By analysing the location of the interview it allowed the spatialities of the encounters to become a key focus of analysis of the interview data when the interview process had been completed.

Conducting Interviews

The interviews conducted for this study of military masculinities provided an opportunity for this sample of former military men to talk in depth about their experiences, often for the first time. In order to garner the information required the interviews were generally unstructured, in that although I had predefined set of ideas that I wanted to bring to the discussions, I was generally more interested in observing and listening to the various directions which this cohort of military men wanted to take. Therefore I used mildly leading questions that sought to encourage reflection and memory; at many stages during interviews the narratives raised subjects that I had not previously considered, and I was keen to explore these ideas further to gain a deeper understanding of the intersecting subjectivities of these men. Many of the interviews reached the three-hour mark as I allowed these themes to be explored and articulated in different ways.

An example of this comes from my interview with Aled. Aled was a very proud Welshman who displayed the Welsh flag outside of his house and several images of the Welsh dragon in his living room and kitchen. I asked Aled the reasons why he joined the army, but then the narrative digressed into a discussion of his proud nationalism.

SA: Okay thanks for that. Now I would like to move on to thinking about the reasons that you joined the Army. How did that come about, you know was it through school or was it something that you sought out?

Aled: Blimey that is taking me back now. I think it was my Ma who wanted me to join up.

SA: It was family pressure?

Aled: Well that wasn't pressure it was sort of how can I get a job and it was like yeah okay when my Ma suggested it. I can't remember exactly how it came about now but there were three lads three of us who wanted to join up and I think I remember it was through school in the end. Yeah it was a signing up thing at the school but it wasn't like with the teachers it was them using the school hall I think.

SA: Right and did all three of you join together and go through the training together or did you go your own separate ways?

Aled: Two of us joined the Army and the other lad [name removed] joined the Navy books, we didn't really keep in touch or anything once we joined we didn't need to you know it was like the first we were too busy with all the training and then like we played rugby with our mates in our division you know.

SA: The true Welshman's sport!

Aled: Ah yes I love my rugby. I know that's one of the good things about the Army is that we get all these sports teams and play against each other and not just in Wales you know we went over to England and beat poncey southerners and shown em how to play fucking rugby.

SA: [Laughs] So when you weren't doing all the training you are beating Englishmen at rugby?

Aled: That's right but they don't understand they'll play football. It's our heritage that rugby is. You know I see those lads go out on their red colours representing Wales and it makes your blood run an' your heart thump you're so proud of being a Welshman. Makes me proud you know. I will tell you another thing, us squaddies here in Wales lot more pride in fighting for our country than English lads. They aren't proud. England is a fucking mess at the moment I wouldn't wanna be in the Army there because I had to go to Aldershot on training and Jesus those lads had no respect for their country no respect at all. If they'd come over to Cymru like that they would have got a fucking kicking. More pride in our country you see.

In the interview I introduced the 'semistructured' theme of recruitment by asking Aled the reasons why he joined the army; the interview then digressed into a more unstructured pattern that identified other aspects Aled felt were important for him to articulate. Aled's digression also allowed me to consider the intersection between gender identities and other identities such as nationalism, athleticism and identities associated with playing sports.

All of interviews were taped, with permission from the interviewee, and initially coded according to the key themes of military domestic spaces, military identities, military training, domestication in the military, domestication following discharge, constructions of the home following discharge, support and welfare, and trauma. Once completed the interviews were transcribed at length in a process requiring approximately 850,000 words and several hundred hours of listening. The interviews were then recoded thematically and broken down into further subsections. My recording, transcription and analysis by coding followed established patterns. However, what I wanted to do was find a way of capturing and using in my analysis something of the emotional reactions of my interviewees. I therefore designed a piece of software that allowed me to place the interviews within particular contexts. There are commercial programs available that enable the researcher to code and categorise interviews; however, none of these has the level of flexibility in playing back audio files that I required. The software I created allowed me to tag the audio from the coded transcription and to then play back that particular piece of audio so that I could listen closely to aspects of the interviews that are too often neglected by researchers; that is, issues such as tone and pitch of voice, emotion displayed vocally and where the respondent paused during the interview. Without this further level of detailed analysis, writing about the expression of emotional geographies would have been difficult. Of course, to go one stage further would have been to video interviews and to tag the video file in order to analyse embodied displays of emotion. For me, however, this would walk a fine line ethically insofar as it would have been an obvious intrusion, particularly as most interviews were conducted within homes and workplaces. I did, however, jot down my observations about certain bodily behaviours as they occurred, such as head in hands, looking at and holding photographs. The following excerpt comes from my notes jotted down after an interview with Rhys:

> Rhys took great pleasure in showing me through some of the photographs and press cuttings from the Falklands conflict. It was clear that these photographs in particular held a great deal of meaning for Rhys as they were displayed in a protective cover which he would not remove them from. The manner in which he thrust the photographs into my hand before we had even started the interview characterized his immense pride at the role that he had performed in the Falklands. When showing these photographs his voice became excited and the pace quickened and highlighted his general keenness to display these images to me. These displays of emotion quickly changed however when we analysed some of the newspaper cuttings. Rhys did not want to be interviewed while we discussed the newspaper cuttings but it was clear that one headline in particular had caused a great deal of distress and he later articulated that one of his friends had been killed in an attack on a British ship, which made the headlines in several British newspapers.

For me, the in-depth nature of these interviews was particularly important as I wanted to understand something of the complex subjectivities of military men. But also, I wanted to position these reflective, complex subjectivities in relation to archival histories produced by the army and to statistical information reported by the Ministry of Defence (MoD). My interviews were buttressed by a close reading of military archive material that provided a historical overview of MoD and government policy changes. The National Archives in Kew, the Imperial War Museum and the National Army Museum were visited. The National Archives proved most fruitful for this particular research project, as they hold records of social and economic policy that relate to the transition from soldier to civilian. The latter two hold information about military campaigns and service records, some of which was useful in contextualising policy and analysing the productions of military masculinities; however they had little information about what happened to soldiers once they came out of the army. Many of these documents relate to a time previous to when this sample of military men undertook their duties; however, they are useful in contextualising the experiences of these men historically and identifying the relative changes in approach to masculinities over the twentieth century.

Emotion and the Research Process

In the preceding section, I mentioned that I undertook many interviews in the homes of those I researched, and how this yielded interesting results as I became not only an interviewer but also a part of the place where I was conducting the interviews. Here, I want to pursue this in more depth, responding to Bennett's (2009) challenge to geographers to think not only about their own emotions and the emotions of others, but also to actively engage with the question of how their feelings connect with the feelings of others.

Prior to conducting a formal interview I had engaged in three telephone conversations with Mike and knew that he had been severely injured and could no longer perform paid work. Therefore this initial interview was one that I had particularly mixed feelings about. I had doubts as to the approach that I should take, as I did not want to stir negative emotions by recalling past events, yet I wanted to gain a sense of Mike's experiences of both his time in the army and his subsequent demobilisation through injury. Equally, I did not want to be patronising and avoid difficult questions because of Mike's physical disabilities. In consequence, this initial interview was perhaps the most difficult to prepare for. These feelings of discomfort were exacerbated as a result of my own positionality as a civilian. I was required to be reflexive within the research due to my awareness of a personal lack of experience of military service, which impacted on the relationship that I initially entered into with the sample of former military men.

This feeling of discomfort continued throughout the interview process with other men who were part of this sample. It was particularly prominent around men who had been injured or disabled, both physically and mentally. What follows is an extract from my field notebook after a particularly difficult and emotionally challenging interview with Bob:

> About half way through the interview [Bob] rolled up his sleeve to show me the scars on his wrists. He felt quite comfortable doing so, I think. We were in his home and there was no one else around and so I think that is why he was able to do so. The comfort of his own home allowed him to show me this very personal and emotional scarring. I was not prepared for this, it shook me badly. I was uncomfortable in his home and I was uncomfortable with what I had been confronted with.

It is evident that my discomfort here was challenging on two levels; first, the place of the research was the participant's own home. It was not neutral ground, and although in many cases interviewing in respondents' own homes yielded interesting and unexpected results, here the power relationship was very much in favour of the researched. These feelings of anxiety were very much centred on being outside of my own comfort zone and my relative inexperience of conducting interviews with this sample of men. This anxiety was exacerbated by a feeling of helplessness in not having the correct training or preparation to deal with the more complex emotional situations. In hindsight it is clear that my screening for post-traumatic stress disorder (PTSD, outlined later in this chapter) should also have included some method of screening for wider emotional and psychological issues so that I could be better prepared to deal with these types of situations. This experience was also an intensely emotional moment for me as a researcher and for the person displaying these literal, emotional scars from the past. My own feelings of discomfort were a contrast with the evident comfort felt by this individual to share these deep emotional and personal experiences. I believe that through the interview process a certain amount of trust had developed between myself and Bob.

For most, the interview process was a means for them to express emotions that they were perhaps not able to in their everyday lives. Roger, for example, suffered a traumatic experience during his service in the army and was not adequately diagnosed upon demobilisation. Roger's narrative was explicitly and visually highly emotional and had a distinct impact on how I felt and acted towards him. What follows is a section from the interview with Roger and my thoughts expressed in my research diary after the interview:

Roger: I get upset a lot now, I'm holding together alright at the moment I think because I was prepared for this [interview]. It's hard I'm not the man I was, I cry I get depressed.

[Research diary: The guy poured his heart out to me. I think I was more like a counsellor than a researcher here. I listened to all of his problems and experienced his hurt, anguish and fear. I didn't know what to say other than he wasn't alone and I had met others with his problems. I think in many ways I was proud to have been there for him and maybe to give him a little hope or at least to know he wasn't alone. He didn't cry but he was so upset I could see it in his eyes.]

This example highlights my own emotional responses to the interview, yet also identifies a comfortable situation for Roger to talk about his emotions. It reminds me to a degree of Dias's (2003) research with anorexia sufferers, where she suggests that when positioned as an 'impartial researcher' she was able to provide a means for these sufferers to talk about their illness away from family and doctors. I believe that by having someone to talk with, Roger felt able to expand upon these emotions, where he perhaps was not able to with other people. In relation to Bennett's (2009) arguments regarding the importance of researchers understanding how their feelings connect with the feelings of others, I believe that Roger and I shared an emotional experience; his recall and my listening impacted on both of us. During his emotional outburst I felt the hurt he was expressing, yet I also felt that he experienced some relief through the process. This experience moved beyond the researched impacting on the researcher to a mutual exchange of emotion. Reflecting on this experience five years on, I still feel particularly troubled by this interview. Roger was very much in need of counselling and further support and I appeared to be the first person he had really opened up to. I did provide Roger with contact details of support charities following our interview, but upon reflection, were I to conduct interviews in this manner again, I would have these details ready to distribute in the interviews.

Ethical Considerations

Because this research project was partly funded by the Economic and Social Research Council (ESRC), I was required to work within the ESRC's guidelines on research ethics (ESRC, 2010). Significantly, my own sense of ethics and morality also influenced the ways in which I discussed, cited and disseminated the information I gained through interviews. As noted earlier, for example, many of the interviews were conducted in different places that had very different meanings both for me and for the people I was interviewing. In particular, there are ethical considerations that need to be borne in mind when conducting interviews in respondents' own homes which went beyond the ESRC framework. For example, there are particular issues with private spaces within the home that I would not have felt comfortable entering without invitation. Research council guidelines cannot fully account for all of the nuances regarding the varied locations of interviews other than on health and safety grounds. With the nature of the interviews as part of my study there were also more localised, moral issues that I had to use my best judgement to deal with.

All of the men interviewed as part of this research were promised and have been given anonymity on several levels including changing all names. Following the interviews I also gave respondents the opportunity to opt out of the research project by withdrawing their interview recording and transcript from the data. I also provided an opportunity for all of the men who were interviewed to read through interview transcripts if desired, and to remove any information that they did not wish to enter into my discussions. As I discuss in the thesis to which this research contributed (see Atherton, 2009), many of the people interviewed felt guilty about actions they had performed while in the army. These feelings of guilt manifested themselves in different ways and often respondents referred to specific places where they had fought, or served on peacekeeping missions. With this in mind I offered respondents the opportunity to remove the names of specific places where they had served on their tour of duty. Around half of the respondents desired the specific place names to be removed. In some cases I removed complete paragraphs from interview transcripts as the information itself provided specific details about places without directly naming them. I also removed specific details of certain military careers such as communications, strategic policy and missions that were not in the public domain. This was a time-consuming process that did entail removing some interesting data. Nonetheless, the process was important for maintaining the anonymity and trust of the men involved in the study.

Elsewhere (see Atherton, 2009, 2014) I also consider the problematic negotiations of identities in the home for men who have suffered some form of trauma during their time in the army. For some of the men this trauma had resulted in physical and/or mental disability. In these interviews a great deal of care was taken to ensure that this topic was broached by the interviewees themselves, and that they had the opportunity to stop the interview at any moment they felt uncomfortable. I was aware of how in-depth interviews on their military experience would mean that these men were reliving often harrowing experiences. The possibility of interviewing men suffering from PTSD was rejected on the grounds that this reiteration of their experiences would be both insensitive and potentially harmful. This was achieved through a self-declaration form that asked men to disclose if they were suffering from or had received treatment for PTSD. In hindsight it is clear that there were perhaps undiagnosed cases of PTSD, and other severe complications, that manifested in some of the interviews referred to earlier in this chapter.

Concluding Remarks

By utilising a mixed methods approach I was able to draw out key themes pertaining to the production of masculinities within military space, and particularly the manner in which domesticity enters into everyday practices and reflections. Much of the research utilised both primary and secondary material, and I hope that while the former provides some idea of the shifting institutional expectations around military life, the latter will prove a useful and telling counterpoint to these army narratives, as soldiers proceed to tell their own stories. In this chapter I have outlined the significance of the interactive nature of the interviews in terms of the verbal information contained in interview responses, but also the nonverbal information that was recorded within my research diary, and how the two approaches allowed a more detailed examination of the productions of masculinities among this sample of former military men. Furthermore, I have outlined the importance of physical space in regards to the interview process. This examination of the spatialities of the interview allowed me to raise further analytical issues regarding the production of military masculinities located within the domestic sphere.

Reflecting back on this process five years on, there are features of the research design that I would adapt. I would maintain the snowball approach to recruiting participants, but I would also work more closely with the armed forces community covenant to identify further organisations providing support and information to ex-service personnel. Were I to engage in this project again I would be in a better position in regards to mental health issues that arose in the interviews. Although I had screening in place for PTSD, I would develop a more rigorous system for identifying other possible issues that may arise in the project, which may have overcome some of the more uncomfortable situations outlined in this chapter. However, the importance of emotions within the research process cannot be dismissed and this aspect would feature prominently were this research to be conducted again.

References

Atherton, S. (2009) Domesticating Military Masculinities: Home, Performance and the Negotiation of Identity. *Social and Cultural Geography* 10(8), 821–836.

Atherton, S. (2014) The Geographies of Military Inculcation and Domesticity: Reconceptualising Masculinities in the Home. In Hopkins, P. and Gorman-Murray, A. (Eds.), *Masculinities and Place*. Surrey: Ashgate, pp. 143–159.

Barrett, F. (1996) The Organizational Construction of Hegemonic Masculinity: The Case of the US Navy. *Gender, Work & Organization* 3(3), 129–142.

Barrett, F. (2001) The Organizational Construction of Hegemonic Masculinity: The Case of the US Navy. In Whitehead, S. and Barrett, F. (Eds.), *The Masculinities Reader*. Cambridge: Polity Press, pp. 77–100.

Baxter, J. and Eyles, B. (1997) Evaluating Qualitative Research in Social Geography: Establishing 'Rigour' in Interview Analysis. *Transactions of the Institute of British Geographers* 22(4), 505–525.

Bennett, K. (2009) Challenging Emotions. *Area* 41(3), 244–251.

Coker, C. (2001) The United States and the Ethics of Post-Modern War. In Smith, K. E. and Light, M. (Eds.), *Ethics and Foreign Policy*. Cambridge: Cambridge University Press, pp. 147–166.

Collier, R. (1998) *Masculinities, Crime and Criminology: Men, Heterosexuality and the Criminal(ised) Other*: London: Sage.

Collinson, D. and Hearn, J. (1994) Naming Men as Men: Implications for Work, Organization and Management. *Gender, Work & Organization* 1(1), 2–22.

Colville, Q. (2009) Corporate Domesticity and Idealised Masculinity: Royal Naval Officers and Their Shipboard Homes, 1918–39. *Gender & History* 21(3), 499–519.

Connell, R. (2006) *Masculinities*. Berkeley: University of California Press.

Dandeker, C., French, M., Birtles, M. and Wessely, S. (2006) *Deployment Experiences of British Army Wives Before, During and After Deployment: Satisfaction With Military Life and Use of Support Networks*. In *Human Dimensions in Military Operations – Military Leaders' Strategies for Addressing Stress and Psychological Support* (pp. 38-1–38-20). Meeting Proceedings RTO-MP-HFM-134, Paper 38. Neuilly-sur-Seine, France: RTO. Available at: http://www.rto.nato.int/abstracts.asp.

Dias, K. (2003) The Ana Sanctuary: Women's Pro-Anorexia Narratives in Cyberspace. *Journal of International Women's Studies*, 4, 31–45.

Economic and Social Research Council (2010) *ESRC Framework for Research Ethics (FRE) 2010*. Available at: http://www.esrc.ac.uk/_images/framework-for-research-ethics-09–12_tcm8–4586.pdf

Elwood, S. and Martin, D. (2000) 'Placing' Interviews: Location and Scales of Power in Qualitative Research. *Professional Geographer* 52(4), 649–657.

Flick, U. (2002) *An Introduction to Qualitative Research*. London: Sage.

Gagen, E. (2009) Homespun Manhood and the War Against Masculinity: Community Leisure on the US Home Front, 1917–19. *Gender, Place & Culture* 16(1), 23–42.

Gaskell, G. (2002) Individual and Group Interviewing. In Bauer, M. and Gaskell, G. (Eds.), *Qualitative Researching With Text Image and Sound*. London: Sage, pp. 38–57.

Gibson-Graham, J. (1997) "Stuffed I Know" Reflections on Post-Modern Feminist Social Research. In McDowell, L. and Sharp, J. P. (Eds.), *Space, Gender Knowledge: Feminist Readings*. London: Arnold, pp. 124–146.

Herzog, H. (2005) On Home Turf: Interview Location and Its Social Meaning. *Qualitative Sociology* 28(1), 25–47.

Hockey, J. (1986) *Squaddies: Portrait of a Subculture*. Exeter: University of Exeter Press.

Hoggart, K., Lees, L. and Davies, A. (2002) *Researching Human Geography*. London: Arnold.

Horn, R. (1997) Not 'One of the Boys': Women Researching the Police. *Journal of Gender Studies* 6(3), 297–308.

Jackson, P. (1991) The Cultural Politics of Masculinity: Towards a Social Geography. *Transactions of the Institute of British Geographers* 16, 199–213.

Kronsell, A. (2005) Gendered Practices in Institutions of Hegemonic Masculinity. *International Feminist Journal of Politics* 7(2), 280–298.

McDowell, L. (1997) Introduction: Homeplace. In McDowell, L. (Ed.), *Undoing Place: A Geographical Reader*. London: Arnold, pp. 13–21.

Monaghan, L. (2002a) Hard Men, Shop Boys and Others: Embodying Competence in a Masculinist Occupation. *Sociological Review* 50(3), 334–355.

Monaghan, L. F. (2002b) Regulating 'Unruly' Bodies: Work Tasks, Conflict and Violence in Britain's Night Time Economy. *British Journal of Sociology* 53(3), 403–429.

Pini, B. (2005) Interviewing Men: Gender and the Collection and Interpretation of Qualitative Data. *Journal of Sociology* 41, 201–216.

Soeters, J., Winslow, D. and Weibull, A. (2006) Military Culture. In Carforio, G. (Eds.), *Handbook of the Sociology of the Military*. New York: Springer, pp. 237–254.

Stacey, J. (1997) Can There Be Feminist Ethnography? In McDowell, L. and Sharp, J. P. (Eds.), *Space, Gender Knowledge: Feminist Readings*. London: Arnold, pp. 115–123.

Wetherell, M. and Edley, N. (1999) Negotiating Hegemonic Masculinity: Imaginary Positions and Psycho-Discursive Practices. *Feminism and Psychology* 9, 335–356.

Winchester, H.P.M. (1996) Ethical Issues in Interviewing as a Research Method in Human Geography. *Australian Geographer* 2(1), 117–131.

Woodward, R. and Winter, T. (2007) *Sexing the Soldier: The Politics of Gender and the Contemporary British Army*: London: Routledge.

20

PUTTING 'INSIDER-NESS' TO WORK

Researching Identity Narratives of Career Soldiers About to Leave the Army

David Walker

Military research in the social sciences is often carried out by individuals with past or ongoing military service as well as by those who have no prior experiences of military life. Regardless of the kind of researcher you are, there are well-known challenges associated with researching the military. In this chapter, I shall draw upon my recent research on career soldiers who were leaving the British Army to explore some of the important methodological dynamics that I encountered (Walker, 2010, 2013). This qualitative research project began only months after I had ended my own lengthy Army career. As the work progressed, my professional relationship to the data and the topic of Army exit shifted from one of proximity to one of critical distance – or at least as critically distant as an ex-soldier-turned-researcher can possibly be. My aim in the chapter is to show that military insider-ness may be put to good research use if a professional and reflexive approach is adopted. During the course of this particular project, I found that the passage of time was a significant resource that I could use to the advantage of the work at key stages of the research process. For example, I wrote the research questions from a position of significant early proximity; then as I became increasingly distant from my Army career, and more critical of the process of Army exit, I found that I had new perspectives from which I could analyse the data and my own prior views – a process that reverses that of the traditional anthropologist who seeks to immerse herself in a foreign culture in order to report back to the academic community. Matters of positionality and reflexivity for researchers are certainly not new considerations, but as I go on to discuss they are less likely to attract the attention they deserve among researchers of the military, although this is beginning to change (cf. Castro and Carreiras, 2013; Soeters et al., 2014). In this chapter, my own contribution to this important body of work is made in the spirit of Amanda Coffey's notion of the researcher as an 'ethnographic self'. For Coffey, the researcher is 'thoroughly implicated in the way we collect, understand, and analyse [. . .] data such that the researching self is often presented as a kind of "medium through which fieldwork is conducted"' (1999: 122).

I didn't know it at the time, but an important relationship had been set in motion when I began this PhD research in 2007. This was between my ongoing experiences as a newly exited soldier and the focus of the research that I had designed. The work examined identity narratives among career soldiers during their last year in the British Army. It explored identity transition among soldiers and officers who, although still very much caught up in Army relations, were nevertheless relating differently to them as they anticipated new horizons in their future civilian lives. I designed the research project as a serving soldier in my last year of Army service and began the work a month after my retirement. In designing the project I had not fully foreseen my own routine and ongoing reaction to leaving the Army, nor had I anticipated how this would shape my relationship with the data in different ways and at different stages. Interestingly, one important finding of the research concerns the difficulty exiting career soldiers have in accurately projecting themselves forward into future lives devoid of the military social relations that were supporting the person they thought themselves to be. In this chapter I explore a number of issues associated with researching the military when a researcher has personal knowledge and experience of the service being investigated. In the next section, I review some relevant literature about insider-ness in qualitative research, and then in subsequent sections I discuss how insider-ness may be put to work in military contexts in general, and how in particular this happened in my own research on Army identity transitions in particular.

Insider-ness in Qualitative Research

No longer can the researcher's place in all stages of research be ignored (Atkinson, 1992; Mauthner and Doucet, 2003) since in qualitative research s/he is a 'research instrument *par excellence*' (Hammersley and Atkinson, 1983: 19) whose influence is pervasive and most pronounced during phases of analysis. But a loss of faith in objectivity and value-neutrality among qualitative researchers should not represent a retreat from systematic and stringent research strategies. Instead, well-designed and consistent methodologies ought to coexist with concern also to offer 'transparency, honesty, and openness' (Higate and Cameron, 2006: 223) about personal, artful and elusive aspects of research. One of the most important developments in qualitative research occurred in the late 1980s in what has been termed a 'crisis of representation'. This marked a point when a number of authors (Geertz, 1980; Hammersley and Atkinson, 1983; Clifford and Marcus, 1986; Marcus and Fischer, 1986; Geertz, 1988; Haraway, 1988; Denzin, 1989; Atkinson, 1992; Hobbs and May, 1993; Denzin, 1995, 1997) challenged the possibility of completely 'voicing' (Hobbs, 1993; Pearson, 1993: xviii) the experiences of the researched. These authors took issue with a tendency to play-down – or worse – to conceal the effects of the researcher in a haze of professional mystique. For Mauthner and Doucet (2003: 416) 'the problem arises through recognition that as social researchers we are integral to the social world we study,' and this translates into a situation where the researcher's presence can be problematic or useful in all sorts of ways.

Insider-ness is a key feature of this, and researchers are encouraged both to embrace and resist it. They might embrace their access to 'local and esoteric knowledges' (Coffey, 1999: 27) which for the 'standard fieldwork model' requires a research journey from 'ethnographer-as-stranger, progressing towards a familiarity and eventual enlightenment' (Coffey, 1999: 20). But the insider has an 'initial proximity' (Hodkinson, 2005) that resonates with Coffey's assertion that 'fieldwork always starts from where we are' (Coffey, 1999: 158).

This is a statement about identity, and for the insider the starting position is one of personal immersion that brings for many a 'definite advantage' (Ohnuki-Tierney, 1984). In reviewing this literature Labaree has divided perceived advantages into four areas: 'the value of shared experiences; the value of greater access; the value of cultural interpretation; and the value of deeper understanding and clarity of thought' (Labaree, 2002: 103). The bonuses of these areas, while apparently quite clear at one level, are also potentially problematic in a number of important ways. Consequently, making insider-ness work seems to depend on an adequate treatment of a range of well-documented pitfalls. For many, researcher reflexivity is the preferred means for achieving this since it 'expresses researchers' awareness of the necessary connection to the researcher situation and hence their effect on it' (Davies, 1999: 7). Insider 'cognitive [. . .] predispositions' (Gergen and Gergen, 1991: 77) that affect how the world is apprehended can escape notice, and may limit what we might experience and the questions we ask. This is why the need to continually induce levels of 'strangeness' is sometimes advocated (Hammersley and Atkinson, 1983) where familiarity either preexisted or has set in. In educational settings, researchers as teachers are often insiders, and Delamont (1992) calls for strategies to deal with this to establish a workable tension between 'strangeness and over-identification' (Coffey, 1999: 23) to avoid total absorption and maintain some degree of professional distance. For Hammersley and Atkinson (1983: 115) this is because 'feeling at home' must be avoided for a critical and analytical perspective to flourish. For these authors, too much familiarity can hinder analysis and inquiry. At worst it can produce uncritical work based on flawed analysis. Coffey (1999: 31) argues that this was the case for Willis's classic work *Learning to Labour* (1993), where he did not adequately consider his 'ethnographic self' and so failed to reflect upon his overidentification with the boys who were the focus of his study, undermining the work somewhat. This kind of unreflexive overidentification is likely to disqualify a researcher from an area of work, since a nonreflexive stance rather emulates members of the subject group, producing work that is noncritical and perhaps based upon common-sense observations. Indeed, as Coffey notes, 'a researcher who is no longer able to stand back from the esoteric knowledge they have acquired, and whose perspective becomes indistinguishable from that of the host culture, may face analytic problems' (Coffey, 1999: 23).

A more political criticism of interpretive research — and by implication, of insider-ness — is considered by Hammersley. This is the critical argument that 'ideological common sense' (Hammersley, 1992: 103) is reproduced by the approach and that this neglects 'the effects of macro-social factors on people's behaviour' (Hammersley, 1992: 103). For these authors, values ought to be explicit, especially those that motivate areas of study, since if value neutrality is unachievable, there must always be political or ideological implications of the chosen work. Inevitably, the position of an insider reflects a certain balance of power that may not be directly addressed. Other common insider pitfalls include self-indulgence and narcissism (Davies, 1999: 179) if they are given free reign. This seems more likely with autobiographical-style research which carries related perils of emotionality. This is a delicate and difficult matter, especially for researchers who approach projects from a position of 'knowing' (Coffey, 1999: 33) that draws to some extent on biographical experiences. Even so, 'emotional connectedness to the processes and practices of fieldwork, to analysis and writing, is normal and appropriate' (Coffey, 1999: 158). More than that, if done well it can produce excellent work such as John Hockey's *Squaddies*, an ethnography that draws on his own Army experiences for mediation (Hockey, 1986; see also Hockey, 1996). Coffey urges advancement beyond polarised accounts of familiarity and strangeness to embrace continually changing characteristics of self and identity, and to think about the researcher in terms of positionality. This allows for a range of researcher selves and interactional performances, including (hopefully fleeting) moments of self-indulgence or too much emotional attachment. Positionality draws attention to the process of research from phase to phase – even from person to person – and this has been especially pertinent to my research about identity at the point of Army exit. This requires from the researcher an awareness of the self as researcher and a willingness to 'critically [. . .] engage with the range of possibilities of position, place and identity' (Coffey, 1999: 36). Insider researchers are part of the research process, and their sense of self is drawn into the work in different ways across the entire process. It is this process of being drawn into the research work in different ways that constitutes the focus of this chapter.

Putting Insider-ness to Work

Putting insider-ness to work in qualitative military research involves making the most of specialist knowledge, but crucially at the same time striving for critical analysis. Charles Kirke gives us an example of insider-ness in military research (2013). He constructs three anthropological types: 'anthropology of the other' (the researcher can only very loosely be termed an insider); 'anthropology from within' (the researcher is a full member of the group being studied); and 'anthropology of the familiar' (researchers who are familiar but not full members). He locates his own research of the British Army in the middle type. Even though this kind of reflexivity among researchers of the military is becoming more common, there remains reluctance among many such researchers to recognise their own place in the research as a consequence of a number of factors. For one thing, the military is a challenging field of study largely because its component parts are closed institutions with unique roles and cultural differences, creating issues of access and understanding in a variety of different ways. Moreover, a preference for statistical 'fact' still influences the norms for certain academic and policy-related outlets speaking to military audiences and readerships. At the level of individual research work, the chosen methodology and theoretical framework, together with the precise details of the research being undertaken, will determine the extent to which military researchers *do* incorporate or recognise their place in the generation of meaning and the extent to which they believe they *should* do so. Professional answers to these two key questions will vary, and the divide between research that uses qualitative or quantitative methods is noticeable because the validity and reliability of these methods are justified in very different ways (cf. Golafshani, 2003). As a researcher of the military who has used both qualitative (Walker, 2013) and quantitative methods (Walker et al., 2014), I can see the very different ways in which credibility for that work has been justified. Now that I have outlined the kinds of issues that are at stake for matters of reflexivity and insider-ness among qualitative researchers I will focus in the remainder of this chapter on the context of my own research project that I introduced earlier. My hope is to ground some of the earlier more conceptual points about insider-ness in the context of this specific research situation and this specific researcher.

Theoretical Framework

As noted before, the theoretical orientation of the work will influence a researcher's stance on whether or not their place is acknowledged in the generation of meaning, and there will be differences of opinion about this. My own research about identity at Army exit took place from within a relativist ontology. This ontology emphasises local and specific co-constructed realities (Denzin and Lincoln, 2008) as part of a wider constructivist paradigm. From this paradigm, knowledge is partial, generated, and 'situational' (Haraway, 1988). As Finlay (2002) notes, research does not generate objective truth but instead something different emerges from the 'intersubjective relation' of research as an important dynamic. I did not take this to mean that there is not a lived reality; in trying to understand this, I drew theoretically on a combination of the work of Paul Ricoeur and George Herbert Mead. Ricoeur claims that the experience of being and acting in the world comes first (Ricoeur and Valdés, 1991) and that this should be the starting point for the analysis of identity. Although events and experience must come first, however, meaning was to be the only currency in a project such as mine. Drawing on Ricoeur's narrative theory of self that operates in the space between event and meaning, and accounting for the more general notion that life has an already storied quality (Somers, 1994: 613), my work could only explore the meaning side of this distinction – between what might be termed the soldiers' lived reality and what we or they can know, think or say about it in a research or interview context. Temporality was key for my work in all sorts of ways. Not only is temporality a central feature of identity, but for soldiers facing a horizon of exit, the relationship between time and meaning seemed especially prominent. This dynamic brought together in the present moment, for the soldiers, changing constructions of their past and anticipated futures – their identities of becoming. During lengthy interviews about their impending Army exit, soldiers chose in the moments of the interview, and in the contexts of their present lives, what to talk about in relation to the questions I asked, the themes pursued either in my analysis or in our interactions with each other.

Negotiating the 'Swamp'

What to do about reflexivity for this study? Some advocate a thorough exploration of every twist and turn in the research process in a kind of 'confessional' act (VanMaanen, 1988) intended to lay bare decisions made and routes taken so as to expose its dynamic. Finlay (2002) suggests five different ways to negotiate the 'swamp' of reflexivity (introspection, intersubjective reflection, mutual collaboration, social critique and discursive deconstruction) that are implicated in the researchers' aims. In a similar vein, Lynch (2000) develops an even more complex inventory of reflexivity (mechanical, substantive, metatheoretical, interpretive, ethnomethodological, methodological) to expose the endless diversity of the term. Both authors, however, warn against the simplistic idea that reflexivity is a process that alone can be conducive to good research. They seem to imply that there can be no final sense of getting reflexivity 'right' and that the endless pursuit of this is no guarantee of successful research. Indeed, for Finlay (2002: 227), researchers 'are damned if they do damned if they don't'. The challenge, it seemed, was for me as the researcher to negotiate the swamp in a way that best employs the undeniable processes of reflexivity for the project at hand.

As a new ex-soldier and postgraduate researcher, I straddled at least two communities and had to communicate in both. Being reflexive meant grasping this dynamic. I belonged neither to the Army nor the academic community. Dick Hobbs makes the same point about the working classes of the East End of London that he had come to call home. His challenge was in getting to know the academic community into which his work would be received, but he discovered he could not present a world he knew so well into an appropriate academic format. He goes on to say: 'in my attempt to perform an ethnographic ventriloquist act (Geertz, 1988: 145) I was using two dummies, and the voices were getting mixed up' (Hobbs and May, 1993: 56). This happened to me, and was also related to my ongoing attempts to adjust to a new role completely at odds with my last. My initial proximity to soldiering was soon consumed by an ongoing need to adjust to a university environment. Not only did this affect my relations with the data but it also significantly shifted interactions with my own prior career and disturbed my interpretive capacities, shattering what I came to see as a prior comfortable sense of belonging. This is why Hobbs's work resonates; I believe that by the time my data reached written form the voices of the soon-to-be-leaving soldier and the veteran-researcher had become thoroughly mixed up, and I think this is a good thing, for it entwined the leavers' own personal experiences of the pre-exit period with my experiences of postexit life, together with a professional and sociological critical analysis informed by the literature. The idea that something new and different emerges from the 'intersubjective relation' (Finlay, 2002) of the research offers one passage through the swamp. In the same way, Corbin and Strauss suggest that 'the constructivist viewpoint that concepts and theories are *constructed* by researchers out of stories that are constructed by research participants who are trying to explain and makes sense out of their experiences and/or lives, both to the researcher and themselves' (2008: 10) is a useful one.

Positionality and Reflexivity at Different Phases of the Research

In addition to a general and ongoing interplay between my veteran researching self and the data, there are quite specific features of the work that might have been different had it not been for insider-ness. During the period of proximity many such matters were beyond my notice and took on a different significance as the work progressed. Some of these kinds of processes are discussed later in relation to key phases of the research.

Topic and Sample Selection

It is possible that I intuitively knew the difficulties I might cause for myself by researching Army identity because I was very tempted by nonmilitary topics. But sometimes the 'selection of a setting for study hardly arises at all because an opportunity presents itself' (Hammersley and Atkinson, 1983: 36), and this was the case for me: I realised that something of sociological importance was occurring in the pre-exit phase because I was experiencing it myself, and I could see that friends and colleagues were similarly affected. Paradoxically, my topic selection resolved my own attempt to leave the Army and was for me at least the surest way to attract funding. In terms of sample selection, the most noticeable insider benefit I had was access: I knew key people who could help. In gathering my sample, I used internal Army categories (for example, corporal, Royal Artillery, twenty-two years' service) as a basis for differentiating people, but later realised how unquestionably

I had divided individuals in this way. Paul Higate notes that in military settings, research subjects might be 'captive' (Higate and Cameron, 2006: 222), unable to resist the researcher who is perceived as more powerful. I had little problem with access, but again with hindsight I could see that I was using without question the 'wilco' (will-cooperate) attitude soldiers have towards each other. Researchers I knew in other fields had a much harder time negotiating access.

Face-to-Face Interviews

On the whole insider-ness brought me easy rapport with the leavers, but this cannot be guaranteed, as Paul Higate found when his Royal Air Force background afforded little purchase with a veteran sample (Higate and Cameron, 2006: 228). Moreover, the manner or demeanour of the interviewer will shape the image created of him by respondents who will 'use that image as a basis of response' (McCall and Simmons, 1969: 80). Along with other factors, differences of gender and rank might influence face-to-face interactions in ways unnoticed by the researcher, perhaps sustaining local power relations of gender and subordination (Higate and Cameron, 2006: 222). For McCall and Simmons, 'observer's data are conditioned by the basis upon which subjects respond to him' (1969: 82); interviewers, too, may find themselves reacting in unforeseen ways. For the last eight years of my own Army service I interviewed (or managed those who interviewed) hundreds of soldiers and officers as part of my role. Before this, I worked close to senior commissioned officers. This meant that I had established ways of communicating with all types of soldiers and officers, compatible with the various roles I had. During the subsequent research interviews, I capitalised on my background and presented myself with a 'just-left-the-Army' story. Indeed, on one occasion this went too far when I was placed into my previous Army role by a passing captain who recognised me and requested advice about an ongoing personnel situation.

Connection was eased by my knowledge of local personalities, places and events. Most of my sample seemed able to place me in Army terms. One or two knew me a little. Most were bursting to talk about leaving the Army, but a few had little to say, and I noted at the time that I found interviewing commissioned officers most challenging – a factor perhaps implicated in my more limited analysis of this group. I connected most easily with full-career noncommissioned leavers who like me joined the Army in the mid-1980s. But, soon I was beginning to see how many of my questions had been framed from within the Army community, motivated by my shared concern about exit. From the analysis phase of the research I looked back on my interviewing self, thoroughly at home there. In early writing, I contemplated an interview with Don. Don occupied a senior commissioned role that had been significant to me as an 18-year-old soldier, but I struggled to connect this and my familiar sense of what a major might be to the person periodically unravelling before me in interview, under the shadow of exit. After the interview I wrote in my research journal:

> Met Don for interview – very odd experience talking to him – he seemed so military and (I was) strangely nervous about interviewing an officer in a position I remember as important and senior (he had a role in the same headquarters I worked in as a private soldier), and occupied by a man of a very different lifestyle – he kept convincing himself he is a top bloke and that he had lots to offer (civilian life).
>
> *(Research Journal, 10 January 2008)*

The temporal aspects of this are important. There are signs I had disconnected a little from the Army ('he seemed so military'), but overall during this face-to-face encounter I connected to him in keeping with the effects of my past, because following the interview I also wrote:

> Don tells me that humility is central to leadership, and I believe him. As he talks I am in his world. I know what he means. I draw on the same social relations for meaning. I can see that I latch on to his talk of service, his notion of giving-back, and his idealised, embedded speech about community and team.

This was an early veteran researching self reflecting on the face-to-face interview, attempting to understand how embedded knowledge that bound us together during the interview seemed now to be problematic. Paradoxically, too, internal know-how also separated us due to the officer/soldier divide. This was further complicated by his reaction to me. As a public school–educated officer he reacted to my new role as a PhD researcher by saying in a surprised tone: 'well . . ., but that will make you middle class!' Eventually I conceived of this interview and others like it in terms of Wittgenstein's notion of 'seeing-as' (Ricœur, 1978: 251), because increasingly I could see how my biography was implicated in much of my understanding because 'seeing as' is 'half thought and half experience' (Ricœur, 1978: 251). I went on to write: 'I don't receive Don's words as an empty receptacle takes water. I hear and experience Don's words at the same time.' Two years later, I could see that the interviewing period featured for me both deep personal understandings of Army exit and increasing distance, as numerous postexit encounters forced strangeness onto the past. In addition, each time I entered a barracks or location I did so as an outsider. This was often disagreeable, but usefully gave me a sense of incongruity now that my rank and belonging were gone. Denied the interactional comfort I had previously enjoyed, on one occasion my veteran (lack of) status meant that I had to stand outside of the guardroom in a manner reminiscent of my recruit days. I noted this at the time:

> Interview: Nigel – had to meet outside guardroom; long time since I had waited outside a guardroom – lots of young lads in and out of uniform small boy racer type cars passing by and young soldiers mostly. Got on to camp no problem – just had to wait for someone. Car parked across the way – someone (of my prior status) coming into camp in car asked who I was.
>
> *(Research Journal, 19 February 2008)*

At the time, I searched these kinds of incident for insight, but in preparing an account of the work two years later, I was surprised to come across the following journal entry about another interview with a provost sergeant:

> I felt the despondency of the guardroom – provost sergeant polishing pace stick – he will be out (of the Army) mending heating systems in a few months – the pointlessness of polishing a pace stick! [. . .] I hated going there – in the cells.
>
> *(Research Journal, 19 May 2008)*

Separation from Army life increased for me as the interviews ended and I began a lengthy and intensive period of analysis and writing. By now, I was quite envious of those I had interviewed because all the signs were that they had moved on, but I was to remain trapped in the process of Army exit for another two years as a consequence of my selected research topic.

Data Analysis

As already stated, the researcher's influence is most pronounced during specific phases of analysis (Hammersley and Atkinson, 1983: 19), but attending to this should not preclude systematic and stringent research strategies. The structured features of this particular research project are outlined in the original work (Walker, 2010, 2013), including, for example, the organisation of transcribed interviews into codes and categories close to the data. In terms of positionality and insider-ness, however, a number of unforeseen factors combined to influence my approach to the analysis. First, I was reading a good deal of literature about self and identity; second, I adopted a broadly symbolic interactionist approach that emphasises the temporal conditions for identity, especially the 'hypothetical' properties of the future and the revisable properties of the past, apprehended by the person during the present moment. This was an orientation that also made sense for me in terms of post-Army situations in which I found myself, particularly when social relations were at odds with my prior self. Trying to unpick all of these sorts of issues would be a poor way to negotiate the reflexive swamp even if it were possible; but still, I could see that my own pre- and post-Army experiences were feeding into my reading of the literature in different ways and at different stages. They also reinforced my understanding of the temporal and situated dimensions of identity about which I had been reading so much. Similarly, I began to observe connections between identity, self and social relations in other settings – insights that I am sure fed into my treatment of the data. For one thing, the contrast between the two organisations I frequented – Army and university – seemed extreme and granted me special insight into their respective organisational needs for different kinds of personal identity that often passed unnoticed among those involved.

As I developed the analysis, I learned how identities are granted or denied in social relations and that there are clear limits for personally anticipating, noticing or incorporating this into a personal narrative – at least for exiting career soldiers. Overall, throughout the research there was an underpinning process for me as researcher that is best described as a movement of position from Army insider, to partial detachment and alienation, towards critical analysis. I began to acknowledge more fully the force of Army social relations on individual soldiers' identities – something I was prone to underestimate because it challenged my starting emphasis on personal agency and the narrative capacity of the individual as a consequence of my starting position and identity. In these ways, the emergent emphasis of the research project – that identity is a (vulnerable) becoming – was somewhat mediated by my veteran-researching-self, although this was obviously the only route or basis for such a conclusion. Overall, in the thesis, I argue for the presence and importance of a concrete, 'real subject' (Denzin, 1992: 2) compatible with an interactionist position and with the theoretical basis of the work, not to mention the perspective of the soldiers themselves. I argued for a middle way between poststructural and essentialist treatments of the self, and to do this

I read a lot about poststructural treatments of identity. This reading was especially pertinent to the experiences I was having as a newly exited soldier and caused me to think very hard indeed about the work and about identity in general. Although a retreat from my interactionist perspective was unnecessary, I nevertheless steered a middle position that was much closer to contingent views of identity than I had expected or intended. This shift in the work reflects the journey that I suspect most Army leavers personally encounter as they come to recognise the extent to which their sense of self can or cannot be continued in the absence of Army relations; at least, that is, for those who do not seek out and find lives and jobs that closely emulate the ones that they were leaving.

Conclusion

My research focused on leavers' narrative attempts to construct and project a continuance of self across the anticipated social rupture of Army exit. Eventually this brought to the fore during analysis, ethical and moral principles that are given worth in Army relations but not always beyond. Personal attributes encouraged in Army relations – especially those concerned with nonindividualistic attitudes – became more peculiar to me as the work progressed and this enhanced critical analysis of these data. In this chapter, I have argued that if researchers of the military who are also insiders are sufficiently reflexive about their connections to the military then their insider-ness can be put to good research-related use. However, precisely how – and even if – this should be done will depend on many different and often complicated dimensions, especially the kind of research being undertaken. I started the identity research project with a personal determination not to talk about my own relationship to the Army and although this clearly changed I still contend that processes of self-reflection, positionality and insider-ness have serious limits and should never take centre-stage in any research work. I agree with Corbin and Strauss when they claim that 'something occurs when doing analysis that is beyond the ability of a person to articulate or explain' (Corbin and Strauss, 2008: 9). For Paul Ricoeur, our actual and complete lived experience is only known fully by those present at that specific and individual moment, and beyond this the generation of meaning is forever a partial construction. It is true that researcher reflections on the research process benefit significantly from the passage of time, but they are also markedly hindered by it. I believe that I found a pragmatic balance in the original research work for putting insider-ness to work but contrary to what this chapter might imply, I did not spend much time on endless introspection, but instead I tried adequately to acknowledge my own changing place in the generation of meaning and to use this to good critical effect.

References

Atkinson, P. (1992) The Ethnography of a Medical Setting: Reading, Writing and Rhetoric. *Qualitative Health Research* 2, 451–474.

Carreiras, H. and Castro, C. (Eds.) (2013) *Qualitative Methods in Military Studies: Research Experiences and Challenges*. Oxon: Routledge.

Clifford, J. and Marcus, G. (1986) *Writing Culture: The Poetics and Politics of Ethnography*. Berkeley: University of California Press.

Coffey, A. (1999) *The Ethnographic Self: Fieldwork and the Representation of Identity*. London: Sage.

Corbin, J. and Strauss, A. (2008) *Basics of Qualitative Research 3e – Techniques and Procedures for Developing Grounded Theory*. London: Sage.

Davies, C. A. (1999) *Reflexive Ethnography: A Guide to Researching Selves and Others*. London: Routledge.

Delamont, S. (1992) *Fieldwork in Educational Settings*. Lewes: Falmer.

Denzin, N. K. (1989) *Interpretive Biography*. Thousand Oaks, CA: Sage.

Denzin, N. K. (1992) *Symbolic Interactionism and Cultural Studies: The Politics of Interpretation*. Oxford: Blackwell.

Denzin, N. K. (1995) *The Poststructuralist Crisis in the Social Sciences Writing Postmodernism*. Brown, R. H. (Ed.). Urbana: University of Illinois Press.

Denzin, N. K. (1997) *Interpretive Ethnography: Ethnographic Practices for the 21st Century*. London: Sage.

Denzin, N. K. and Lincoln, Y. S. (2008) *The Landscape of Qualitative Research*. Thousand Oaks, CA: Sage.

Finlay, L. (2002) Negotiating the Swamp: The Opportunity and Challenge of Reflexivity in Research Practice. *Qualitative Research* 2(2), 209–230.

Geertz, C. (1980) Blurred Genres. *American Scholar* 49, 165–179.

Geertz, C. (1988) *Works and Lives: The Anthropologist as Author*. Stanford: Stanford University Press.

Gergen, K. and Gergen, M. (1991) From Research to Reflexivity in Research Practice. In Steier, F. (Ed.), Research and Reflexivity. London: Sage, pp. 76–95.

Golafshani, N. (2003) Understanding Reliability and Validity in Qualitative Research. *Qualitative Report* 8(4), 597–607.

Hammersley, M. (1992) *What's Wrong With Ethnography?: Methodological Explorations*. London: Routledge.

Hammersley, M. and Atkinson, P. (1983) *Ethnography: Principles in Practice*. London: Tavistock.

Haraway, D. (1988) Situated Knowledges: The Science Question in Feminism and the Privilege of Partial Perspective. *Feminist Studies* 14, 575–599.

Higate, P. and Cameron, A. (2006) Reflexivity and Researching the Military. *Armed Forces and Society* 32(2), 219–233.

Hobbs, D. (1993) *Peers, Careers, and Academic Fears: Writing as Field-Work. Interpreting the Field*. Hobbs, D. and May, T. (Eds.). Oxford: Clarendon Press.

Hobbs, D. and May, T. (1993) *Interpreting the Field: Accounts of Ethnography*. Oxford: Clarendon Press.

Hockey, J. (1986) *Squaddies: Portrait of a Subculture*. Exeter: Exeter University Press.

Hockey, J. (1996) Putting Down Smoke: Emotion and Engagement in Participant Observation. In Carter, K. and Delamont, S. (Eds.), *Qualitative Research: The Emotional Dimension*. Aldershot: Avebury, pp. 12–27.

Hodkinson, P. (2005) 'Insider Research' in the Study of Youth Cultures. *Journal of Youth Studies* 8(2), 131–149.

Kirke, C. (2013) Insider Anthropology: Theoretical and Empirical Issues for the Researchers. In Carreiras, H. and Castro, C. (Eds.), *Qualitative Methods in Military Studies: Research Experiences and Challenges*. Oxon: Routledge, pp. 17–30.

Labaree, R. V. (2002) The Risk of 'Going Observationalist': Negotiating the Hidden Dilemmas of Being an Insider Participant Observer. *Qualitative Research* 2(1), 97–122.

Lynch, M. (2000) Against Reflexivity as an Academic Virtue and Source of Privileged Knowledge. *Theory Culture Society* 17, 26–54.

Marcus, G. and Fischer, M. (1986) *Anthropology as Cultural Critique: An experimental Moment in the Human Sciences*. Chicago: University of Chicago Press.

Mauthner, N. S. and Doucet, A. (2003) Reflexive Accounts and Accounts of Reflexivity in Qualitative Data Analysis. *Sociology* 37, 413–431.

McCall, G. J. and Simmons, J. L. (1969) *Issues in Participant Observation: A Text and Reader*. Reading, MA: Addison-Wesley.

Ohnuki-Tierney, E. (1984) 'Native' Anthropologist. *American Ethnologist* 11, 584–586.

Pearson, G. (1993) Talking a Good Fight: Authenticity and Distance in the Ethnographer's Craft. In Hobbs, D. and May, T. (Eds.), *Interpreting the Field*. Oxford: Clarendon Press, pp. vi–xx.

Ricœur, P. (1978) *The Rule of Metaphor: the Creation of Meaning in Language*. London: Routledge & Kegan Paul.

Ricoeur, P. and Valdés, M. J. (1991) *A Ricoeur Reader: Reflection and Imagination*. New York: Harvester Wheatsheaf.

Soeters, J., Shields, P. M. and Rietjens, S. (Eds.) (2014) *Routledge Handbook of Research Methods in Military Studies*. New York: Routledge.

Somers, M. R. (1994) The Narrative Constitution of Identity: A Relational and Network Approach. *Theory and Society* 23(5), 605–649.

VanMaanen, L. (1988) *Tales of the Field: On Writing Ethnography*. Chicago: University of Chicago Press.

Walker, D. (2010) *Narrating Identity: Career Soldiers Anticipating Exit from the British Army*. PhD dissertation, Durham University.

Walker, D. (2013) Anticipating Army Exit: Identity Constructions of Final Year UK Career Soldiers. *Armed Forces & Society* 39, 284.

Walker, D. I., Cardin, J. F., Chawla, N., Topp, D., Burton, T. and Wadsworth, S. MacDermid (2014) Effectiveness of a Multimedia Outreach Kit for Families of Wounded Veterans. *Disability and Health Journal* 7(2), 216–225.

Willis, P. E. (1993) *Learning to Labour: How Working Class Kids Get Working Class Jobs*. Aldershot: Ashgate.

21

RESEARCHING AT MILITARY AIRSHOWS

A Dialogue About Ethnography and Autoethnography

Matthew F. Rech and Alison J. Williams

Air Forces Monthly magazine, the self-proclaimed 'world's number one military aviation magazine', included in its September 2010 issue a one-page report on the RAF Waddington International Airshow which had taken place on 3–4 July. Beginning with a comment about the reporter's enjoyment of the event – 'It has such a friendly feeling about it' (Warnes, 2010: 76) – the article went on to detail a number of the key attractions available to visitors, including flying displays from a Vulcan bomber, a Royal Air Force (RAF) Typhoon, and the Battle of Britain Memorial Flight. The reporter concluded that 'if you want a show that you can take your kids to without it costing the earth to keep them entertained, then head for RAF Waddington' (Warnes, 2010: 76).

Airshows have, since their rise to prominence as cultural occasions in the early years of the twentieth century, provided the stage for a politics made aesthetic, and for nationalism as expressed 'through the art of the spectacular . . . technological modernism and various modes of voyeurism, comfort and spectatorship' (Adey, 2010: 57). Airshows are also premier tourist events. The UK's annual airshow calendar includes approximately 100 shows with the larger of these, like Fairford's Royal International Air Tattoo (hereafter RIAT), attracting crowds of 130,000. Host to aerobatic displays and stirring evocations of airpower and wartime heritage, airshows, after Kong and Yeoh (1997: 216), are instances where a spectacle (the display of aircraft) 'is used to inspire positive feelings of admiration and wonder' in the military and the military adventures of states. This admiration is 'attained through the deliberate use of ceremony; the conscious construction of pomp; the creation of occasion and circumstances for celebration' (Kong and Yeoh, 1997: 216). But vitally, it is attained through 'visual effects', a 'spectating airshow-goer and a willing audience' (Adey, 2010: 61). As a space and set of techniques for spectacular observation (Crary, 1992), the airshow is, thus, a 'means of situating the . . . citizen within the political world of the state' (MacDonald, 2006: 57) and of inculcating senses of the borders, boundaries, dangers and differences integral to modern geopolitical imaginations.

Both of us have attended many airshows and have watched aircraft demonstrations, talked with military aviators, and visited a host of stalls, military reenactment zones and military recruiting presences. However, our experiences of Waddington airshow (and others) were quite different to those of the *Air Forces Monthly* reporter quoted earlier, and our motivations for attending were not the same as those looking for an entertaining day out for the kids. Our focus was on the spectacle of the airshow and its role and place in processes of militarisation. Yet, our individual motivations also differ from each other, one focusing on collecting data for PhD research on military recruitment strategies, the other thinking through what researching airshows might actually mean methodologically and in terms of researcher positionality as part of wider analysis of the geographies of airspaces (see Williams, 2011a, 2011b; Rech, 2012, 2015). These separate yet overlapping interests have meant that over the last five years we have both visited a number of airshows across the UK for academic research, although we have never attended one together. We have clocked up approximately ten different airshows, held at RAF stations (RAF Waddington, RAF Fairford), on beach fronts (Sunderland, Bournemouth), and at county shows which often include appearances by the RAF's Red Arrows or the Battle of Britain Memorial Flight display teams. We have paid money, bought tickets and immersed ourselves in the official airshow experience. But for the purposes of research, we have also sat at the end of an airport runway with a small number of aviation photography enthusiasts watching aircraft taking off en route to a public airshow being held elsewhere, and have experienced unplanned, surprising encounters with aircraft flying to and from airshow displays. We have often spent a portion of our airshow visits sitting making notes, surreptitiously taking research photographs or eavesdropping on conversations between young show-goers and military recruiters.

Taking the decision to position ourselves as researchers at airshows has required us to consider what being a researcher means in this context, both in terms of methodology and strategies of scholarly engagement. In this chapter we focus on how the concepts of ethnography and autoethnography offer us ways of engaging with these events, their discourses, spatialities and materialities as critically informed scholars. We also consider what it might mean when we position ourselves in relation to our own interests beyond academia. In order to achieve this, we provide a brief overview of the history of airshows in the UK. We go on to discuss our positionality and how autoethnographic practices help us consider what it means to be a researcher at an airshow. We then consider the utility of adopting specific methodologies for data collection at airshows, and discuss a range of techniques we have used to both acquire and generate research materials. Because of our differing experiences of doing airshow research, and our differing formative positions, we decided to write each of these subsections separately, so when we consider our personal encounters, we write in the first person rather than attempting to speak for each other. In concluding, we bring our reflections together in dialogue, and provide some final thoughts on the effectiveness of these approaches for engaging with public military events and spaces. Alongside this, we offer a vision of critical military methodologies as about a working-through of the relationship between enchantment and 'criticality'.

A Brief History of the Airshow

The history of the military airshow stretches back to the early years of aviation in the first decade of the twentieth century. The first aeroplane flight in the UK took place in 1908 at Farnborough, when renowned adventurer Samuel Cody took control of a copy of the Wright Brothers' famous Flyer I. In his discussion of the 1909 airshow at Brescia, Italy, which involved a number of record attempts by aircrews in these early aircraft, Demetz (2002) considers how the success of this event came from its spectacle, with crowds celebrating the achievements of the pioneering crews and their planes. After the First World War, countries across Western Europe and North America embarked on military reduction plans, resulting in a glut of aircraft and aviators. At the same time military and government bodies realised that the advent of aviation could be put to good use, and the quest to promote air-mindedness emerged.[1] As Adey (2010) and others (see Corn, 1983) have noted, this sense of being aware of aviation was encouraged in part through airshows, as they offered opportunities for the public to recognise that the aerial dimension was one that could now be controlled and used by the state. This desire to promote air-mindedness dovetailed with the abundance of skilled aviators and aircraft, resulting in a host of airshow-type events being held throughout the interwar period. In the UK there were a number of air races and air pageants which offered opportunities for the general public to witness aircraft being flown (Adey, 2010; Watson, 2010). The most popular of these were the Hendon Air Pageants, held at the north London aerodrome during the 1920s and 1930s. These boasted a huge following with the final pageant in 1937 attracting approximately 186,000 visitors (Watson, 2010: 16).

In the aftermath of the Second World War there were few, in the UK at least, who were unaware of the effects of military aviation both as a tool of war-making and geopolitical rhetoric. Yet the RAF began a programme of airshows as soon as 1945, with Battle of Britain commemoration events held only weeks after the end of the war. By the 1950s the RAF had instituted a series of 'At Home' days, where RAF stations were opened to the public for one Saturday in September as part of war commemorations (Watson, 2010). Although no longer needing to so strongly instil air-mindedness in the populace, these events served to offer an insight into the postwar operations of what was seen as a victorious RAF. As such, airshows, while maintaining a Battle of Britain commemoration element, increasingly came to include displays from the RAF's front line aircraft. As the Cold War set in, the number of these events was reduced (for financial and operational reasons) to be replaced by fewer, larger shows, offering the public a day-long spectacle of British military aviation capabilities.

By the 1980s only four RAF stations continued to hold airshows, as pressures of time and money and the deployment requirements of the Cold War and other UK overseas commitments led to their reduction (Watson, 2010). However, this decline in RAF airshows was, and continues to be, mitigated by an increasing number of private or local council-organised events, and by the ability of organisations to book the RAF Red Arrows and Battle of Britain Memorial Flight display teams, for example, to perform at these events. Thus, as the UK Airshow Review website (http://www.airshows.org.uk) reveals, a total of 105 airshows were planned across the UK in 2014; these ranged from official RAF and Royal Navy events through to historical reenactments with air displays, country fairs, beach front displays, festivals of flight, speed and history, and myriad other events at which the public can watch historical and current military aircraft in flight (Adey, 2014). The preponderance of opportunities to encounter these spectacles, and what this might mean in terms of cultural militarism, is what drives our research. In the following section we discuss our approaches to doing research at and on airshows. We focus on our experiences of attending an airshow at RAF Waddington, how we positioned ourselves at the event and how we recorded information for later analysis.

Being at the Airshow

The airshow offers a unique opportunity to get close to the military. As political geographers interested in military spaces and the projection of state power, these occasions offer us chances to encounter military personnel and materiel somewhat on their own terms. Airshows have also become important recruiting and awareness-raising outlets for our armed forces: at RAF Waddington, there were recruitment and information stalls not only for various branches of the RAF, but also for the Royal Navy, British Army, and Royal Marines, as well as their associated reserve forces. Furthermore, large airshows such as Waddington include a 'plethora of stalls selling everything military and aviation related, and somewhat bizarrely lots of things that had no relevance to the military at all' (Alison diary) (see Figure 21.1).

Thus, the airshow offers us unrivalled opportunities to consider issues of militarism and militarisation, their spatialities and materialities. It also offers us ways to investigate the blurring of civil-military boundaries and relations. As noted earlier, many of the UK's airshows now take place in public spaces, on beach fronts or as displays at fairs and other gatherings.

Figure 21.1　Military ephemera on sale at Waddington 2010 airshow.

Source: Alison Williams.

These events require us to consider what the inclusion of military aircraft means for the spatialities of military power and the militarisation of civil space. Much of this is beyond the scope of this chapter, although we wish to highlight these issues to illustrate the multiple and varied research avenues that researching airshows offers. This section, however, focuses on the issues inherent in negotiating our position as researchers at these events. The two subsections that follow are written by each of us in turn to offer an insight into the individuality of these negotiations and their operation, and to illustrate the utility of autoethnography, ethnography and observant participation as ways to work through the complex and sometimes contradictory practice of being at airshows.

Alison at the Airshow

For me, being a researcher at an airshow is a complex position to adopt, as I have a long association with them which significantly predates my academic interest (cf. DeLyser, 2001). I attended my first airshow as a child for a friend's birthday party. I had already started to develop an interest in military aviation by that point, and during my teenage years I attended numerous airshows across the UK, sometimes going with my parents to relatively local events, sometimes travelling by coach, on trips run by a travel agency in my home town, to large international airshows at air bases hours away from my home. Although I moved away from these interests as a student, as an academic I have begun to reengage with airshows as spaces of aerial power projection (my key research focus). Of course this is not unusual: researching something of interest naturally enables us to be more enthusiastic about our subject matter (see Woodyer and Geoghegan, 2012). When researching the military, being cognisant of our motivations and underlying biases seems even more significant because critical social science has tended to shy away from military research, and thus this enthusiasm can lead to misunderstandings about positionality and objectivity (as Matthew later notes). Thus, bringing an extant interest in airshows to conducting research at airshows produces a complex positionality, yet one that I hope to show here, can offer useful and insightful material and experiences from which to draw.

Engaging with the airshow as a data collection site and process is different to other forms of data collection, such as conducting semistructured interviews with military personnel, and this approach is influenced by my personal perspectives more than using any other data collection methods has been. Here I gesture to work by Wylie (2009) on walking in landscapes, as he argues that we are imbued with and immersed within the spaces and places of our analysis rather than simply being observers. Thus, my airshow experiences and my aviation interests have shaped my perspective on how I conduct research at these events, and have required me to think carefully about how I engage with them.

In order to negotiate these complex positionalities I have sought to adopt something of an autoethnographic approach. As Reed-Danahay (1997: 9) notes, autoethnography simply describes 'a form of self-narrative that places the self within a social context' (quoted in Butz, 2010: 138). However, as Butz and Besio illustrate, there are many ways of doing autoethnographic research. Importantly, although they note that autoethnography 'may be understood as the practice of doing . . . identity work self-consciously' (Butz and Besio, 2009, 1660). As such, this approach, as I will illustrate, enables me to recognise how my experiences of airshows are bound up in how I engage with them inclusive of my past experiences and interests.

This approach embodies and enacts a broadly feminist geopolitical position (see, for example, Dowler and Sharp, 2001; Pain, 2009; Massaro and Williams, 2013) in which the focus of my inquiry are my personal and emotional experiences of the militarised airshow, recognising rather than seeking to ignore my own positionality and its implications for my work. Rech (2015) has recently written on the spectacle of the airshow (and elucidates on this in more detail later), yet my perspective is different because I frame my research within the wider context of my personal affectual experiences and aviation interests rather than solely my academic and intellectual knowledge, which comes second and sometimes only later upon reflection. As Butz and Besio (2009, 1662) note, autoethnography 'radically foregrounds the emotions and experiences of the researcher as a way to acknowledge the inevitably subjective nature of knowledge'. This approach illustrates that rather than my scholarship being absent from my work on airshows, allowing my innate interest in aviation to dominate over my intellectual framing in the moment of being at the airshow, it enables me to conceptually frame my experiences in order to more fully understand their significance through a process of critical reflection. Thus my experience is not explicitly bounded by intellectual positioning, but rather I reflect upon and consider how my implicit academic framings and my enthusiasm and interest in the airshow are drawn together in messy and often untidy, unsettling ways. Critical reflection after the event, contra to Matthew's position, revels in this messiness of the conceptual and empirical, intellectual and emotional, producing an engagement with the military airshow that for me is more fruitful.

My research practice at 2010 Waddington airshow, for example, revolved around an unstructured approach. I experienced the airshow as an interested participant, allowing my prior knowledge of and fascination with military aviation to be my dominant emotional driver. Of course it would be erroneous to suggest that my extant academic and intellectual knowledge was not also bundled into those emotional responses – something which continues to cause a jarring and discordant sensation when attending such events. At Waddington, as much as possible I actively sought to not intellectualise my experience at the point of its event, although of course this does not mean that I shed my academic sensibilities entirely. Instead I revelled in the affectual resonances the airshow brought, and drew upon these to fulfil my intellectual analyses. These included all manner of emotional responses, from joy at seeing a particular aircraft to disappointment that another was not in attendance. I engaged all my senses in these responses, the sound of an aircraft often eliciting the strongest feelings of expectation and excitement.

These affects and experiences could lead to accusations that I have 'gone native', that I have lost my researcher objectivity. And, it could be hard to disagree. But this is the very reason why I employ an autoethnographic methodology to airshows. I am aware that any attempt on my part to play the role of an objective academic researcher, only interested in the airshow for its intellectual aspects (as Matthew discusses later), would lead to unsatisfactory data collection for me given that my fascination with the airshow predates my academic career. Thus, my approach to being at the airshow is to experience it as a relatively well-informed military aviation enthusiast. Autoethnography enables me to do this while also collecting worthwhile data (as Butz and Besio, 2009 encourage) because it explicitly enables me to foreground my nonacademic, emotional engagement with the experience over my more abstracted scholarly positioning while not seeking to expunge either.

I do this through two data collection methods. One involves taking lots of photographs of things that interest me. I purposely do not compile a list of required images, or even subject matter, before the event, in order to leave the capturing of sights and sites open to chance and moments of interest piqued. The other is through completing a narrative diary of the event, as soon as possible after leaving the airshow. This offers me two important and interlinked opportunities. The first is to simply narrate the day in a form that I can return to as an aide-memoire long after the event. The second is to begin the process of rebalancing what I might a little starkly describe as 'the aviation geek' with 'the geopolitical scholar'. This process enables me to contemplate my experiences and emotions and consider the photographs I have taken from a more critical perspective. In what follows I discuss some of the comments I made in my Waddington diary. Following this, Matthew will discuss his experience and reflections from his visits to Fairford's Royal International Air Tattoo and the Waddington and Sunderland airshows. We do this to illustrate the differences in our approaches to engaging with our experiences of being at the airshow and recording those experiences for subsequent academic use.

My diary for the 2010 RAF Waddington airshow was written by hand soon after I left the show and details the day from arrival at the air station at 9.30am to leaving during the Red Arrows display at 3.30pm. It takes an unstructured narrative form and does not seek to answer any research questions or stratify the events of the day according to any sort of academic framing. Of course it is beyond the scope of this chapter to discuss the entirety of my record of the airshow. Instead I have chosen to focus on one moment of the day, discussing how I recorded one experience and using that to reflect more broadly on the utility of employing this autoethnographic method for this research.

The part of the day at the Waddington, 2010 airshow that I have chosen to focus on relates to my experience of and emotional reaction to the UAV (unmanned aerial vehicles) hangar there. I recorded the moment when

> we turned left from our Blades [aerial display team] vantage point towards the hangar [and I was] completely overwhelmed by what I saw. In front of me were three huge UAVs, one of which was a Global Hawk . . . I couldn't believe my eyes and was so excited a tear came to my eyes. I began taking photos.
>
> *(Alison diary)*

I have chosen to reflect on this passage as this clearly illustrates the complexities I encounter when attempting to conduct research at airshows, and why an autoethnographic approach is the only one I feel ethically at ease with. In the moment of turning the corner and seeing the Global Hawk I was completely unable to perform the role of objective researcher, even if that had been my intention. My emotional response to seeing this aircraft is still vivid in my memory (although my diary enables me to be clear on how I felt at the time, without the fuzziness of memory altering it). As such, it illustrates how autoethnography enables me to engage with the airshow event fully, performing myself honestly as someone interested in military aviation rather than simply, and less coherently, as an objective military researcher (for more on autoethnographic performance see Spry, 2001).

Further down the diary's page I record how even while talking with one of the BAE Systems people on their UAV stand I was 'still in a state of astonishment' at what I had seen (i.e. the Global Hawk) (Alison diary). However, I go on to note immediately after this:

> It also struck me that the other people walking past had no idea of how exciting it was to be seeing these aircraft. Few people were stopping to take photos, I on the other hand was taking lots.
>
> *(Alison diary)*

Here the autoethnographic approach illustrates my concern to consider not only my own positionality in this situation but to try to contextualise it in relation to other people's perspectives, as a way to judge my own reaction to the presence of the UAVs. At the airshow I seem to do this via the camera – not just considering how many photos I have taken, but also whether or not other people were taking photos and ascribing the level of their interest in, or reaction to, aircraft or other materiel and events at the airshow through their photo-taking activities (see Figure 21.2).

My diary continues to describe my actions and emotional state as I walked around the UAV hangar (the large building in the top right of Figure 21.2), talking to military personnel and civilians from various defence industries. It is interspersed with a range of more prosaic comments. For example, in relation to the Watchkeeper UAV I comment in the diary that 'currently in trials, this will be the UK Army's largest and most capable UAV to date when it enters service' (Alison diary). My notes on this UAV are more restrained than for Global

Figure 21.2 Not taking photos of UAVs.

Source: Alison Williams.

Hawk, and in reading my diary back they have caused me to think about the reasons for this. I conclude that it is the level of my emotional reaction that influences the type of comments I make, and that these reactions are often related to my research foci, as I discuss later.

In 2010, when I visited the Waddington airshow, I had active research interests in US and UK drone operations in Afghanistan, and a long-standing research focus on US military air power. I had recently visited RAF Waddington to interview serving aircrew, including a drone pilot, as part of research on military airspace use. Thus, I suggest that the Global Hawk evoked a stronger reaction than the Watchkeeper simply because it connected more strongly to the work I had been doing. Seeing the Global Hawk caused an emotional reaction because my research on drones had often been hampered by the secrecy surrounding their operations and the difficulties I had had gaining access to drone pilots and other official material on their activities. To simply walk around a corner at a public airshow and come face to face with something I had been 'chasing' for months from my desk resulted in shock and surprise that manifested itself in a visceral reaction.

This brief example enables me to think about the complexities of conducting autoethnographic research in an environment that is well known. Of course, in this my engagements with airshows is not unique. (DeLyser's (2001) discussions of her research on and in Bodie, California, for example, offers a fascinating discussion of autoethnographic research in a location well known to her). I suggest this approach offers an important way to think through how we actively recognise the variety of ways in which and through which we engage with military events and/or the process of militarisation not simply as military researchers but in our daily lives (which may include visiting military airshows or other military events). My diary of the Waddington, 2010 airshow was written in a first-person narrative style, without paragraphs, references or editing, and I have not sought to rewrite it in a more structured form since. As such, it offers an insight that is more rounded and unscripted than other methods might offer. This is important simply because the event I am researching is a military airshow. It is a space and place of explicit militarism and militarisation. As a self-defined critical military researcher, I recognise that my emotional attachment and reaction to military aircraft places me in what for many may be considered a difficult and problematic position. There is a clear difference between being knowledgeable about military materiel (as my diary comment on the Watchkeeper UAV illustrates) and having an emotional reaction to it (as in the case of the Global Hawk).

This is exactly why an autoethnographic approach offers such a useful methodological approach for engagements with military events. Rather than having to negotiate around personal feelings, and knowledge of the subject matter, through ignoring them or intellectualising them, it enables us to recognise and take ownership of them in a critically reflexive way that facilitates an exploration of their impact on our data collection experiences. This approach thus offers military researchers an opportunity to actively consider their emotional reaction to their personal experiences of military events. I contend that this offers us the ability to write more honestly about how we intellectualise these occasions rather than feeling a requirement to always prioritise the critical intellectual perspective at the expense of the affectual or experiential (cf. Sidaway, 2009). It is possible, thus, to remain analytical yet interested in military issues and events, and to explicitly recognise that interest as a factor affecting positionality. Autoethnographic methods enable us to be honest about our experiences, yet to still be able to analyse them critically, something which is of vital importance when conducting military research.

Matthew at the Airshow

Contrary to Alison's experience of research at military airshows – and particularly her positionality as someone interested in military aviation – my methodological approach emerges from a specific intellectual agenda, and has been grounded in an attempt to reveal the everyday geographies of British militarism. Central to my research has been a critique of military recruitment and the extent to which the RAF uses 'experiential' marketing and careers promotion at events such as airshows. Whether held in civilian or military spaces, airshows form part of a wider military publicity effort which, as I have argued elsewhere (see Rech, 2014, 2015), draws upon dominant geopolitical imaginaries and forms an important node in contemporary popular militarism. Thus, my methodological approach has been informed – outside any interest in and enjoyment of the airshow as a cultural event per se – by an effort to theorise airshows as geopolitical and militaristic. Approaching the airshow has meant for me, therefore, considering it a priori an instance of politics-made-aesthetic, whereby formal recruiting strategies are accompanied by experiences of a dazzling and persuasive spectacle of aerobatic displays, mock bombing runs and formation flying. My methods are ethnographic in nature and have included photographic documentation, note-taking and the collection of material objects. In discussing these methods and my position as an 'observant participant', I aim to offer an insight into the problems which inhere to critical research of this sort.[2] Specifically I will speak to issues related to the determinism of methods designed to document military presences, and the difficulties in maintaining a critical distance at the airshows where these events are designed, inherently, to enthral and to suppress critical reflection.

Being at airshows, for me, has been guided by a straightforward attempt at ethnographic research where ethnography is meant as a

> sojourn amongst a group of people where the researcher immerses himself or herself in daily life, continuously reflecting on meticulously-kept field notes, to learn the social understandings of the group [in this case, the show-going public] in its own terms.
>
> *(Megoran, 2006: 256)*

The use of ethnography provides a crucial rejoinder to normative methodological approaches to war because, as Lutz (1999: 617–618) argues, 'while battle has beckoned as the central place of war to many observers, it [is] more important to see the crisis [of global militarization] in the mundane, the everyday.' Added to this, ethnography and qualitative fieldwork is pivotal where we aim to gain an understanding of the 'sinews that connect individual actors [show-goers] to larger trends [such as militarisation]' (Allen, 2009). The sinews which connect show-goers to militarism, at least in the case of British military airshows, are in part military promotional and recruiting presences. But there are also crucial links in the range of instances where show-goers are enrolled into certain practices of observation (see MacDonald, 2006) and spectacular consumption.

My methodological approach to the airshow is an attempt to capture both of these phenomena, and as ethnography, involves partaking in the usual activities (watching air displays, making walking tours of the airfield, visiting stalls and hangar exhibitions) and so 'becoming the phenomenon' (Laurier, 2003: 134) of a typical show-goer. In terms of data collection, I keep a field notebook which allows room for frequent personal reflections on presences at airshows, on the reactions of show-goers and my own responses to air displays and a range of other phenomena in and around the showground (Figure 21.3). Data collection at the airshow also involves first asking questions of recruiters about the importance of airshows to the military publicity effort. Second, it involves photographic documentation. Drawing on Suchar's (1997: 35; see also Rose, 2007) method of 'shooting scripts', I attempt a 'strategic organization of field photography in order to establish a base of photographic information'. As part of shooting scripts, I produce a set of questions before the airshow which reflect the research concerns/questions and which relate to the conceptual framing. Photographs are then taken *in answer* to these questions. Where recruitment at the airshow is twinned to the show as a broader experience and aesthetic, the questions I set for my documentation were:

- How is the RAF present at airshows in a recruiting capacity?
- How is the RAF recruiting done at airshows?
- How does the broader culture of the show (both as a visual and material experience) enable recruiting?

Figure 21.3 Researcher in the field.

Source: Matthew Rech.

Documenting the presence of the RAF entails photographing the broad span of marquees, tents, stalls and other temporary spaces which differ widely according to the type and size of airshow. The Sunderland show in the northeast of England, for example, gives up a large proportion of its beachfront space to recruiting presences, whereas the Farnborough or Paris shows afford recruiting much less importance. These differences are significant and often reveal something about where recruiting efforts are targeted nationally and also the political economy of the show's organisation. Documenting how recruiting is done entails photographing the range of techniques and strategies used to engage the show-going public and to bring them into proximity to recruiters. This sort of engagement is often passive, with recruiters relying on a display of weaponry or other hardware to attract attention. But it is also often active, with the RAF offering the potential recruit elaborate aircraft simulator experiences, or more formal sit-down meetings where personnel will discuss with a show-goer their options for possible entry into the service. Perhaps most importantly, photography is used to document how the aesthetic experience (sights, sounds, sensations) of the show might bolster recruitment. Recruiting at the airshow is an opportunity to directly engage the potential recruit via a host of senses, but only where engagement entrains a particular vision of the world (Rech, 2015). For example, unlike many other recruiting contexts, the show allows for weaponry and other military equipment to be engaged with directly, with show-goers being encouraged to handle rifles or mortar shells, to sit in vehicles or to try on military clothing (Figure 21.4). But in turn, these sorts of experiences are framed as geo-politically significant: the show-goer pictured in Figure 21.4 was encouraged to imagine the weight of a Bergen (rucksack), rifle and flak jacket while under fire and 'at 50 degrees Celsius in Iraq'. As this brief discussion undoubtedly implies, even a close focus just on recruitment at airshows can result in a huge span of possibly significant observations. Shooting photographic scripts, in turn, allows the researcher to systematise these observations and to tie methodological practice as closely as possible to the aims of the research.

A third method of data collection was the collecting of documents (recruiting pamphlets, corporate arms company brochures) and objects (pens, keyrings, stickers, etc.). The airshow is inherently materially profuse, and in 'becoming the phenomenon' – in attempting to emulate as closely as possible the experience of the show-goer – the collection of these types of things was a necessity. Based on the assumption that 'objects may not merely be used to refer to a given social group, but may themselves be constitutive of a certain social relation' (Miller, 1987), the collection of 'stuff' at the airshow is set within a small literature on material ethnography (cf. Geismar and Horst, 2004 and related special edition). The argument here is that while material objects (such as recruiting pamphlets) may be used as 'interpretative tools to understand the nature of society [one should rather consider them] an active agent within the social relations' which are being studied (Geismar and Horst, 2004: 6). Engagements with the materials and material cultures of the airshow was not, then, only about amassing evidence for subsequent analysis. Rather, engagement with objects and the taking-home of mundane things like RAF mouse mats, posters and pencils become worthy of analysis in-and-of-themselves, and should be considered a pivotal site in the constitution of military and geopolitical imaginations, and importantly, central to the process of recruitment.

In aiming to offer a dialogue between two different methodological approaches to the airshow, and to think through how critical military researchers might intellectualise their

Figure 21.4   The weight of battle, Fairford Airshow.

Source: Matthew Rech.

engagement at such events, I offer two points of reflection. First, approaches to popular militarised spaces which a priori script events like the airshow as inherently worthy of critique (as my approach does) don't necessarily afford the opportunity for 'enchantment'. Enchantment, suggest Woodyer and Geoghegan (2012), opens up the possibility for negotiating unintelligibility, the sublime, and for positively negotiating experiences which otherwise might fit outside normative strictures of research (in this this instance, the strictures of the terminologies of militarism or recruitment). In terms of the airshow, an approach confining research praxis to these terms only risks inuring the researcher to the broader phenomenology of the airshow. In a similar vein, an approach to the airshow or any large military event which has defined itself as an exploration *only* of state militaries and the role of the airshow as a crucible for statecraft risks sidelining aspects of airshow culture *not* formally associated with the state – for example the prevalence of corporate flight and promotional teams, civilian or national (but not state) aircraft displays or the fundamental role of arms manufacturers and the aerospace industry.

A second point of reflection (one which points to perhaps the key contradiction of airshow research) is researcher positionality and the dangers of becoming *too* enchanted with military spectacle. Having visited a number of airshows with a straightforward ethnographic and documentary motive, my personal interest in military aviation has grown – by stealth it seems. I find myself now being able to identify various types and classes of fast jet and reconnaissance aircraft, and am quite conscious of the presence of a rare or particularly vaunted aerial spectacle (like that of the Avro Vulcan bomber). Moreover, having taken a material-ethnographic approach to the airshow, the growing range of posters and other ephemera in my workspace has often prompted colleagues to question my criticality as a military researcher.[3] Thus, although I've positioned myself as a critical military researcher, this has often gone alongside being and finding myself enthralled at the military spectacle. This, as I've argued (see Rech, 2015), is *precisely* the point of the airshow: that through a range of practices of observation, 'appropriate' means of consumption, only certain means of movement and the control of the means of the inhabitation of space, the airshow is designed to discourage critical reflection of any sort. Thus, the military researcher at airshows or other military spectacles treads a fine line between criticality and enchantment. Although a level of enchantment is entirely necessary for engaged research, military research in particular risks engendering a fascination for the very object of critique: this is the very essence of popular militarism: a nearly unconscious reverence for military things which is inseparable from aesthetic experience.

Concluding Reflections

In offering reflections on our experiences of researching at military airshows, we've attempted to open up a dialogue between two different approaches which, in themselves, demonstrate some of the crucial aspects of studying contemporary popular militarism. We offer the following concluding thoughts. First, airshows have been, and are, a crucial part of aviation cultures where aviation from its earliest moments was scripted as a spectacle. This spectacle, however, has also always been inseparable from militarism because aviation was quickly militarised in the early years of the twentieth century. Today, military airshows are a key location where civilians are tied into broader currents of militarism, either directly through military recruitment, or more subtly through a range of embodied and aesthetic experiences. As researchers we cannot ignore the implicit and explicit spectacle of the airshow. In this chapter, through discussing our differing methodologies we have illustrated how we can recognise this and critically reflect on its importance as a tool of militarisation and its impact on individual show-goers.

Second, airshows are unique, being far more prevalent than Navy Days or terrestrial forces equivalents, and are internationally prevalent events which draw hundreds of thousands of spectators annually. The access they grant to military personnel and materiel, often on military sites, offers an opportunity to undertake research that explicitly engages with how the military interacts with and relates to civilian show-goers. Adopting a research methodology that recognises the experiential and emotional aspects of these interactions enables us to more fully understand the politics and practices of these events. Rather than being the objective scholar, part of this approach involves us recognising and critiquing our own experiences of being at the airshow and offers us opportunities to be more nuanced in our analyses of how these events work. That airshows pose this positional challenge has broader relevance to military researchers. In providing a dialogue between our two differing approaches we hope to have illustrated that retaining criticality in these situations does not need to jar against the often enthralling and persuasive nature of the airshow. For our part, we argue that there must be room in military research into airshows for enchantment. However, 'such an affirmative approach should not be equated with a naive optimism and a permanently pleasurable disposition' (Woodyer and Geoghegan, 2012: 16), for as we hope to reveal in our work, there is little to be optimistic about regarding popular militarisms. Rather, there must be room in critical military research to channel our outlook as scholars of the military to 'attend to the simultaneous experience of charm *and* uncanniness' of the airshow experience (Woodyer and Geoghegan, 2012: 17). Indeed, with its stark contrasts, clashes and its status as both space of recreation and stage for the flexing of lethal military muscle, the military airshow demands, we argue, a methodological sensitivity to the uncanniness of popular militarism.[4] In this sense, the very nature of 'becoming' enchanted with the airshow, for military researchers, should be part of our work. We do not think it anathema, therefore, that committing to being open about our experiences of military events might include being honest about the uncanniness of being enchanted by militarism and that recognising that this can be a strength for critical military research.

Notes

1 As the Oxford English Dictionary suggests, air-mindedness can be defined as a state of being 'interested in or enthusiastic for the use and development of aircraft'. A broad and ongoing examination of the histories and geography of this term can be found at http://airminded.org, which is maintained by Brett Holman.
2 'Observant participation' is associated here with Moeran (2007: 14) who suggests that, as opposed to participant observation, this approach implies an 'involved detachment'. As we go on to describe, the airshow requires of the critical military researcher an effort to remain both involved *and* detached, or 'enchanted' while also 'critical'.
3 Indeed, a common question I received from my fellow doctoral colleagues was whether it was my intention to join the RAF after my PhD (it wasn't).
4 In line with this argument, we point to the range of contemporary artists who document airshows and who often make stylistic use of juxtaposition in an attempt to reveal the uncanny. See for example the websites of Nina Berman (http://www.ninaberman.com/homeland) and Melanie Friend (http://www.melaniefriend.com/thehomefront/).

References

Adey, P. (2010) *Aerial Life: Spaces, Mobilities, Affects.* Oxford: Wiley-Blackwell.

Airshows.org.uk (2014) Available at: http://airshows.org.uk/

Allen, R. (2009) *The Army Rolls the Indianapolis: Fieldwork at the Virtual Army Experience.* Available at: http://journal.transformativeworks.org/index.php/twc/article/view/80/97 (Accessed 26 July 2012).

Butz, D. (2010) Autoethnography as Sensibility. In DeLyser, D., Aitken, S., Herbert, S., Crang, M. and McDowell, L. (Eds.), *The Sage Handbook of Qualitative Geography.* London: Sage, pp. 138–155.

Butz, D. and Besio, K. (2009) Autoethnography. *Geography Compass* 3(5), 1660–1674.

Corn, J. (1983) *The Winged Gospel: America's Romance With Aviation.* Baltimore: Johns Hopkins University Press.

Crary, J. (1992) *The Techniques of the Observer.* London: MIT Press.

DeLyser, D. (2001) "Do You Really Live Here?" Thoughts in Insider Research. *Geographical Review* 91, 441–453.

Demetz, P. (2002) *The Air Show at Brescia, 1909.* London: Farrar, Straus and Giroux.

Dowler, L. and Sharp, J. (2001) A Feminist Geopolitics? *Space and Polity* 5(3), 165–176.

Geismar, H. and Horst, H.A. (2004) Materializing Ethnography. *Journal of Material Culture* 9(5), 5–10.

Kong, L. and Yeoh, S.A. (1997) The Construction of National Identity Through the Production of Ritual and Spectacle. *Political Geography* 16(3), 213–239.

Laurier, E. (2003) Participant Observation. In Clifford, N. J. and Valentine, G. (Eds.), *Key Methods in Geography.* London: Sage, pp. 133–148.

Lutz, C. (1999) Ethnography at the War Century's End. *Journal of Contemporary Ethnography* 28(6), 610–619.

MacDonald, F. (2006) Geopolitics and 'The Vision Thing': Regarding Britain and America's First Nuclear Missile. *Transactions of the Institute of British Geographers* 31, 53–71.

Massaro, V. and Williams, J. (2013) Feminist Geopolitics. *Geography Compass* 7, 567–577.

Megoran, N. (2006) For Ethnography in Political Geography: Experiencing and Re-Imagining Ferghana Valley Boundary Closures. *Political Geography* 25, 622–640.

Moeran, B. (2007) From Participant Observation to Observant Participation: Anthropology, Fieldwork and Organizational Ethnography. *Creative Encounters* Working Paper #2, Copenhagen Business School. Available at: http://openarchive.cbs.dk/bitstream/handle/10398/7038/wp%202007–2.pdf?sequence=1 (Accessed 7 April 2015).

Miller, D. (1987) *Material Culture and Mass Consumption.* Oxford: Blackwell.

Pain, R. (2009) Globalized Fear? Towards an Emotional Geopolitics. *Progress in Human Geography* 33, 466–486.

Rech, M. F. (2012) *A Critical Geopolitics of RAF Recruitment.* PhD thesis, Newcastle University.

Rech, M. F. (2014) Be Part of the Story: A Popular Geopolitics of War Comics Aesthetics and Royal Air Force Recruitment. *Political Geography* 39, 36–47.

Rech, M. F. (2015) A Critical Geopolitics of Observant Practice at British Military Airshows. *Transactions of the Institute of British Geographers*, 15(4), 536–548.

Reed-Danahay, D. (1997) Introduction. In Reed-Danahay, D. (Ed.), *Auto/Ethnography: Rewriting the Self and the Social.* Oxford: Berg, pp. 1–20.

Rose, G. (2007) *Visual Methodologies: An Introduction to the Interpretation of Visual Methods.* London: Sage.

Sidaway, J. (2009) Shadows on the Path: Negotiating Geopolitics on an Urban Section of Britain's South West Coast Path. *Environment and Planning D: Society and Space* 27(6), 1091–1116.

Spry, T. (2001) Performing Autoethnography: An Embodied Methodological Praxis. *Qualitative Inquiry* 7, 706–731.

Suchar, C. S. (1997) Grounding Visual Sociology Research in Shooting Scripts. *Qualitative Sociology* 20(1), 33–55.

Warnes, A. (2010) The Friendly Show. *Air Forces Monthly*, September.

Watson, I. S. (2010) *The Royal Air Force at Home: The History of RAF Air Displays from 1920.* Barnsley: Pen and Sword.

Williams, A. J. (2011a) Enabling Persistent Presence? Performing the Embodied Geopolitics of the Unmanned Aerial Vehicle Assemblage. *Political Geography* 30, 381–390.

Williams, A. J. (2011b) Reconceptualising the Spaces of the Air: Performing the Multiple Spatialities of UK Military Airspaces, *Transactions of the Institute of British Geographers* 36, 253–267.

Woodyer, T. and Geoghegan, H. (2012) (Re)enchanting Geography? The Nature of Being Critical and the Character of Critique in Human Geography. *Progress in Human Geography* 37(2), 195–214.

Wylie, J. (2009) Landscape, Absence and the Geographies of Love. *Transactions of the Institute of British Geographers* 34(3), 275–289.

22

PERCEPTIONS OF PAST CONFLICT

Researching Modern Understandings of Historic Battlefields

Justin Sikora

Sites of conflict can be overt and poignant '*lieux de mémoire*' (Nora, 1989), foci of collective identities and national shrines of memorialisation. There are countless evocative, visceral and timeless worldwide representations to the acts of war, from monuments and memorials to arboretums and parades. Yet battlefields where armies met and fought stand out in two conspicuous, interrelated ways. First, although they are only fleeting moments in the history of a place, often lasting little more than minutes or hours, battlefields are sometimes the only tangible vestige of ephemeral clashes with substantial repercussions stretching across centuries. Second, particularly in the case of British battlefields, they are often difficult to positively identify and map because tangible evidence is often lacking or completely missing. Consequently, there are instances where a battle may be seminal in a nation's history, but its site not yet positively identified, such as the Scots' 1314 victory over the English at the Battle of Bannockburn in Scotland (Foard and Partida, 2005: 8, Pollard and Banks, 2009: xiii), or (until the archaeological discoveries announced in 2009) the 1485 Battle of Bosworth in England, where Richard III was killed and the usurper Henry Tudor proclaimed king (Battlefields Trust, 2009a).

Therefore, the first point to consider when researching historic battlefields is whether the nature of the inquiry is about uncovering knowledge of the physical *place*, or the conceptualisation of the *event*. In academic research, the vast majority of scholarly time and energy has been expended in answering the former rather than the latter. Accordingly, research on historic battlefields in recent decades has been concentrated in the related – yet sometimes epistemologically separate – disciplines of military history, battlefield archaeology and war tourism. Innumerable resources and scholarly works have been produced in understanding where battles were fought; why and how they were fought, both militarily and politically; and visitation in the direct aftermath of the event, usually by veterans. In enormous contrast, far less attention has been paid to what values these historic sites have for people today. Why are huge sums of money spent every year acquiring seemingly undistinguishable farmland and building visitor centres for millions of visitors from around the globe?

Indeed, it is a challenging inquiry to gather what subjective values a piece of land – which may or may not be correctly identified as the 'real' battlefield – might have for people today, how those values are shaped and by whom. These questions became the research focus of my doctoral thesis concerning the role of on-site interpretation in conceptualising values at historic battlefields (Sikora, 2013). The nexus of this inquiry centred on how modern inter-pretive information – or lack of it – presented at historic battlefields (such as information panels, tour guides, visitor centres and mobile apps) shapes the heritage values of those spaces to current visitors. The purpose of this chapter is to investigate the research methods and practices which were used to explore questions about visitors' experiences and values at historic battlefield sites.

Before elaborating on the research approach and methodologies of this project, it is important to clarify the terminology around what is meant by 'historic' battlefield. There are no clear-cut distinctions in the literature as to what is meant by this imprecise appellation; when does a battle go from being modern to historic? The semantic discrepancies can come down to different academic disciplines, as well as which aspects from a conflict one wishes to focus on, such as era of political rule, tactics or weaponry. It can be equally difficult to distinguish between a battle and other military engagements such as a skirmish or a siege, particularly when considering the time period, although this is typically an inconsequential designation in regards to how these events are valued today. For the purposes of this chapter, 'historic battlefields' are understood as sites of battles which feature little or no perceptible remains of conflict and are outside of living memory.

The distinction of memory connected to tangibility is an important one since, in stark contrast to more modern eras of conflict or sites of continued military occupation, historic battlefields often contain little to no recognisable signs or symbols of a fleeting military pres-ence in what is often now peaceful, bucolic open space. Indeed, one of the great paradoxes of battlefields is the inherent lack of visible militarisation present within these bounded spaces, and their relative calm today (Carman and Carman, 2006: 155, Prideaux, 2007: 17). It is important to bear in mind that they were frequently occupied by military forces for a com-paratively short amount of time relative to the rich and diverse histories of established sites of military occupation stretching back many hundreds and sometimes thousands of years of human occupation. This distinction between past conflict events and current appearances of peace is important to make, as it influences how those sites are perceived and managed, and thus how visitors engage with them.

Yet even with temporal distance, in some instances historical battlefields may have still provided military personnel with a practical classroom to study first-hand lessons of war in terms of strategy and tactics. For instance, the site of the American Battle of Antietam in 1862 still features an observation tower built by the then War Department in the 1890s to assist with study and analysis of America's bloodiest day. At Culloden, the disastrous night march by the Jacobite army and aborted surprise attack on the Hanoverian encampment the night before the battle has been reenacted in recent years by members of the Territorial Army to gain first-hand experience of complicated, fatiguing night manoeuvres and how they can fail. At Flodden, there has been an increase in military personnel using the site to study the use of defilade fortifications and manoeuvring against a flanking opposing force (Hallam-Baker, 2014). Yet despite some continued interest from military personnel, historic battlefields are not manifestly militarised spaces, often devoid of any clear signs to the battle event. Consequently, they cannot be regarded in the same light as sites of long-term military presence or even explicit past militarised spaces, such as trenches from the First World War or bunkers from the Second World War where physical evidence still remains.

The situation at historic battlefields is further complicated by the fact that the majority of battlefields in the United Kingdom have not been positively identified, a major reason why some are not included in nonstatutory listings in England and Scotland (Sutherland and Holst, 2005: 1; Battlefields Trust, 2009b; Historic Scotland, 2010). Indeed, countless battlefields contain no marker, monument or interpretive element, so it is difficult to know where a battlefield is without specialised knowledge. Those sites where a battle is known or presumed to have taken place may feature on-site interpretation which acts to flag and designate an area as a historic site. In contrast to other types of military locales, having to distinguish between empty fields is a unique aspect of historic battlefields which feature no visible outward signs of militarisation. Historic battlefields contrast substantially enough from overtly militarised spaces that the practice of researching them requires a differing methodology within the context of a broad conceptualisation of military research methods.

Paradoxically, they are not very similar to other types of traditional heritage sites either, so framing them epistemologically within heritage studies has been equally challenging. Even cultural landscapes, which are quite possibly the most apt comparison, are not entirely suitable since they are associated with areas of long-term human occupation and adaption of nature featuring clearly defined tangible spaces, as opposed to the ephemeral military activity at battlefields. Recognising the need for a clearer distinction, Garden (2006, 2009) coined the term 'heritagescape' to describe landscapes with significant heritage elements. Referencing Culloden, she elaborated that 'whilst the heritagescape has been applied most often to built sites, it also offers potential for sites that possess few or no built remains but which are recognised spaces' (Garden, 2006: 404). Specifically referring to battlefields, Garden has aided in defining a nebulous particularity in heritage definitions, which further shows the uniqueness of these historic sites and thus the utility of considering how one might proceed in researching them.

Research Approach and Methodology

Despite the great interest in historic battlefield research among historians, archaeologists, and tourism researchers, there has been comparatively little scholarly activity in this area from heritage studies. Although some commentators have interspersed examples of military heritage in their works (Uzzell, 1989; Moore, 1997; Beck and Cable, 2002; Howard, 2003; Tilden, 2007), studies which have focused on the military from a heritage perspective have instead looked at objects in museums (Pearce, 1992, 1994) or have focused primarily on conflicts of the twentieth and twenty-first centuries (Gegner and Ziino, 2012). Academics publishing in the *Journal of War & Culture Studies* have written almost exclusively on conflict from the last hundred years – an absence of scholarship which sits at odds with the widespread popularity of visitation to sites of conflict outside of living memory. This overall lack of investigations into historic battlefields ostensibly excludes from academic consideration the continued impact these sites hold for people today, and disregards their popularity as sought destinations with resonant international importance.

Although military infrastructure has featured in global heritage designations for many years – castles other erstwhile fortifications and defence structures feature significantly on UNESCO's World Heritage List – there has been a lack of distinction for battlefield protection and listing outside of the United States. In the United Kingdom – where I was based for my doctoral studies – only non-statutory legislation has been passed, by English Heritage in 1996 (English Heritage, 2010) and Historic Scotland in 2011 (Historic Scotland, 2011). Accordingly, this research was not only concerned with a critical inquiry around these military landscapes, but also with how organisations in the heritage industry evaluate and manage them, and more importantly, how they have been influenced by the ideas of leading researchers in history and archaeology.

In deciding how to proceed with my research on visitor negotiations of on-site interpretation at historic battlefields, it was determined that a significant factor in the lack of more protective legislation was the fact that solely historical and tangible factors (such as the size of the armies which fought, or areas with known archaeological finds or strong potential for their discovery) were used to determine sites' heritage value and protection. Further investigations strongly suggested that modern, intangible values were largely neglected due to the pervading influence of historians and archaeologists who – particularly with their emphasis on physical remains from period documentation and artefacts – have almost singlehandedly provided the justification for values in heritage assessments. In contrast, it became clear from my fieldwork, juxtaposed with other parallel studies, that nonacademics place much greater significance on the more general themes these battles represent, as well as what occurred in the aftermath. This omission of popular valuations most certainly stems from the dominance of the research values historians and archaeologists hold for battlefields, and not from the broader and more representative sets of ideas held by a public with its own diverse interests.

Since it was ambiguous which groups' values were represented in the formal assessments of battlefields in Britain, and that there clearly were excluded paradigms, another key factor was determining which group(s) should be included in the data collection. There was ample data on the valuations from historians and archaeologists, and more generally on historic tourism in Britain and abroad. However, the largest unrepresented and underresearched group with a clear stake in determining value at historic battlefields were modern-day visitors. Visitors were specifically selected since in terms of numbers, they are the largest and most frequent group to engage with these sites, and yet my literature review suggested them to be the least consulted for formal designation under nonstatutory legislation. The answers they elicited could also be extrapolated to form a more general understanding of the larger public's evaluation and understanding of battlefields. Moreover, there has been a large number of comparable studies of visitation at battlefields – albeit historic tourism or to sites from modern conflict – so cross comparisons could be made to triangulate and validate new findings.

The term 'visitors' was explicitly employed rather than 'tourists', since understanding engagement with the interpretive displays did not require having come a long distance to visit, but could include local residents as well. It would be an entirely separate inquiry to solely investigate local community involvement and values at historic battlefields, since discussions with site staff and participant observations concurred that locals residents tend not to visit these sites in the same way as tourists, opting instead to utilise the open green space for recreation and outdoor pursuits (see Sikora, 2016, on community involvement and engagement with historic battlefields). However, for the purposes of this study, local residents were included in order to capture views which might otherwise be excluded.

When information is provided to visitors to historic battlefields – in the form of, for example, an informational panel, a guide, an audio tour, a visitor centre or related medium – it is inevitably imbued with the values and biases of site managers and owners, and of the researchers providing the interpretive content. For these reasons, the interpretative presentations were the focus of my study of how values are transmitted to visitors, and how visitors constructed their own idea of a site's values. The values and experiences obtained on-site by visitors were also juxtaposed against what they knew about the battle before coming to the site and where they had received previous information.

The research focused on three case study sites in the United Kingdom for fieldwork: Bosworth (1485) and Flodden (1513) in England, and Culloden (1746) in Scotland. These battles were chosen for the type of interpretive presentations available at each site and their relational representativeness to analogous battlefields. Bosworth and Culloden are 'comparative case studies' (Pearce, 1993: 28–29), compared and contrasted due to the large amount of interpretation featured on-site. They both have very sophisticated battlefield visitor centres, reflecting substantial financial and political capital investment. The new visitor centre at Culloden was opened in 2007 at a cost of nearly £10 million (Gareth Hoskins Architects, 2012), and the 2009 renovated visitor centre at Bosworth featuring new exhibits cost about £750,000 (YouTube, 2010; Heritage Lottery Fund, 2011). Flodden, in contrast, was used as a 'negative case' (Punch, 2005: 146), since it features markedly less on-site interpretation and was included to present nascent, low-budget interpretation strategies. With a stone cross monument from 1910 which does not even mention the battle, and a walking trail with interpretation panels erected by a resident of the local village with £23,300 from the Heritage Lottery Fund (Heritage Lottery Fund, 2014), the battlefield feels almost forgotten in comparison to Bosworth and Culloden. This is symbolised by the 'visitor centre' in the nearby village of Branxton, in an old phone box purchased for £1 and claimed as the smallest such attraction in the world (see Figure 22.1; BBC, 2014). There have been efforts, however, to increase the profile of the battle. Flodden has been the site of an ecomuseum since 2013 (see Corsane et al., 2009; Davis, 2011 on ecomuseology), which brings potential for new methods of interpreting heritagescapes.

To generate the appropriate data for this study, a broadly ethnographic qualitative approach was selected. I was interested in the nuances in visitor responses, and in what could be elucidated through visitors' own words through semistructured interviewing rather than by using predetermined questionnaire criteria. A pilot study was conducted at Culloden in order to determine if qualitative interviewing would be possible with visitors who may have been concerned about committing too much time speaking to me about their experiences. In contrast to previous studies of battlefield tourism, and similar 'dark' (Foley and Lennon, 1996; Lennon and Foley, 2007) or 'thanatouristic' sites (Seaton, 1996, 1999), which have either used predominately quantitative methodologies – or little to no empirical research (Biran et al., 2011; Stone, 2011: 327) – a qualitative approach allowed for a greater flexibility in visitor responses. Surprisingly, once consenting to an interview, visitors were generally unconcerned with time restrictions, in some cases speaking with me for an hour.

Figure 22.1   Flodden battlefield phone box 'visitor centre' at Branxton village.
Source: Remembering Flodden (2012).

Participant observations were also a fundamental aspect of the fieldwork. Individuals were not selected and followed, but rather I positioned myself in a specific area of a visitor centre or of the battlefield and made notes of general trends, a form of 'unobtrusive (nonreactive) observation' (Angrosino, 2008: 166). This included the length of time spent in certain areas, questions asked to other visitors or staff members and noting which areas attracted more attention than others. Along with extensive formal and informal discussions with site staff and volunteers, this allowed for a 'thicker description' (Geertz, 1973) of the interpretive experience. Interview and observational data were then analysed following a grounded theory approach (Glaser and Strauss, 1967).

Analysis and Reflection

Unlike areas with a continued military presence which feature tight security checks or special permission to enter, the majority of British battlefields are either on public land with open access, or owned by private individuals who allow variable rights of entry. Largely lacking barriers may allow for a greater degree of accessibility, but battlefields are still places which are either individually or collectively owned (e.g. by a regional or national entity such as Leicestershire County Council, English Heritage, or the National Trust for Scotland). There is little in their nature which is secretive or sensitive, requiring literal or figurative camouflage and restrictions, a most obvious point of comparison with contemporary military landscapes.

Due to this, it was a simple process to gain access to do fieldwork at each site, through contact with an individual involved in education and interpretation. Equally, as the pilot study revealed, it was simpler than originally envisaged to have visitors speak with me in-depth about their experiences with the interpretive information, although some did mistake me as an authorised agent of the site, despite my thoroughly explaining that I was not. It was also beneficial that I speak French and German. At Culloden about 75% of visitors are from abroad, and the two largest groups of nonnative English speakers come from France and Germany. Thus, using my language skills I could gather more nuanced appraisals of values from those who could not speak fluent English. Before starting the fieldwork, my only concern was that as an American I might not be taken seriously enough as a researcher interested in discussing Britain's heritage with visitors from the United Kingdom. Yet this was never an issue during the fieldwork, and may have even helped in gaining interviews with people who were curious why I was taking an interest in their history and heritage.

The fieldwork process differed greatly from site to site, and even day to day. At Culloden, where for various reasons I spent most time collecting data, I built a stronger rapport with the various staff who work there than the other two case study sites: those from the gift shop, the front desk workers selling tickets, the live interpreters, the various managers, the cafe workers, the staff responsible for the handheld interpretive devices used on the field and volunteers. By shadowing tours and presentations, to conducting formal interviews, as well as eating lunch in the staff break room, I was able to penetrate and explore all areas of the site's interpretation, management and daily operations. In spending time discussing how the site runs with people whose professional lives revolve around it throughout the year, I was able to gain a level of knowledge about the site which my limited time on the ground conducting observation did not allow. The same was largely true at Bosworth, and although I did not spend as much time there, I was able to speak with the various staff there and draw on their detailed experiences generated over their time working at the site, which assisted in understanding the data I collected and how representational it was in a wider context.

There were very few moments during the fieldwork at those two sites when I was not busy with some data collection task; the time I spent at Culloden and Bosworth were both fascinating and exhilarating. The vast majority of my time was spent observing participants and how they interacted with the multiple forms of interpretation on display, of which there was a multitude. Due to all the things to see and do, visitors often planned in advance to spend significant periods of time at these two sites. There were therefore large numbers of visitors who were generally amicable and very willing to devote time to speak with me. Although the semistructured interviews were key in fleshing out visitors' valuations and feelings at these sites, without the many hours spent combing through the interpretative displays and observing the general flow of how they were set up, alongside my discussions with staff, I would never have been able to dive as deep as I was able to with visitors and get to the very heart of why these places matter to visitors, and how a site's interpretation influences that valuation.

The fieldwork experience at Flodden differed almost completely in every respect from the other two case study locations. Lacking any building or cover from the elements, the battlefield's most conspicuous absence was a complete lack of amenities – not even a toilet. Moreover, there were no staff there, only the local resident who installed the interpretive signage and occasionally gives tours. Today's site, like most battlefields through Britain and indeed the world, is a peaceful, calm place. While it was a fine locale for a quiet holiday, my attempts to conduct research with visitors was a lonely, frustrating experience. It was the only site where people were largely uninterested in being interviewed, and the only place where I was denied interviews; someone even walked out in the middle of one without warning. It rained the majority of one of my days there, which limited visitors to the site, and made for a paradoxically soggy and loud interview with one hardy photographer under a golf umbrella.

It was a broad challenge at every site to pick apart and analyse what drew visitors to these sites, and what values they held for them in the present. Although the events of these battles occurred centuries in the past, many visitors and staff members alike still held passionate feelings about the people who fought, and more importantly, the resulting aftermath of each battle's residual effects. There was even evidence of 'prosthetic memory' (Landsberg, 2004), whereby visitors spoke about historic events as if they occurred more recently, or they had happened to people they knew intimately. Much of this had to do with the continued impact of the aftermath of these conflicts, and how some of the same issues originally fought over are pertinent to them today.

This was particularly pronounced in how broadly visitors spoke of the intangible values of these sites, such as how they represent ideas of identity and nationhood, even if they came from another country. The nexus of national identity and politics was a fixture of the commentary at each site, especially at Culloden and Flodden. It was at these sites where notions of what it means to be Scottish – or even British – were spoken of in heady, passionate terms. My fieldwork and research took place during the heated build-up towards what was ultimately the rejection for Scottish independence from the United Kingdom, a point which was not lost upon visitors who spoke frequently and ardently about how these battles represented a continued, or parallel, struggle in Britain today. This illustrates in turn the contextual nature of the values and meanings which visitors confer on sites. For example, one Scottish visitor from the Outer Hebrides discussed this at great length, and with a clear bias in favour of the Jacobite (or Scottish, in his mind) cause. He suggested that knowing his own family history, and whether they were Jacobites or Hanoverians, might change his mind about the conflict. However, when prompted as to his thoughts if he discovered his relatives were Hanoverians fighting with the British government troops, he still rationalised a pro-Jacobite perspective:

> Then you would, I suppose you would need to know why and you try to understand that, and I think you would pay, you pay more attention to the Government's cause . . . But I think the reason you sympathise with Bonnie Prince Charlie, and the reason I think he got a lot of followers is because he was, the Stuarts were the heir to the, the rightful heir to the throne in Scotland, at the time. So I think that's why they gather a lot of sympathy, because if you're Scottish and you don't know a lot about it you think that the Scots should be in charge of their own country.

Without a doubt, battlefields are a focal point of national pride, and representative of values which may not directly relate to the event itself. In the case of this Scottish visitor, the conflict represented his modern values and perspectives, justifying them with a historical case of being 'wronged' by the Hanoverians, which he considered to be exclusively English, despite a historical reality that roughly half the Hanoverian troops at Culloden were Scottish. He twisted this incorrect history from the past tense to the present by proclaiming the Scottish right to political sovereignty today.

Whether or not he firmly believed this before coming, or if his perspective was reinforced by the site's presentation, is difficult to determine. However, in some cases, it is clearer what the site's intentions are, and how that can influence visitors' values. This is clearly demonstrated by who funds and operates these battlefields, such as the recently opened visitor centre at Bannockburn. The centre is run in conjunction by Historic Scotland and the National Trust for Scotland; it cost a total of £9 million, of which only £4 million was paid by the Heritage Lottery Fund, and the rest directly by the Scottish government (Heritage Lottery Fund, 2012). Although the actual battlefield has not been physically identified, there has still been a concerted effort to present a narrative of Scotland winning independence in 1314, and the possibility of that freedom in the modern day.

More unexpectedly, my research also revealed that those interviewed focused on the importance of the aftermath of a battle as a principal fixture of its value. In fact, the events after a battle were frequently stated as being *more* important than the actual battle which precipitated them. At Bosworth, Richard III being killed and Henry Tudor being crowned king became secondary to the fact that the later Tudors Henry VIII and Elizabeth I deeply impacted English, British and global history in ways which would not have happened without victory at Bosworth in 1485. At Culloden, it was the effects of the Acts of Proscription – which included banning the wearing of the tartan and the dismantling of the clan system, exacerbated by the Highland Clearances – which visitors gravitated to in the interviews. Although these events are directly connected to a battle, it was not inevitable that they were destined to occur because of a battle's result. There are many instances where substantially larger numbers of troops clashed – an important factor for both English and Scottish non-statutory designation – yet the aftermath was inconsequential or neutral; most notably the Battle of Pinkie, which despite being Scotland's largest battle in terms of numbers, resulted in almost no change to the status quo. The fact that visitors were very clearly able to understand battles as separate events to their aftermath was notable, although it could argued that they did so independent of the site's interpretation from general knowledge learned elsewhere.

Indeed, one of the major obstacles to unravelling visitors' evaluations of historic battlefields was revealing where information was received about the battle – whether through authorised sources like academic histories, the site interpretation itself, or through fictional representations – and how that factored into visitors' assessments of value. It was clear from visitors' comments during the interviews that Hollywood films, such as Mel Gibson's *Braveheart* and Laurence Olivier's *Richard III*, have had a pervasive influence on people's understandings of these centuries-old battles – even if, as is the case with *Braveheart*, it has nothing to do with the Battle of Culloden, or even the same time period (see also McLean et al., 2007: 221–222; Pollard and Banks, 2010: 418). Yet the film has become a profoundly effective symbol of Scotland, and therefore Culloden, in the minds of a large number of visitors who referenced the film frequently.

When visitors mentioned these popular representations, or made statements that I was clearly able to deduce came from these sources, more often than not I asked for confirmation of how they had learned about a particular element, and how they compared it to their experience at the site that day. Sometimes visitors recognised that the books and films they had seen or read were fictionalised accounts and were able to separate them from what they learned on-site. Other times, depending on the context, I suggested that what they had previously heard might not be reliable, evidence-based historic facts. The only time I told a visitor that something he believed was in fact false was an Australian man who had told his family all about the plot to *Braveheart*, since he was convinced Culloden was the site where the film's key events were located.

Another source of fictional misinformation was Diana Gabaldon's popular *Outlander* series which chronicles a contemporary time-travelling woman back to the eighteenth century, who meets a Jacobite highlander who fights at Culloden. Although highly fictionalised, and an ire to the site staff attempting to reverse falsehoods, the books are well stocked in the gift shop and were referenced by several visitors in interviews. In turn, this could be interpreted as explicitly conflicting with what the site is attempting to represent via its interpretation and the latter's appeals to historical accuracy.

Remarkably, many visitors referred back to information they had acquired through popular fictional representations, such as these films and novels, even if they read information to the contrary at the site. In some cases, visitors either still believed the fiction over the facts presented, or at least sternly questioned if the site had presented the information fully or even correctly. A nearly identical case was noted at Gettysburg, with visitors citing Jeff Shaara's popular American Civil War fiction *The Killer Angels* (Chronis and Hampton, 2008: 116).

While initially surprising, upon reflection it made sense that visitors trusted these popularised images and ideas which they spent infinitely more time with than the perhaps one or two hours – or an average of fifteen minutes, in the case of Flodden – at a battlefield's visitor centre or interpretation trail. Battling a Hollywood epic made on a massive budget almost seems like an unfair fight to sites attempting to grasp the attention of and teach visitors from every age group, socioeconomic demographic, and cultural origin in a fraction of the time. Moreover, these conflicts were not easily definable affairs with clear delineations of who was 'right' and 'wrong', and explanations as to why people took the sides they did. It is therefore unsurprising that a film's simplified, understandable narrative with easily identifiable characters is easier to grasp than the complicated historical realities presented at battlefields, no matter how entertaining or informative the latter's presentation.

However, there might also have been an issue in *what* information was being presented through the 'official' message at each site. Since these displays have been developed and researched by historians and archaeologists, it makes sense that values which they have for those sites came through the strongest. Certainly, scholarly research on historic battlefields in the disciplines of history and archaeology has yielded fascinating insights into the nature of warfare and has contributed a great deal in understanding seminal events in world history. However, the ways in which battlefields have been researched and then presented on-site leads one to the impression that the event and place is trapped in a bygone bubble, analysed with a clinical distance which has largely separated our present values from the past (for a critical assessment of archaeology's role in heritage valuations see Waterton and Smith, 2009).

This conclusion was also substantiated in the interviews with visitors who rarely, and usually only with pointed questioning, discussed the importance of any original objects – or replicas, in the case of most of the objects at Bosworth – which were on display in the visitor centres. The same was true for the graves at Culloden, which visitors seldom commented on without specific questioning, something also observed in another research project there (McLean et al., 2007: 233). Neither did visitors state that they would leave the site thinking about the minutiae of historic exactitude, so often posed in dense panels of text. Although some visitors expressed interest in the military history of a site, or of other sites, the over-whelming majority spoke only about how the decisions made on the battlefield affected history in the aftermath of the battle, and into today. Despite any alternative intentions by site planners (something discussed in interviews), it would be fair to conclude that historic battlefields are almost wholly demilitarised sites. Instead, they are clearly focal points where deeper narratives on more salient contemporary issues, like politics and identity, are drawn out, digested and deconstructed by visitors of all backgrounds.

More than anything, it was striking how little visitors associated these historic battlefields today with the acts of militarism, and detached – at least in their minds – the silent fields today, from brief epicentres of violence and terror which have defined them through history. Quite to the contrary, they were often cited as places of relaxation and reflection, a spot to holiday and spend a fun afternoon with family. It is this dichotomy between the tourist space today, and the military clash of yesteryear, which makes historic battlefields enigmatic and appealing. No matter how much more factual, historical and tangible under-standing is accrued of these fields of conflict, little will change about how they are broadly valued. As sites of research for understanding their modern values, historic battlefields are so firmly disassociated with more traditional militarised spaces that they ultimately must be approached with a methodology which firmly embraces and comprehends this paradoxical circumstance.

References

Angrosino, M.V. (2008) Recontextualizing Observation: Ethnography, Pedagogy, and the Prospects for a Progressive Political Agenda. In Denzin, N.K. and Lincoln, Y.S. (Eds.), *Collecting and Interpreting Qualitative Materials*. London: Sage, pp. 161–183.

Battlefields Trust (2009a) *Bosworth Battlefield Survey*. Available at: http://www.battlefieldstrust.com/resource-centre/warsoftheroses/battlepageview.asp?pageid=824 (Accessed 23 October 2009).

Battlefields Trust (2009b) *The Resource Centre*. Available at: http://www.battlefieldstrust.com/resource-centre/faq/ (Accessed 23 October 2009).

BBC (2014) *The Legacy of Flodden Field*, 4 January. Available at: http://www.bbc.co.uk/programmes/b03mj1y4 (Accessed 16 March 2014).

Beck, L. and Cable, T. (2002) *Interpretation for the 21st Century: Fifteen Guiding Principles for Interpreting Nature and Culture* (2nd edition). Champaign, IL: Sagamore.

Biran, A., Poria, Y. and Oren, G. (2011) Sought Experiences at (Dark) Heritage Sites. *Annals of Tourism Research* 38(3), 820–841.

Carman, J. and Carman, P. (2006) *Bloody Meadows: Investigating Landscapes of Battle*. Stroud: Sutton.

Chronis, A. and Hampton, R. D. (2008) Consuming the Authentic Gettysburg: How a Tourist Landscape Becomes an Authentic Experience. *Journal of Consumer Behaviour* 7, 111–126.

Corsane, G., Murtas, D. and Davis, P. (2009) Place, Local Distinctiveness and Local Identity: Ecomuseum Approaches in Europe and Asia. In Anico, M. and Peralta, E. (Eds.), *Heritage and Identity: Engagement and Demission in the Contemporary World*. London: Routledge, pp. 47–62.

Davis, P. (2011) *Ecomuseums: A Sense of Place*. London: Continuum.

English Heritage (2010) *Battlefields: Definition*. Available at: http://www.english-heritage.org.uk/server/show/ConWebDoc.16550 (Accessed 23 January 2010).

Foard, G. and Partida, T. (2005) *Scotland's Historic Fields of Conflict: An Assessment for Historic Scotland By the Battlefields Trust*. Available at: http://www.battlefieldstrust.com/media/660.pdf (Accessed 23 October 2009).

Foley, M. and Lennon, J. (1996) Editorial: Heart of Darkness. *International Journal of Heritage Studies* 2(4), 195–197.

Garden, M.C.E. (2006) The Heritagescape: Looking at Landscapes of the Past. *International Journal of Heritage Studies* 12(5), 394–411.

Garden, M.C.E. (2009) The Heritagescape: Looking at Heritage Sites. In Sørensen, M.L.S. and Carman, J. (Eds.), *Heritage Studies: Methods and Approaches*. London: Routledge, pp. 270–291.

Gareth Hoskin Architects (2012) *Culloden Battlefield Visitor Centre*. Available at: http://www.gareth hoskinsarchitects.co.uk/projects/arts-and-culture/culloden-battlefield-visitor-centre (Accessed 16 May 2012).

Geertz, C. (1973) *The Interpretation of Cultures*. New York: Basic Books.

Gegner, M. and Ziino, B. (Eds.) (2012) *The Heritage of War*. London: Routledge.

Glaser, B. G. and Strauss, A. (1967) *Discovery of Grounded Theory: Strategies for Qualitative Research*. Chicago: Aldine.

Hallam-Baker, C. (2014) *Personal Communication* (Skype Interview With Director of Remembering Flodden), 24 March 2014.

Heritage Lottery Fund (2011) *Project – Bosworth*. Available at: http://www.hlf.org.uk/ourproject/Pages/Nov2004/df10c27d-8369–4d66-a728–72ea99969b8f.aspx (Accessed 5 March 2011).

Heritage Lottery Fund (2012) *Construction Begins on New Battle of Bannockburn Visitor Centre*. 27 June. Available at: http://www.hlf.org.uk/news/Pages/ConstructionbeginsonnewbattleofBannockburn visitorcentre.aspx#.UQ5UxR0z2cc (Accessed 14 January 2013).

Heritage Lottery Fund (2014) *Remembering Flodden Project, Your Heritage Project*. Available at: http://www.hlf.org.uk/ourproject/Pages/May2012/0575cc83-a5b9–49c9-b994-a577cdbe8dc9.aspx#.U2lBZ_ldUlA (Accessed 22 January 2014).

Historic Scotland (2010) *Valuing our Heritage: Battlefields*. Available at: http://www.historic-scotland.gov.uk/index/heritage/valuingourheritage/battlefields.htm (Accessed 23 January 2010).

Historic Scotland (2011) *A Guide to the Inventory of Historic Battlefields*. Available at: http://www.historic-scotland.gov.uk/guidetobattlefieldinventory.pdf (Accessed 10 November 2011).

Howard, P. (2003) *Heritage: Management, Interpretation, Identity*. London: Continuum.

Landsberg, A. (2004) *Prosthetic Memory: The Transformation of American Remembrance in the Age of Mass Culture*. New York: Columbia University Press.

Lennon, J. and Foley, M. (2007) *Dark Tourism: The Attraction of Death and Disaster*. London: Thomson Learning.

McLean, F., Garden, M. C. and Urquhart, G. (2007) Romanticising Tragedy: Culloden Battle Site in Scotland. In Ryan, C. (Ed.), *Battlefield Tourism: History, Place and Interpretation*. Oxford: Elsevier, pp. 221–234.

Moore, K. (1997) *Museums and Popular Culture*. London: Leicester University Press.

Nora, P. (1989) Between Memory and History: Les Lieux de Mémoire. *Representations* 26, 7–24.

Pearce, D. G. (1993) Comparative Studies in Tourism Research. In Pearce, D. G. and Butler, R. W. (Eds.), *Tourism Research: Critiques and Challenges*. London: Routledge, pp. 28–29.

Pearce, S. M. (1992) *Museums, Objects and Collections: A Cultural Study*. Leicester: Leicester University Press.

Pearce, S. M. (1994) Objects as Meaning; Or Narrating the Past. In Pearce, S. (Ed.), *Interpreting Objects and Collections*. London: Routledge, pp. 19–29.

Pollard, T. and Banks, I. (2009) Editorial. *Journal of Conflict Archaeology* 5(1), xi–xvi.

Pollard, T. and Banks, I. (2010) Now the Wars Are Over: The Past, Present and Future of Scottish Battlefields. *International Journal of Historical Archaeology* 14(3), 414–441.

Prideaux, B. (2007) Echoes of War: Battlefield Tourism. In Ryan, C. (Ed.), *Battlefield Tourism: History, Place and Interpretation*. Oxford: Elsevier, pp. 17–27.

Punch, K. F. (2005) *Introduction to Social Research: Quantitative and Qualitative Approaches*. (2nd edition). London: Sage.

Remembering Flodden (2012) *About the Remembering Flodden Project*. Available at: http://www.flodden.net/about (Accessed 18 November 2012).

Seaton, A.V. (1996) Guided by the Dark: From Thanatopsis to Thanatourism. *Journal of Heritage Studies* 2(4), 234–244.

Seaton, A.V. (1999) War and Thanatourism: Waterloo 1815–1914. *Annals of Tourism Research* 26(1), 130–158.

Sikora, J. (2013) *"This Deathless Field": The Role of On-site Interpretation in Negotiating Heritage Values of Historic Battlefield*. PhD, Newcastle: Newcastle University.

Sikora, J. (2016) Embattled Legacies: Challenges in Community Engagement at Historic Battlefields in Britain. In Onciul, B., Stefano, M. and Hawke, S. (Eds.), *Engaging Communities*. Rochester, NY: Boydell & Brewer.

Stone, P. R. (2011) Dark Tourism: Towards a New Post-Disciplinary Research Agenda. *International Journal of Tourism Anthropology* 1(3–4), 318–332.

Sutherland, T. L. and Holst, M. R. (2005) *Battlefield Archaeology – The Archaeology of Ancient and Historical Conflict*. Available at: http://www.bajr.org/documents/bajrbattleguide.pdf (Accessed 23 October 2009).

Tilden, F. (2007) *Interpreting Our Heritage* (4th edition). Chapel Hill: University of North Carolina Press.

Uzzell, D. (Ed.) (1989) *Heritage Interpretation Volume 1: The Natural and Built Environment*. London: Belhaven Press.

Waterton, E. and Smith, L. (Eds.) (2009) *Taking Archaeology Out of Heritage*. Newcastle Upon Tyne: Cambridge Scholars.

YouTube (2010) *The Best in Heritage 2010: Project No. 3 Bosworth Battlefield Heritage Centre*-filmed in Dubrovnik, Croatia. Available at: http://www.youtube.com/watch?feature=player_embedded&v=xdCQb-ULKbQ (Accessed 27 May 2011).

SECTION 4

Senses

23

A VISUAL AND MATERIAL CULTURE APPROACH TO RESEARCHING WAR AND CONFLICT

Jane Tynan

Despite the growth of military-based research in the arts and humanities, few scholars take a visual or material perspective on war and conflict. The work of the soldier might be far removed from the gentler pursuits of artists, designers and photographers, but there is a point of connection. More and more, news of military conflict reaches us through images. Here, I set out to argue that visual and material research methods are of significant value to military studies. A comparatively narrow field, academic research in art and design has traditionally been linked to museums and primarily concerned with attribution; only recently have artworks and designed objects been taken seriously as agents of change. This pattern to view objects within their wider social contexts has, however, stimulated new interest in the part the visual and material play in military life.

In this chapter, I explore what relationships between spaces, objects and images reveal about various military conflicts. This perspective owes much to cultural studies as an academic discipline. Raymond Williams, often considered its leading light, saw cultural activity as more than merely a reflection of economic and political actions. For Williams, it was a productive force in its own right, which inspired his 1958 essay 'Culture Is Ordinary' (Williams, 1989). Stuart Hall, director of the Birmingham Centre for Contemporary Cultural Studies in the 1970s, continued in this vein, making connections between culture and power. His work became particularly influential to those interested in *representation* as a site of power (Hall, 1997). By the 1980s, many academics in art and design sought to highlight the power of the visual, giving birth to the subdiscipline of visual culture.

By the end of the 1990s Nicholas Mirzoeff observed that visual culture had become a new means of doing interdisciplinary work (Mirzoeff, 1998: 4). There was a suspicion among many that the truth claims of the discipline of art history were based on intellectually lazy assumptions. Such criticism reflected the significant number of scholars breaking out of art history to contribute to wider debates within the arts and humanities. This new intellectual project engaged with questions of power, it went beyond images and artworks to take in the whole visual world. Paul Virilio was one of the first thinkers to apply visual analysis to military practices, and in doing so he asked new kinds of questions about the relationship between the military and civil society (Virilio, 1989). For Virilio, it was no coincidence that military and cinematic techniques developed alongside one another in the early twentieth century. His logistics of perception described prosthetic vision but also explored the visuality inherent in the coordination of complex military operations. Thus, Virilio called the field of battle a 'field of perception' (Virilio, 1989: 26) to demonstrate that war is both visualized and embodied.

How are events constructed for our eyes? How is war and conflict mediated for us, and by whom? How are the senses enrolled to shape our understanding of particular conflicts? These are all questions of power that require a close analysis of forms of representation. Alongside the emergence of this project to explore visual culture, a similar reassessment was taking place concerning research into designed objects. Material culture emerged within archaeology and anthropology to challenge the 'heroic' structures of narrowly defined disciplines such as design history. Material culture, just like visual culture, asked how objects of all kinds created identities and mediated power. Both the emergence of material culture and visual culture were critical to breaking down distinctions between high and low culture to highlight instead structures and systems of power and belief. Now there is much more fluidity, and material culture draws on archaeology, anthropology, art history, museum studies and design history. Historians, sociologists, geographers and those working in politics and media studies departments are increasingly alert to the significance of the 'stuff' of conflict. Clearly images and things are critical to the process of militarization, which is why, more than ever, we must to be attentive to how the military puts them to work.

My focus on military culture does not follow a social science model, with empirical techniques such as ethnographic field research, direct and participant observation. Instead, I examine images, objects and texts to determine how they mediate modes of seeing, and thus knowing, military institutions and their activities. Researchers using visual and material approaches are reflexive in their engagement with military topics. There are various ways to approach military-based projects, as exemplified by methods employed in visual sociology or new forms of field research around social media use. There is a drive among a new generation of thinkers to develop a methodological toolkit that takes account of how the senses are enrolled to communicate ideas about military thought and action. This might mean reflecting on how specific conflicts are represented in, say, the popular media, but it might equally involve analysis of how civilians transform their bodies to be recruited into the military. A key question might be, how do civilians 'learn' about military life, or perhaps, how do civilians come to be militarized? Interdisciplinary methods informed by cultural studies habitually focus on the most ordinary, everyday cultural practices.

In this chapter, I suggest three sites of 'sense' that the researcher of military practices might utilize: first, the visual image; then fields of perception; and finally the body and materiality. My work is concerned, among other things, with the organizational culture of the military; the surface effects of clothing, visual images and media on one hand, and on the other, the objects and cultural production that underpin the military institution. In both there is scope to examine the role of the senses in constructing military cultures, whether on a symbolic or a material level. How the body at war is visualized is critical to understanding the social and cultural meaning of war and conflict. Is this because the visual might be one of the first senses we engage to perceive the meaning of a new conflict? Nicholas Mirzoeff (2013) argues that visuality is itself a 'militarized technique' of control and domination. For him, the visual is an activity rather than an after-effect, not just a trace of the event. It is, as Virilio suggests, a military practice in itself.

In my research, I treat visual images and designed objects as cultural practices playing a part in shaping military conflicts. The British historian Jeremy Black (2012) recently observed a cultural turn in military history which has given rise to various studies that explicitly critique *militarism* as a cultural formation. Concepts of culture are central to the approach I take to researching the military, to explore its dynamic, ever-changing processes and transformations. If the visual is an activity rather than an after-effect, then could it be considered part of the conduct of warfare? Indeed, across the arts and humanities there is increased interest in the significance of visual perception to military operations, to the internal culture of military institutions and to its external representation.

For instance, Suzannah Biernoff, in her research on the visual representation of facial mutilation, adopts a visual culture perspective to examine material from the First World War (Biernoff, 2011). Exploring the visual record enables her to incorporate facial injury into a cultural history of the war, rather than leaving this work to medical and social historians. The result is a distinctive perspective that considers the perception of facial wounds as a loss of humanity and identity for wounded men, and a crisis of representation. By employing the concept of culture, Biernoff and others acknowledge that the military cannot be set apart from the rest of social life. As her work highlights, we can usefully study cultures of militarism to scrutinize the images, objects and texts that construct popular myths and memories of war. This is one of the ways in which research on military culture might employ humanities-based and arts-based methods to test representation against experience. Here, Biernoff's exploration of war wounding finds a distinctive rhetoric emerging to cope with facial injury. The work of surgeons may have been aesthetic reconstruction, but what was at stake was the identity and humanity of the patient. The interdisciplinary nature of this research and its concern with relationships between spaces, objects and images makes for a revealing study of the cultural meaning of wounding in First World War Britain.

A History of the Discipline

Within the arts and humanities, research on the military has been dominated by concerns with the politics and conduct of war, but also with its representation in literature and art. While social and cultural histories of conflict are gaining ground, research focusing on the objects and textures of everyday military life are less apparent. Scepticism about the reliability of representations among scholars of visual and material culture is a feature of this type of methodology. Images are inherently unreliable and must be subjected to close scrutiny and tested against other types of sources. My work on the trench coat illustrates how these surface qualities can draw out new meanings from wartime objects and images (Tynan, 2011). In this case study, I read the trench coat as a symbol of the militarizing of the home front during the First World War. A range of firms, including Burberry, sold weatherproof coats during wartime; most were called trench coats. My research examined images that promoted the trench coat in the context of a changing tailoring trade and increased social mobility in the British Army.

The approach I took was interdisciplinary. First, the masculine body is read through media images, but their reliability is tested against sources on the tailoring trade and social class participation during the war (Tynan, 2011). Thus, designed objects and the images they generate can enhance understanding of the politics and conduct of war, but are also critical to interpret changing military masculinities. It became apparent that the trench coat was a thoroughly modern solution on the field of battle, despite its image as a traditional garment. Light fabric gave soldiers mobility, while water-repellent material protected them from wet weather on the Western Front. Developments in technology during the war saw clothing become more significant to survival. A focus on everyday military life revealed that objects and images are not incidental to warfare, but hold clues that can enhance our understanding of the meaning of the conflict.

How did the art and design historians of the past go about interpreting military images, objects and trappings? In contrast to the new cultural studies of war there was instead a desire to view uniforms, for instance, in isolation from the structures of military or civilian life. Research on the design and use of clothing worn by combatants had traditionally been the preserve of dress historians, who explained change in military uniform design through the internal principles of dress history. As an example, James Laver (1948) saw transitions in the appearance of British military uniform as an evolutionary process. For Laver, changes in military costume were governed by tensions between what he terms a utility principle and a seduction principle. Michael Barthrop's (1982) study of British infantry uniforms lists changes in uniform codes, highlighting the functionality of military costume for combat requirements. While Barthrop and Laver are clearly interested in the reasons why military uniforms look as they do, they are reluctant to view them through a wider lens. These traditional approaches to researching military uniform offer little in the way of social history and context; they dissociate military design from social processes, despite the potential for research on military clothing to illuminate the relationship between conflict and civil society.

This might be because design has rarely been given a place in serious discussions of the military. For instance, the military uniform can be ignored because it does not follow the standard narrative in design history. A nonstandard commodity such as these are useful for attribution but fail to question how design decisions shape military institutions. We cannot assume that objects are rational and progressive because they appear so, a mistake Philip Steadman (1979) observes is made far too often in the analysis of functionalist design. Utilitarian objects, such as those made for military purposes, offer insights into the institutional rules and regulations encoded in the designs. Traditional studies of military uniform often fail to situate them within a broader social or cultural context, instead tracing aesthetic themes and listing facts and figures (Barnes, 1950; Carman, 1957; Barthrop, 1982). Connoisseurial studies such as these are useful for attribution but fail to question how design decisions shape military institutions.

Increasingly, scholars concerned with military design are placing primary sources within a wider social framework. Scott Hughes Myerly (1996) uses the concept of military spectacle to describe how the visual appearance of the British Army drives its management, discipline and morale. Nathan Joseph's sociological study describes uniform as a special form of clothing that 'makes the wearer's position or status much more visible than do other types of dress' (Joseph, 1986: 67). Both studies interpret the visual appearance of army clothing as encoding hierarchy and military status. Military uniform may be of interest to fashion and art historians, as well as social scientists, but there are few studies within the existing literature that sustain a deep social or cultural analysis of specific types of uniforms. Neither does the study of military uniform fit comfortably within histories of design. The heroic structures of design history make no sense without designers. State involvement with its production and supply places military uniform design outside the disciplines normally concerned with aesthetics, fashion and design. As such, research on military uniform demands interdisciplinary methods that uncover a range of social, political and economic concerns. If questions of power and public truth are critical to research on war and conflict, then it is a mistake to exclude the softer forms of militarism; these are the cultural forms most likely reach a wider public.

New Directions

An interdisciplinary approach is thus useful to researchers engaging with military design. Design history has by now embraced the political spirit of the cultural studies project. Fashion history has undergone a transformation that is breathing new life into research on style, clothing and the body. Much of the new work is interdisciplinary, making efforts to place dress practices within wider social formations. As Joanne Entwistle (2000) argues, fashion is a social process that creates the body through dress, making it social in modern society. Clothing can tell us much about the history, experience and representation of the body, but for my purposes, military uniform is revealing of how war is lived and embodied. For instance, histories of military and naval uniform have traced the move from a privatized to a publicly owned body.

Why would a focus on the military uniform be useful to the researcher? While uniforms physically direct and coerce the activities of wearers, they also signal authority. Uniform, an embodied social practice that reinforces or undermines wider social hierarchies, also fits with the needs of the nation-state, its forms of standardization, bureaucracy and centralized organization systems. Regulation clothing embodies specific forms of citizenship, and in various regimes its transformative power was highly valued. For instance, fascist projects deployed clothing to signify and embody ideals of masculinity, patriotism and action. There are few studies in fashion or design history that examine military uniform in this way, but Wendy Parkins (2002) has explored the interconnected concerns of dress, gender and citizenship by examining the role of uniforms in the formation of collective disciplines.

Important also are the productive tensions between official regulation and social processes in the design and making of army uniform. By focusing on the relationship between styles of presentation and institutional structures, researchers can determine the symbolic force of the image, or consider how bodies are fashioned for specific military tasks. The consumption of officially controlled goods, such as military uniform, constitutes a rich source for research on how design and material culture function in everyday life. Consumer practices are shaped by perceived needs and are constructed through specific social, economic and political structures. Penny Sparke (1986) links the social significance of design with processes of mass production and consumption. It is significant that academic studies on consumption have neglected military uniforms, due in part to the perception, observed by Ben Fine, that the state is somehow anticonsumption (Fine, 2002: 176–186). However, the reality is that the state is a primary producer of commodities, particularly for war. Thus, the visual appearance of the uniform becomes even more significant in the context of the systems of its provision.

To isolate military goods from wider debates on consumption ignores the cultural and political significance of some of the most ambitious projects that have mobilized the means of state production. Military institutions may not like fashion, but they do value the power of clothing, which is critical to body discipline and troop morale in the army. Examining the military uniform, in all of its various manifestations, might reveal how the military body is envisaged within modern warfare. Daniel Roche (1994) argues that the discipline of appearances, which is the purpose of modern uniform, is primarily concerned with the formation and training of bodies for combat. For him, the perfect military body is achieved through the instrumentality that standardized clothing gives 'to shape the physique and the bearing of a combative individual, whose autonomy conditions his docility and whose obedience transforms individual strength into collective power' (Roche, 1994: 229). His analysis highlights the considerable power of uniforms to express military values and to shape military cultures.

That collective power has many dimensions: visual, psychological, social and material. During the First and Second World Wars, the British Army produced commodities, including uniforms, managed by a field of official documentation regulating their design and indicating how they should be worn. The relationship between state and trade can be traced through processes that design, make and supply uniform clothing to the troops. The fact that uniforms had to be made is often overlooked in studies of war and conflict, but these huge projects drew on the resources of the state and civil society. Consistent with standardizing nationalist projects, the supply of military uniform requires investment in scientific innovation and access to the means of industrial production. To ask what kinds of images and objects are made to support war highlights how military institutions have set out to design bodies for conflict conditions.

Scholars from various academic disciplines examine cultures of militarism, but there is growing interest in the politics of war images in art history, media studies and visual culture. The image has become central to questions of legitimacy in warfare, particularly concerning contemporary forms of armed conflict (Michalski and Gow, 2007). Dora Apel (2012) argues that the contest of images has normalized war across popular culture. Spectacle has become a key concept through which to interpret war images, particularly those that advance a 'military neo-liberalism', or constitute a continuous image of war to underpin the state apparatus (Boal et al., 2005: 72).

Despite much debate on the status of the war image over the past few decades, the photographic image has not lost its potency in the field of battle. If visuality, as Mirzoeff (2013) argues, is a 'militarized technique' of control and domination, then these various optical weapons of war operate within a whole system of military spectacle and surveillance. There are many optical weapons in time of war, but theorists such as Allen Feldman argue that we have gone beyond simply using visual technologies to extract information; they are now deployed for more grotesque forms of 'optical penetration,' exemplified by the Abu Ghraib torture photographs (Feldman, 2013). We have moved on from thinking about visual technologies manipulating reality to consider modes of visuality as weapons of capture and torture in and of themselves. Now, with increased academic interest in the role of senses in perceiving conflict, we see the various ways in which civilians are mobilized in support of war.

The realm of the visual is not confined to the analysis of static images, but to other modes of seeing, what one theorist describes as a 'vigilant visuality', mobilized on the US home front to wage the so-called war on terror (Amoore, 2007). The recent surge of interest in the role of the image in warfare marks a new analytic sophistication in the debate, particularly around visual technologies and practices of looking (Boal et al., 2005; Michalski and Gow, 2007; Apel, 2012; Feldman, 2013; Mirzoeff, 2013; Stallybrass, 2013). There is also a growing emphasis on the image as surveillant rather than as representation. An interdisciplinary approach offers insights into how images or objects mediate and participate in military conflict. In particular, the extra-artistic concerns of these studies indicate a decisive shift in scholarship on the visual culture of war. Far removed from the gentle concerns of connoisseurship and attribution that once shaped the disciplines of art and design history, new approaches and methods have brought this kind of research into the realm of the geopolitical.

Camouflage and Visibility

If the visual is about all modes of seeing and perceiving, then the politics of vision is critical to understanding what constitutes the military body. For my work, this is about exploring the body's construction and visual impact. By tracing the historical emergence of military designs, we can reach a better understanding of changing requirements of the body on the battlefield. The historian Thomas Abler (1999), for instance, observes how certain aspects of British military uniform design came into being. Both khaki and puttees were staple forms of army dress during the First World War, but as Abler shows, they emerged first in nineteenth-century India (1999). As military procedures changed, clothing came to be regarded as part of the technology of warfare. This decisively affected the visual appearance of soldiers, in particular the colour of their uniform for combat. Thus, new military surveillance technologies shaped uniform design in the second half of the nineteenth century.

Weapon design was also bound up with visualizing techniques. Smokeless magazine rifles, used from the 1890s, altered behaviour in battle; black powder no longer obscured the soldier's field of vision (Spiers, 1980: 210). By rationalizing the field of visual perception, the traditional military spectacle of display made way for camouflage, which became a decisive tactic on the battlefields of the First World War. Surveillance shaped the design of uniforms and weapons to transform the conduct of warfare thereafter.

Understanding how illusions are created for combatants and civilians brings us closer to deciphering what counts as truth in war and conflict. It also offers insights into the images and illusions preserved to memorialize war. Tracing the history of the body at war might increase our awareness of how military values are embodied. If former military garb was designed to fit notions of military honour, the new visuality of the modern battlefield made these strong colours redundant. By the early twentieth century, in response to surveillance technologies and dispersal tactics, armies were designing uniforms in camouflage colours (Abler, 1999; Newark and Miller, 2007). My own research explores the significance of visuality as a technique in the battlefields of the First World War. In trench warfare, with very little time available to safely investigate no man's land and enemy defences, trench-level observation became critical to military success, when camouflaged apertures in the parapet caught a glimpse of the enemy line (Barton, 2005: 44). It was rare to glimpse the enemy, and too dangerous to look over the parapet, with the result that aerial photography became the most valuable reconnaissance tool available to the army during the war. Attention to specific styles of image and the modes of visuality they employ are tasks for the visual researcher.

Since Virilio's first pronouncements concerning war and cinema, many connections have been made between transformations in social experience and the history of visual technologies. Camouflage is one such visual form, which was the result of a series of significant changes in the conduct of modern warfare. This was when clothing for combatants altered in direct response to military technology. Art historians have been particularly interested in researching camouflage in the context of links between art and science within modernity. Created through the collaboration of artists and scientists, the meaning of the word camouflage comes from the French verb *camoufleur*: 'to blind or veil' or 'to disguise' (Goodden,

Figure 23.1   British armed forces in Germany (1945–1975).

Source: © Imperial War Museums (CT 1223).

2007: 10). Established in France during the First World War, the Battalion of Royal Engineers set up the first British camouflage service in 1916, when hiding equipment and ammunition was the most urgent concern. Only snipers required full camouflage, and the work of these services also gave rise to the use of dummy soldiers, but it was the Second World War that saw the mass production of camouflaged uniforms (Feldman, 2009). Camouflage itself was an invention that traversed the arts and the sciences.

Camouflage was a cultural form designed for strategic military purposes. This art of deception was invented to disguise whatever the enemy might want to attack. Since artists were deployed to create dazzle-painted ships in the First World War, art historians dominated subsequent research on these creative camouflage schemes (Behrens, 1999; Covert, 2007). It was Gertrude Stein who observed that camouflage had echoes of Cubism, which she suggested was reminiscent of the work of the artist Pablo Picasso. The American writer Wylie Sypher made the same connection: 'It is said that cubism invented camouflage, and indeed the First World War used camouflage in a cubist way' (Sypher, 1962: 85). Thus, art and literary theorists amplified the role of aesthetics in camouflage design. Research into camouflage and social histories of its development reveal that key changes came through psychological research into human perception, but in terms of research it was art historians who took it up with the most enthusiasm.

Despite the fact that art historians were keen on camouflage, it is a topic that could benefit from various perspectives: scientific, artistic and social. What is clear from the literature on visuality and war is that art, illusion, geography, military uniform and visual technology are interconnected. As Rachel Woodward (2004: 3) argues, 'militarism's geographies are about the control of space, about creating the necessary preconditions for military activities.' This is suggestive of links between the body and the picturing of landscape, which were critical to the invention of camouflage uniform design. If a visual culture approach asks how war zones are imagined, constructed and experienced, then the senses are critical to gaining insights into the cultures of military spaces. Eyal Weizman (2012), in his investigation of the transformation of the Occupied Palestinian Territories, draws on architectural concepts to explore 'political issues as constructed realities' by conceptualizing the military, politicians and other activists as 'architects' of the chaotic frontier (2012: 6–7). What might seem far removed from art historical work on camouflage, or media studies of war, is in fact probably the most inventive use of visuality as a tool for academic research on military conflict. Weizman reveals visuality as an analytic tool by demonstrating the interconnectedness of visuality, politics and geography. He has perhaps done most to highlight the urgency of using visuality and design as critical concepts in the study of war and conflict.

Weizman's interdisciplinary approach demonstrates the usefulness of visual images, when subjected to a political, social and economic analysis. His work within visual culture studies reveals the practical application of theories of visuality to the analysis of military conflicts. As his Centre for Research Architecture asks, 'Might the notion of architecture be expanded to engage with questions of culture, politics, conflict and human rights?' (Centre for Research Architecture, 2015). Thus, for Weizman the disciplines of design theory, architecture and visual culture are enlisted to resolve practical questions and problems within politics and international relations. How war is embodied is key to understanding its impact on human beings. The archaeologist, Nicholas Saunders, observes that technological modern warfare destroys but also recreates forms, as he puts it, 'the transformation of matter through the agency of destruction' (Saunders, 2004: 5). He sees the material culture of conflict embracing a variety of disciplines, improving our capacity to reflect on how human experience is embodied in relationships between people and objects. This is where designed objects and images mediate and act as go-betweens. Design, when removed from consumer culture, offers interesting critical concepts to study war and conflict. The military institution is, after all, in the business of designing bodies and spaces for war. Exploring the geography of military activity raises questions about how bodies and objects interact, asks why modes of seeing characterize military activity, and considers how visual technologies are deployed as weapons of war.

Making sense of military conflicts involves asking how war zones are imagined, constructed and perceived are critical to military activity. Thus, visuality is key to understanding how the geography of disputed territories and frontiers are created and transformed. This is how I engage the notion of design, by asking how spaces and bodies are designed for war. Military uniform design is shaped by geographical space; after all, what soldiers wear on the battlefield counts. It can be a matter of life or death. Traditionally, discussions of art and war were limited to the preoccupations of discrete disciplines. New interdisciplinary approaches that explore the shape of the military body or the transformation of war zones illustrate how useful the study of aesthetics and materiality can be those researching military topics.

Questions of public truth are central to research on war and conflict, but the visible world is full of illusions. Thus, as Weizman and Saunders demonstrate, visual and material approaches offer useful theoretical models through which to explore the politics of images and the transformation of bodies. To imagine place involves picturing landscape. So too the visual was a significant technique on the battlefields of the First World War, in response to the sheer violence of technological warfare. Both sides invented smart machines to 'scope' the enemy and measure the scale of risk. Control over the field of vision gave tactical advantage, lending visual practices, such as mapping and camouflage, strategic value in this and later conflicts.

What makes research into visual cultures of war distinct is its attention to practices of looking and how they might amplify questions of power. Images created to map out a strategic military task take on an instrumental mode of visuality to discipline, standardize and order reality. As Hanna Shell observes, 'the forms and media of camouflage keep changing as environments and surveillance technologies evolve' (Shell, 2012: 21). Thus, the images created for warfare and the visual appearance of the military body offers scope for those researching conflict, to trace changes in geography, military tactics and the weapons of war. W. J. T. Mitchell argues that we should treat 'visual images as "go-betweens" in social transactions, as a repertoire of screen images or templates that structure our encounters with other human beings' (Mitchell, 2002: 243). For him, visual analysis offers a language to explore the social meaning of images and objects, to uncover their political role in everyday life. Thus, an interdisciplinary approach to the study of military conflict draws in the concerns of geography, politics, social theory, design and media studies to determine how visuality has been deployed to mediate reality and to structure the conduct of warfare.

Bodies at War

With so much attention to the textures and surfaces of military life, increased academic interest in the minutiae of the body at war is hardly a surprising development. Recent edited volumes on the body have given over space to the analysis of the clothing worn by combatants and war prisoners (McSorley, 2013; Cornish and Saunders, 2014). Many scholars are starting to acknowledge clothing as part of the social experience of the body in war and conflict. A number of studies link the visual and the material in an effort to understand how clothing both creates the body as an image, but also becomes part of the occupational habitus of the soldier, prisoner, nurse or transport worker.

My research adopts methodologies in visual and material culture that go beyond the concerns of art and design history. I challenge long-held assumptions that aesthetics and design are incidental to warfare, to argue that war and conflict are among other things cultural formations. Here, the materiality of the body is critical to understanding how war is perceived by the combatant or the civilian. How clothing touches and shapes the body should be a central concern for researchers of the military. Bodies shape the material legacy of conflicts in interesting ways, and are much more tangible than abstract accounts of military operations. Much can be learned from the deployment of cultural practices such as masquerade, mimicry, camouflage, disguise and bodily transformation in the field of battle. A material culture approach issues a useful challenge to the official memory of military conflicts. Often forgotten is the violence, the chaos and the various ways in which people are enlisted. All of these are inscribed on the military body.

An aspect of my research on the uniforms worn by soldiers in the First World War involved a study of what colonial troops wore (Tynan, 2013). It struck me that few studies on race and ethnicity in the British army use images as primary sources. I was interested in the role of photographs to define images of the war, particularly through the iconic image of the man in khaki. I was curious about how popular culture mobilized certain kinds of feelings among the British public about what made an ideal soldier, and what that meant for the experience of colonial soldiers. During the First World War, news of the participation of colonial troops often reached the public through images, but the question for me was what kinds of feelings did the social appearance of colonial troops mobilize among the British public during wartime? Their presence was hardly hidden: the visual evidence shows that nonwhite soldiers on the Western Front generated a lot of interest, and they were remarkably visible throughout wartime popular culture. For instance, in 1914 images of Indian lancers and Sikh infantry appeared on the front page of the *Daily Express*. The newspaper published photographs of Indian soldiers on their way to battle in France. Images of turbaned men gave picture papers an exciting and novel image of war. Their exotic appearance constructed a narrative that Britain had access to the best in global military manpower.

This is where a visual culture approach becomes illuminating. Distinctive forms of military dress for colonial soldiers heightened ethnic differences and created potent forms of military spectacle. Postcolonial theory was critical to making sense of these images. Through the study of images and press stories I identified a discourse that formed around the construction of the image of colonial troops. Regulation clothing maintains hegemonic masculinities in the army by embodying power and hierarchy. Benign views of empire might suggest that variation in uniform designs were a concession to ethnic minorities, a means by which colonial troops could express alternative cultural or ethnic allegiances, but distinctive clothing also marked out colonial troops. The work of Cynthia Enloe (1980) was useful to question how the images

Figure 23.2   The Indian Army during the First World War.

Source: © Imperial War Museums (Q 70214).

functioned during and after the war. She argues that military planners do not reflect ethnic conditions but shape them for their own purposes when they 'deliberately foster new ethnic identities for the sake of achieving military goals' (Enloe, 1980: 25–26). For Enloe, ethnic identities are fostered by the military in the pursuit of more strategic objectives.

It was only by first identifying a distinctive standardizing military masculinity that I could locate something very different in the treatment of colonial soldiers. This is where it became clear that a visual analysis of the military had particular value, to create space to reflect on what is brought into view and why. How we come to conceptualize military activity is highly reliant upon how it enrols the senses. Returning to those who analyse military modes of visuality, the maintenance of a continual image of war that identifies 'insiders' and 'outsiders' is critical to the state apparatus. Mirzoeff's (2013) notion that visuality is a 'militarized technique' has resonance with modes of seeing mobilized in First World War popular culture, whereby military uniform became part of the social project to design civilian men for war.

Khaki formed an idealized image of wartime masculinity; it also had the material force to transform men from civilians to soldiers. However, I also found that many images were circulating during wartime that reflected the ethnic and racial diversity of British Army soldiers on the Western Front, which were notably absent from the collective memory. Whatever currency images of Sikh and black soldiers had during the war, they were less than desirable in the postwar context. This is where questions of visibility become deeply political. Imperialist discourse, in particular its racist images, maintained and justified unequal power relations between men in a range of institutions, including the British Army. Up until the Second World War, the regular British Army found its officers from a narrow social class, which excluded men who were not white. Clothing was a powerful visible marker that distinguished colonial troops from other regiments during the war. By encoding social inequalities, distinct forms of military dress were performing a function far beyond the expression of cultural allegiances. Racism reproduces unequal power relations, but images offer a unique insight into how racial inequality is structured. First World War images of Indian and black soldiers suggest a visual economy that maintained and justified the low social status of colonial troops.

An *Illustrated London News* feature clearly traded in colonial spectacle, describing the war on the Western Front in terms of the 'exotic' troops enlisted to reinforce the British Army. Pictures of Indian soldiers in 'strange' exotic dress were a huge attraction to British readers; adopting a visual presentation that scrutinized individual Indian soldiers' bodies gave dress prominence in illustrating ethnic differences. What I found was that while the colonial relationship subordinated Indian men within the army structure, propaganda images exploited the peculiarities of their exotic dress to advertise their fighting efficiency for the British war effort. Their appearance was 'military' in the subordinate sense; Indian troops were valued for their labour power, not their leadership. When the situation demanded it, Indian troops were used for combat roles. But the unusual appearance of Indian troops on the Western Front was exploited as military spectacle within popular culture. Indian soldiers wore a uniform that reflected aspects of their ethnic difference, not necessarily something that met the soldiers' personal needs on the Western Front. Rather than reflect a proud military tradition, the clothing that Indian soldiers wore symbolized their lowly rank.

Many colonial soldiers were confined to noncombatant roles. The directive from the War Office was that black men should either serve in tropical campaigns or be confined to noncombatant duties in temperate zones; these rules led to the British West Indies Regiment and the South African Native Labour Contingent, raised in 1916, to be employed on the Western Front. Racist ideas that shaped recruitment policy during the war were also prominent in many regiments. Black soldiers were either ignored or dismissed, were often distrusted in their military roles and thought to be a danger to women. Once black soldiers made an appearance on the Western Front, their movements were monitored. In an apparent effort to protect working-class British women, the mobility of black soldiers was tightly controlled when they were away from the battlefield; such were the fears about their rampant sexuality. This view was also reflected in descriptions of black men in popular culture, which was where their bodies were put on display and scrutinized.

When black men wore the uniform they attracted accusations of false pride and duplicity. Black soldiers on the Western Front were considered a threat, controlled on the one hand by a popular media that trivialized the image of black men in uniform, and on the other hand by an army that deployed them in labour units and excluded them from military or frontline duties. At first clothing may seem to innocently convey distinct military traditions, but further research revealed uniform as a marker of ethnic and racial differences in the army.

I found was that the appearance of colonial troops, and in particular, their military uniforms, mobilized certain kinds of feelings among the British public and helped to maintain the racial exclusivity of the British army. The uniform was an object and an image that sought to stimulate certain ideas and feelings among the British public during wartime. This involves examining images and testing them against other kinds of sources, to uncover the relationship between what is seen and what is known.

Conclusion

Research that takes a visual or material culture perspective on conflict can offer insights into how the battlefields of the future might be configured. Indeed, the growing emphasis on the image as surveillant rather than representation has paved the way for significant developments, such as the recent work on digital militarism. Visual analysis offers real insights into the social meaning of things that mediate military life for civilians. Visuality, a militarized technique of control and domination, is thus an activity as well as an after-effect. If the power of photography is most apparent in military contexts, then this is clearly a potent area for visual culture research.

Researching material things with a strong visual impact, such as military uniform, might at first seem trivial, particularly considering the gravity of war and conflict, but the neglect of serious historical research on what soldiers wore in the past ignores a critical part of war experience. Research on military culture employing visual methods is primarily about testing representation against experience, to engage questions of power and public truth. Methodologies that enrol the senses utilize visual images and spatial practices; they are key to the militarizing of civilian bodies. My focus on how, why and where discourses about the military body are constructed involves an exploration of aesthetic and material practices that underpin war and conflict. In particular, my research on military clothing in twentieth-century warfare considers how images and objects reveal the structure of public projects that fashion civilian bodies for war. This might concern highly regulated institutions such as the British Army, but also has potential for research on insurgents who improvise a military appearance. All these examples, however, demonstrate a relationship between the presentation of the military body and the shape of events and technologies of warfare.

Studying military design, particularly for the body, is not about aesthetic beauty. As this discussion has shown, aesthetics and design are not incidental to warfare, but are part of the cultural formation of the military. If we are to deepen our understanding of cultures of militarism, then there is much to learn from the material structure of military life. This involves taking even the most ordinary, everyday objects and images seriously.

References

Abler, Thomas (1999) *Hinterland Warriors and Military Dress: European Empires and Exotic Uniforms.* Oxford: Berg.

Amoore, Louise (2007) Vigilant Visualities: The Watchful Politics of the War on Terror. *Security Dialogue* 38(2), 215–232.

Apel, Dora (2012) *War, Culture and the Contest of Images.* New Brunswick, NJ: Rutgers University Press.

Barnes, M. (1950) *A History of the Regiments and Uniforms of the British Army*. London: Seeley.

Barthrop, Michael (1982) *British Infantry Uniform Since 1660*. Dorset: Blandford Press.

Barton, P. (2005) *The Battlefields of the First World War: The Unseen Panoramas of the Western Front*, Volume 1. London: Constable/Imperial War Museum.

Behrens, Roy (1999) The Role of Artists in Ship Camouflage During World War I. *Leonardo* 32(1), 53–59.

Biernoff, Suzannah (2011) The Rhetoric of Disfigurement in First World War Britain. *Social History of Medicine* 24(3), 666–685.

Black, Jeremy (2012) *War and the Cultural Turn*. Cambridge: Polity Press.

Boal, Iain, Clark, T. J., Matthews, Joseph and Watts, Michael (2005) *Afflicted Powers: Capital and Spectacle in a New Age of War*. London: Verso.

Carman, W. (1957) *British Military Uniforms from Contemporary Pictures, Henry VII to the Present Day*. London: Leonard Hill.

Centre for Research Architecture, Goldsmiths, University of London at URL. Available at: http://roundtable. kein.org (Accessed 30 March 2015).

Cornish, Paul and Saunders, Nicholas (Eds.) (2014) *Bodies in Conflict: Corporeality, Materiality and Transformation*. London: Routledge.

Covert, Claudia (2007) Art at War: Dazzle Camouflage. *Art Documentation: Journal of the Art Libraries Society of North America* 26(2), 50–56.

Enloe, Cynthia H. (1980) *Ethnic Soldiers: State Security in a Divided Society*. Middlesex: Penguin.

Entwistle, Joanne (2000) *The Fashioned Body: Fashion, Dress and Modern Social Theory*. Cambridge: Polity.

Feldman, Allen (2013) On the Actuarial Gaze: From 9/11 to Abu Ghraib. In Mirzoeff, N. (Ed.), *The Visual Culture Reader* (3rd edition). Oxford: Routledge, pp. 163–180.

Fine, Ben (2002) *The World of Consumption* (2nd edition). London: Routledge.

Goodden, Henrietta (2007) *Camouflage and Art: Design for Deception in World War 2*. London: Unicorn Press.

Hall, Stuart (Ed.) (1997) *Representation: Cultural Representations and Signifying Practices*. London: Sage/OU.

Joseph, Nathan (1986) *Uniforms and Nonuniforms: Communication Through Clothing*. Westport, CT: Greenwood Press.

Laver, James (1948) *British Military Uniforms*. Middlesex: King Penguin Books.

McSorley, Kevin (Ed.) (2013) *War and the Body: Militarisation, Practice, Experience*. London: Routledge.

Mirzoeff, Nicholas (1998) *The Visual Culture Reader*. London: Routledge.

Mirzoeff, Nicholas (2013) *The Visual Culture Reader* (3rd edition). London: Routledge.

Michalski, Milena and Gow, James (2007) *War, Image and Legitimacy: Viewing Contemporary Conflict*. London: Routledge.

Mitchell, W.J.T. (2002) Showing Seeing: A Critique of Visual Culture. In Moxey, Keith and Holly, Michael Ann (Eds.), *Art History, Aesthetics and Visual Studies*. New Haven: Yale University Press, pp. 231–250.

Myerly, Scott Hughes (1996) *British Military Spectacle: From the Napoleonic Wars Through the Crimea*. Cambridge, MA: Harvard University Press.

Newark, Tim and Miller, Jonathan (Eds.) (2007) *Camouflage*. London: Thames and Hudson.

Parkins, Wendy (2002) *Fashioning the Body Politic: Dress, Gender, Citizenship*. Oxford: Berg.

Roche, Daniel (1994) *The Culture of Clothing: Dress and Fashion in the 'Ancien Regime'*. Cambridge: Cambridge University Press.

Saunders, Nicholas (Ed.) (2004) *Matters of Conflict: Material Culture, Memory and the First World War*. London: Routledge.

Shell, Hanna Rose (2012) *Hide and Seek: Camouflage, Photography and the Media of Reconnaissance*. New York: Zone Books.

Sparke, Penny (1986) *An Introduction to Design and Culture in the Twentieth Century*. London: Routledge.

Spiers, E. M. (1980) *The Army and Society 1815–1914*. London: Longman.

Steadman, Philip (1979) *The Evolution of Designs*. Cambridge: Cambridge University Press.

Stallybrass, Julian (Ed.) (2013) *Memory of Fire: Images of War and the War of Images*. Brighton: Photoworks.

Sypher, Wylie (1962) *The Loss of Self in Modern Literature and Art*. New York: Random House.

Tynan, Jane (2011) Military Dress and Men's Outdoor Leisurewear: Burberry's Trench Coat in First World War Britain. *Journal of Design History* 24(2), 139–156.

Tynan, Jane (2013) *British Army Uniform and the First World War: Men in Khaki*. Basingstoke: Palgrave Macmillan.

Virilio, Paul (1989) *War and Cinema: The Logistics of Perception*. London: Verso.

Weizman, Eyal (2012) *Hollow Land: Israel's Architecture of Occupation*. London: Verso.

Williams, Raymond (1989) Culture is Ordinary. In *Resources of Hope: Culture, Democracy, Socialism*. London: Verso, pp. 3–14.

Woodward, R. (2004) *Military Geographies*. Oxford: Blackwell.

24

STUDYING MILITARY IMAGE BANKS

A Social Semiotic Approach

Ian Roderick

In this chapter I will outline a social semiotic-based method to studying military image banks. Rose (2012: 16–17) offers three criteria that serve as a litmus test for a critical visual methodology: (1) it must take images seriously and "look very carefully at visual images . . . because they are not entirely reducible to their context . . . and have their own effects"; (2) it must explicitly attend to "the social conditions and effects of visual objects"; and (3) it must invite consideration of the researchers own practices of looking at images. Together these three criteria are suggestive of an approach to studying images that is simultaneously empirical, material and reflexive. A social semiotic approach meets these criteria because it emphasizes systemic description, the social nature of communication and engagement through a critical and historical approach to semiosis.

Accordingly, the chapter begins by detailing the increasing influence of image banks in public communication. Drawing from the work of Frosh (2003) and Machin (2004) to characterize the ways in which image banks shape their content and the kinds of messages that can be communicated, certain qualities associated with stock images are then identified. Having established the significance of image banks and their role in creating a visual style for representing complex ideas and issues in a manner that is positive, upbeat and largely apolitical, the essay then turns to the new military image banks and how they have come to adopt this style. While on the one hand, the images curated on sites like DefenseImagery.mil[1] and in Flickr photostreams seem to serve a documenting function, I contend that these images need to be understood in relation to the growing visual dominance of image banks and stock photography as well as the post–Cold War expansion of military involvement into civil affairs.

So while the images can be treated as part of an internal practice of organizational storytelling (Boje, 2008), my focus instead is on the role of military image banks in external forms of communication. In particular, what concerns me is how the images can be understood as an attempt to shape, in addition to serving as "a primary site for negotiation and articulation of civilian discourses seeking to ascribe meaning to soldiers and their activities" (Woodward et al., 2009: 211).

To adequately address the emergence of this stock style of military imagery, I present a social semiotic approach to studying images that follows from Machin's (2013) proposed two-step method of systemic description and analysis. Finally, as an illustration of how to apply this method, I briefly detail the photographic representation of US Department of Defense (DoD) personnel engaged in humanitarian exercises and how these particular images semiotize what can be termed humanitarian militarism.

Military Image Banks

Image banks have come to be the go-to source for visual content when creating public communications in the form of advertising, marketing, public relations and journalism (see Machin and Jaworski, 2006; Hansen and Machin, 2008). It has been estimated that 70% of images in advertising, marketing and design are sourced from what has come to be known as 'the visual content industry' (Frosh, 2003: 2). Dominated by a small handful of multinational licensing agencies – with Getty Images, Corbis and Sipa Press being the largest – the commercial stock image industry is an industry worth more than a billion dollars per year. Getty alone licenses thirty to forty million image uses each year. For David Machin (2004: 317), the ubiquity of the stock image has meant "a shift . . . from emphasis on photography as witness to photography as a symbolic system." This shift is due primarily, of course, to the need for agencies to maximize the profitability of their image banks by ensuring that individual images will be generic enough to have multiple uses. Quite simply, it makes financial sense to only hold an image in stock if it can be applied to myriad contexts (see Frosh, 2003: 4–5). This results, according to Paul Frosh (2003: 196), in "a radical disjunction between the moment of image production and later moments of distribution and reproduction" so typical of the culture industries and the commodification of the cultural form. The upshot is that commercial image banks have come to function as a global visual repertoire (Frosh, 2003: 111) for expressing, borrowing from Machin (2004: 334), work as a laptop, freedom as a jump, ethnicity as the wearing of bright, multicoloured clothing and so on.

Coinciding with the expanding role that PR plays in communicating to publics the significance and outcomes of military activities and operations, western militaries have increasingly sought to offer image banks of their own. Although indexed and searchable, much like their civilian counterparts, it must be acknowledged, of course, that military image banks differ not only in terms of monetization but, by extension, in that there is not such a neat disjuncture between the moment of production and moments of distribution and reproduction. While commercial stock images are fundamentally intended to be used as visual content in other people's messages, military image banks are intended first and foremost to convey concepts, values, and ideas about, as well as document, the activities of the military institution that hosts the bank, although they can and are repurposed as visual content (see Roderick, 2009). In other words, military image banks such as DefenseImagery.mil are primarily focused upon stocking images that will serve to promote a specific, well-defined organizational 'message'. Nevertheless, as I shall discuss later, the aesthetic influence of the commercial image bank as cultural institution has meant that the military images have also come, to some degree, to take on the generic 'visual language' of the visual content industry.

Like World Wide Web presence, image banks have become a normal part of the internal and external communication practices of the world's militaries. All of the NATO member states, for example, offer photo archives on either the websites for their defence ministries and/or their services. Furthermore, as northern militaries have moved to embrace social media, countries such as United States, Canada, the UK and Australia all have established Flickr photostreams to make their image content readily available to the public. For example, with its revised policy on social media in 2009, the US Department of Defense, its various directorates and its service branches all have Flickr accounts where members of the public can more easily browse, favourite, comment on and share approved photos and videos generated by department personnel. In short, the use of image banks serves not simply to document the activities of militaries for their public, but more importantly, image banks need to be understood as playing an important role in creating the branded 'regimes of signification' (Lash, 1990) that legitimate those activities and the budgets necessary to sustain them.

A Social Semiotic Approach

My technique to analyzing military image banks such as the US DoD Flickr photostream is based upon a social semiotic approach to visual communication. Social semiotics is a critical theory of communication practices that has been developed from a wide range of interdisciplinary scholarship. Foundational work in visual communication by Kress and Van Leeuwen (1996), Van Leeuwen (2005) and Machin (2007) draws from both the Parisian school of semiotics and the systemic functional school of linguistics associated most closely with Michael Halliday. Although these authors all draw from the work of Roland Barthes, by adopting the Hallidayan view of language as one of a number of semiotic modes available to us, their work differs from Barthesean conceptions of semiotics by departing from a focus on codes and signs and instead foregrounding the way people draw from available semiotic resources to both produce and interpret communicative artefacts and events.

Semiotic resources, then, are selected by communicators in accordance with the meanings that they afford in particular social circumstances. Following from Halliday (1985), semiotic resources are always tied to social and communicative practices and so are understood as being socially organized to accomplish specific communication tasks. These tasks are realized through what are referred to as metafunctions: the representational, the interactional and the compositional. Semiosis is understood as always being realized through this tripartite system such that each communicative act draws upon semiotic resources to do three things simultaneously: (1) produce representations of the social action, (2) create interactions and constitute relations between social actors, and (3) realize those representations and interactions through the composition of specific kinds of texts. Communication can therefore never be simply reduced to an act of transmitting information; it is always bound up in the ongoing constitution of social relations and organization of semiotic resources. Accordingly, military image banks do not simply document military activities – because the images they supply "are not entirely reducible to their own context" (Rose, 2012: 17) – but are also constitutive of relationships between viewers and the military organization in question.

As Machin and Mayr (2012: 10) propose, one of the key strengths of a social semiotic approach to analyzing discourse is that it "allow(s) us to show more clearly *how* they make meaning as well as *what* they mean." In other words, a social semiotic analysis looks first to the actual affordances of the semiotic resources employed within the text so as to substantiate any claims then made about how the text functions discursively. This is accomplished by engaging in a practice of systematic description, which then serves to support the ensuing process of analysis (see also Machin, 2013).

Discourses are more than models of the world because they not only inform how we interpret and experience the world, but also how we might act in the world. Van Leeuwen (2005: 104) makes clear the link between discourse and practice when he observes that "the discourses we use in representing social practices such as eating are versions of those practices plus the ideas and attitudes that attach to them in the contexts in which we use them." Particular discourses, accordingly, support specific social practices by providing evaluations, purposes and legitimations (van Leeuwen, 2005: 104–105), and thus "discourse is socially constitutive as well as socially conditioned – *it constitutes situations, objects of knowledge, and the social identities of and relationships between people and groups of people*" (Fairclough and Wodak, 1997: 258, emphasis added). A social semiotic discourse analysis, then, will highlight *how* semiotic resources are selected and combined in texts so as to allow the analyst to specify *what* situations, objects, identities and social relationships are being constituted.

A Discourse of Humanitarian Militarism

Humanitarian militarism can be characterized as one of a number of closely related discourses on military engagements that have emerged since the Cold War era. One other such discourse is what Machin and van Leeuwen (2005) label 'a Special Operations discourse of war', which van Leeuwen (2005: 94) elsewhere describes as representing war as a unilateral intervention by a democratic northern power which places its elite, professional forces between the undisciplined and oppressive forces of a nonstate actor or failed state and their weak and helpless victims (local populations, UN peacekeepers, food relief agencies, etc.) in need of protection. The operation is characterized by its logistical precision and speed as well as technological and training superiority on the part of the elite forces. Humanitarian militarism also presupposes similar tripartite between a logistically superior highly trained professional northern military, a state that cannot or will not properly aid its own people, and a helpless local population. Nongovernmental organizations (NGOs) and other aid agencies may or may not be partners, but if they are, they follow the northern military's lead. In many ways, this can be understood as an expansion of special operations discourse to what are called military operations other than war (MOOTW).

The expansion of military engagement from military affairs into what could be called 'human affairs' corresponds to the expansion of NATO in the post–Cold War period. Merje Kuus (2009: 546) charges that with the expansion of membership in NATO there has been a corresponding "reconfiguration of where and how military power is exercised and normalized." Kuus's thesis is that traditional geopolitical forms of militarism have been supplanted by what she characterizes as cosmopolitan militarism, which is legitimated by "a story of the whole humanity and the whole global space gently guarded by a beneficent NATO" (Kuus, 2009: 546). As Kuus clarifies, it is not that NATO has become cosmopolitan but, rather, it has come to present itself and its actions as bound up in a spirit of cosmopolitanism. Humanitarian militarism, then, is one facet of the new face of cosmopolitan militarism. It is not only reflective of the changing symbolism employed by northern militaries to represent themselves as good citizens of the global village but of changing practices such as the militarization of aid work (see Lutz, 2002).

Accordingly, in what follows, I will describe specific examples of articulations of semiotic resources so as to better illustrate how the DoD photostream images function representationally, interactively and compositionally to discursively constitute military 'humanitarian' efforts as abstracted and apolitical enterprises. Focusing upon the US DoD image collection hosted on Flickr.com, over 500 images were collected from the DoD photostream using the search term 'humanitarian'. Of those images, 81 images were selected for meeting the criteria of representing DoD personnel interacting with civilians. A number of the humanitarian exercises documented in the collections were joint exercises, and so those photographic images that depict military personnel from other countries interacting with civilians were not included. The photographs were then coded by participant (female military personnel, male military personnel, adult female civilians, adult male civilians and/or local children) and by the instances each type of representational, interactional, and compositional resource as they appear for each of the five types of participant.

Representing the Actors and Actions of Humanitarian Militarism

Every act of communication has specific ways of representing participants according to the kinds of transitive process being realized. Transitivity, as Machin and Mayr (2012: 104) summarize, "is simply the study of what people are depicted as doing and refers, broadly to who does what to whom, and how." Halliday (1985) has proposed that there are six primary types of process: material (processes of doing), mental (processes of sensing), relational (processes of being), behavioural (processes of physiological and psychological behaviour), verbal (processes of saying) and existential (processes of existing or happening). Examining the realization of transitivity structures allows the researcher to highlight not only how the actions of actors come to be represented but also the distributions of agency among participants.

Given that the images collected on the DoD Flickr site can be characterized as action-shots and are intended to tell the viewer a 'story' about US military involvement in humanitarian efforts, not surprisingly, the 81 images of US personnel interacting with civilians are narrative representations (see Kress and van Leeuwen, 2006). This is not to say that narrative images do not communicate concepts – that would presuppose the possibility of being discourse-free – but that conceptual images function to relay attributes of the represented participant(s) rather than actions and relationships.

Since all of the photographs are narrative in structure, behavioural and material processes are by far the most commonly represented. By and large, the photographs depict military personnel engaging with local civilians often in the context of medical, dental or vision examinations or engaging in some sort of play or entertainment with local children. For both male and female personnel, there are slightly more instances of behavioural processes being represented when compared to material (16% more and 13% more, respectively). The larger number of behavioural processes suggests that, ultimately, the subject matter of these photographs is the interaction between local civilians and members of the US military rather than the aid process itself. Indeed, the most common represented behavioural process was smiling with 47 photos of personnel smiling, nine of personnel with a slight smile and only 17 in which personnel appear not smiling. Also striking is how few verbal processes appear in this group of photographs (three in total). Clearly there is a preference for pictures of smiling and positively interacting people.

In addition to examining the types of processes being represented in the images, the transitivity analysis of images also details the kinds of participants that are represented (or not) and whether they are active initiators of (actors) or passively undergoing (goal) the represented process (van Leeuwen, 2008: 32–35). Additionally, when represented as being at the receiving end, either positively or negatively, the actor is said to be beneficialized. Since these photographs are intended to document US military personnel doing humanitarian aid work, it should come as no surprise that the military are invariably represented as activated actors while the civilians are represented as being beneficialized (goals) by the actions of the military. While there are 54 instances of female (27) and male (26) military personnel being

Figure 24.1   An officer exchanges greetings with a Djiboutian woman.

Source: DoD photo by Mass Communication Specialist 1st Class Eric Dietrich, US Navy/Released. https://www.flickr.com/photos/39955793@N07/13870443104/

represented performing material processes, there are only 13 representations of civilians (three female, three male and seven child) performing material actions. Accordingly, the vast majority of processes being performed by the civilians are largely reactive and behavioural (smiling, laughing, looking, etc.). This asymmetry clearly reflects the passivation of civilians as they are put in the role of beneficiary. Indeed, there are only two photographs that depict local civilian populations actually participating in aid work (forming mixed relay lines to move provisions). Likewise, there are only nine photographs that include a local aid worker, and in each case the worker is always backgrounded. On the other hand, because the initiators of represented processes are invariably military personnel, agency is almost entirely delegated to the military. In other words, aid is being represented as a process that is done to/for a group of people and the military humanitarians are the actors while the local civilian populations are the goal of the process.

In addition to functioning as either actors or goals, participants can also be represented either as individuals or as groups. Depicting actors as a group tends to homogenize them whereas individuation obviously depicts the actor as somehow distinctive and unique. Of the 81 photographs being discussed, only five photographs depict military personnel as a group, and in two of the five photos an individual member is identified by name in the accompanying verbal text. Additionally, the specificity of the two verbally individuated members is further modulated by their location in the foreground of the photograph. This, of course, repeats the 'us versus them' way in which Europeans have represented themselves in relation to their colonial Others, but it also affords the opportunity to represent US military personnel as moral agents engaged in humanitarian acts. Since agency, and by extension morality, tends to be understood as a property of individuals, viewers of the photostream see repeated evidences

Figure 24.2   Surgeon examines a patient as part of a cooperative health engagement.

Source: DoD photo by Lance Cpl. Allison DeVries, US Marine Corps/Released. https://www.flickr.com/photos/39955793@N07/14183947575/

of the good character as well as professional behaviour of US military personnel. At the same time, while the beneficialized civilians are more likely to be represented in groups than the military members (20 of the 81 images), there is still a tendency to represent the civilians as individual actors as well, but not as profoundly. This, I would argue, is entirely in keeping with the way in which the photographs reproduce the visual language of humanitarian aid employed by NGOs as well as government agencies (see Manzo, 2008). The individuation of actors essentially frames humanitarian aid as a compassionate act between benefactor and beneficiary, and in doing so draws upon an apolitical 'humanist' discourse of aid as a charitable one-to-one meeting in which the broader structural relations that produced the need of aid in the first place are abstracted away. In this way, while the representations of the civilian particcontinues themselves are not overly generic, the ways in which the photos tend to be composed actually is. This, arguably, affords many of the same familiar stereotypes of the Global South as comprised of passive victims who can only have a relationship of dependence with the affluent North. It is also worth noting that in all but three of the twenty group images, the grouped actors are almost exclusively children. In fact, images of children feature prominently in the photographs (64 of the 81). But also striking is the near absence of adult male civilians. In all, there were eight images that depict local men in the entire photo set. The predominance of children and the near absence of men, consequently, redounds with the way in which aid provider-client relations are being imagined and further reproduces the visual representations that we have come to associate with foreign aid and the appeals of NGO aid groups.

Figure 24.3　Officer hands a stuffed animal to an Afghan orphan.

Source: DoD photo illustration by Lt. Chad A. Dulac, US Navy/Released. https://www.flickr.com/photos/39955793@N07/9449817937/

Constituting the Humanitarian Militarism Relationship

Every communicative act entails two kinds of participant: the represented participants within the image and the interactive participants actually engaged in the act of communication (Kress and van Leeuwen, 2006: 48). This means that there are potentially three forms of interactional relation: between participants in the act of communication, between represented participants within the communicative act, and between the communicating participants and the represented participants. What I am arguing, therefore, is that military stock images semiotize social relationships by orienting the viewer of the images so as to encourage a shared visual field and, by extension, a shared attitudinal orientation between producers and viewers – in this case between the US DoD and visitors to its Flickr site.

One way in which interpersonal relations are realized in images is through modality or claims to truthfulness. Since communication is understood to function as an exchange as well as a representation, when considering how 'truthful' an image or other text might be, the concern is for 'how true is this being represented?' rather than 'how true is this representation?' Modality, as addressed by Halliday (1985: 335), is often not simply expressed in absolutes. As Halliday (1985: 332) puts it, modality realizes the writer's "opinion regarding the probability that his [*sic*] observation is valid." Therefore, modality is first the degree of truth as the interactants understand it, and second the resources they opt to use in order to express that claim to truth.

All of the photographs, as documentary evidence of the US DoD and its members' efforts to provide humanitarian aid, can be said to fall within a naturalistic (photorealistic) modality type. The photographers have with near consistency applied the same modality markers to their images: high degrees of articulation of details and backgrounds, high degrees of colour saturation (without oversaturating), modulation and differentiation, and high degrees of articulation of light, shadow and tone. Essentially, the claim of truth here is founded upon verisimilitude, an accepted fidelity between what is captured in the image and what actually took place. At the same time, however, it is worth noting that despite the naturalistic settings of the photographs where lighting can be a problem, the images are typically bright and colourful. This, I would argue, corresponds with the bright, positive and optimistic message of US humanitarian militarism as 'doing good' in the world and is further established by the way in which viewer orientation is constructed in the photographs.

Viewer-participant orientation also functions to realize interactional meanings by using viewer perspective as a semiotic resource. In the selected photostream images, it is most often the facial expressions that determine the mood of the address. Angle of interaction is defined along two primary axes: horizontal angle, which realizes the degree of involvement, and vertical angle, which realizes power relations. The degree to which the represented participant(s) meets the gaze of the viewer or not determines involvement. A low-angle perspective tends to elevate the represented participant in status, whereas a high-angle perspective tends to diminish the represented participant. Finally, distance refers to the social distance between the participant(s) and the viewer. Distance is largely realized through the size of the framing of the participant(s) from close-up (intimate) to medium shot (social) to long shot (impersonal). Together, these three axes of orientation function as part of a simultaneous system of resources to metaphorically semiotize the degree and kind of interaction between viewer and represented participant (see Machin, 2007: 115).

Because the photographs are of interactions between military personnel and civilians, the tendency is for the photographic subjects to be pictured face-to-face according to the 180-degree rule. This means that very few military personnel are pictured face-on, and so their contact with the civilian(s) takes precedence over contact with the viewer. In all there are only seven instances of viewer-contact gaze for female and male personnel. Otherwise, personnel are typically portrayed from side-on to the viewer. Conversely, for the few images of male adult civilians, gaze is entirely constructed as noncontact (three side shots and five in which the head is completely turned from the viewer). For adult female civilians, there is a greater instance of viewer-contact. There are 10 instances of contact but also 10 side-on and eight in which the face is turned from the viewer. Finally, it is a very different situation for the structuring of gaze with child participants. While the majority of images would still be considered noncontact (eight in which the child faces away from the viewer and 25 where the child is facing side-on), 31 of the images of children are in fact what can be labelled contact gaze. While this could in part be explained by noting that children are more apt to look at the photographer, it should nonetheless be remembered that these are the photos that were chosen for inclusion in the photostreams as best able to tell the stories of military humanitarian aid.

The positioning of military personnel in the photographs affords moderate degrees of viewer involvement approximately three times as often as either high or low involvement. Viewer involvement with adult female civilians is divided almost equally between high,

Figure 24.4   Officer examines patient's ears during a consultation at a medical site in Puerto San Jose, Guatemala.

Source: DoD photo by Staff Sgt. Alesia Goosic, US Air Force/Released. https://www.flickr.com/photos/39955793
@N07/5905853928/

medium and low, with high involvement being slightly more frequent. Interestingly, none of the photographs of adult male civilians is high involvement; instead, they are primarily low involvement. Again, this tends to reproduce the tendency to not represent adult males as beneficiaries in visual representations of foreign aid. Of course, when it comes to the representation of children, viewer involvement tends to be high (30 photos) or moderate (26 photos), with only a few (eight) affording low involvement. This, too, is very much in keeping with the practice of tying First-World aid to the plight of children in foreign lands and the long-standing notion of ensuring that charity only goes to 'the deserving'.

At the same time, while one might expect the bulk of the photographs to be taken with high-angle shots, thus affording high viewer power, this is not actually the case. Instead there were about as many high-angle (32) as there were medium-angle (33) shots, which suggests that while many did afford a perspective of viewer power over the represented participants, just as many (one more, in fact) instead suggest a more egalitarian perspective. Nevertheless, it also needs to be noted that there are only 16 low-angle shots in the set, which would suggest that there is still a sense on the part of those taking the photographs and/or assembling the photo collections that viewer subordination is not desirable.

Finally, the use of distance is also very consistent in the photo set. By far the most common shot is a medium one suggesting a 'social' level of familiarity. This is not surprising since it is a comfortable level of social distance for your typical North American: it is not too intimate and neither is it so distant that it affords a sense of clinical distance or even indifference. Additionally, however, the use of medium shots has another important effect in terms of composition.

Composing Humanitarian Militarism

The third element of visual meaning is realized through what are termed compositional meanings. Here the concern is with how the image functions as a message by integrating the previous two dimensions of meaning, representational and interactive, through the distribution and emphasis placed upon the elements within the image. For the purposes of this analysis, I shall restrict my comments here to the use of framing and information linking.

Framing foregrounds how the individual elements of visual compositions are presented in varying degrees of connectedness to the other components. The stronger the framing, the more isolated or disconnected the element is, and therefore the closer it is to being presented as a separate and distinct element within the composition. The greater the degree of connectedness between elements, the more they will function together as one informational unit. What I want to propose is that the predisposition to using medium to medium close-up shots not only creates strongly framed images but also segregates the immediate interaction from any sort of grounding context. In the photo collection we rarely see images with a meaningful backdrop and thus we are rarely given a sense of the background to the immediate events that are unfolding in the images. All we see, and this is in keeping with stock imagery in general (Machin, 2004; Hansen and Machin, 2008), is the happy and positive encounter between service personnel and civilians in need. In this way, the strong framing of interactants further reinforces the abstraction of history and context that I discussed earlier. We are literally given no vantage point from which to evaluate the image-event.

Figure 24.5   Hospital corpsman practices speaking Kinyarwanda.

Source: DoD photo by Senior Chief Mass Communication Specialist Jon E. McMillan, US Navy/Released. https://www.flickr.com/photos/39955793@N07/3817645006/

The other compositional resource to consider is information linking and, in particular, the linking of text to image. Barthes (1977: 39–41) characterizes two kinds of relationship: on the one hand, an accompanying text or caption may fix the meaning of the image (anchorage), and on the other hand it may simply complement what is already communicated through the image itself (relay). Anchorage, as the term suggests, specifies or has a denominative function: "the text *directs* the reader through the signifieds of the image, causing him [*sic*] to avoid some and receive others" (Barthes, 1977: 40). This is how the captions provided on Flickr to accompany the photographs function. Captions, in keeping with the caption style guide provided by the DoD (http://defenseimagery.mil/dms/dvi-documents/StyleGuide-131028.pdf), focus upon the 'five Ws' so as to describe the image-event in the style of journalistic objectivity. What the images also do is identify service members by name while anonymizing local civilians. So, for example, the caption for Figure 24.5 appears as follows:

> US Navy Hospital Corpsman 2nd Class Porfirio Nino, from Maritime Civil Affairs Team 104, practices speaking Kinyarwanda, one of the official languages of Rwanda, during a civil observation mission in Bunyamanza, Rwanda, Aug. 7, 2009. The team is assigned to Combined Joint Task Force ? Horn of Africa and is in Rwanda to assess civil-military operations being conducted by the Rwandan Defense Force.

Likewise, the caption for Figure 24.1 clearly specifies the officer in some detail while the use of the indefinite article reduces the anonymous Iraqi woman to a generic class of attributes:

An Iraqi woman thanks US Army Capt. Jan Rose for care received during a combined humanitarian mission in Tahrir, Iraq, June 30, 2007. Rose is from Charlie Company, 431st Civil Affairs Battalion. DoD photo by Airman 1st Class Christopher Hubenthal, US Air Force.

And again there is the same use of the indefinite article for captioning the Afghan girl in Figure 24.3, but also note the detail afforded the mission, which is actually tangential to the activity captured in the image:

US Navy Lt. j.g. Meghan Burns, with Provincial Reconstruction Team (PRT) Farah, hands a stuffed animal to an Afghan orphan during a key leader engagement at the Farah Orphanage in Farah Province, Afghanistan, Aug. 4, 2013. PRT Farah's mission is to train, advise and assist Afghan government leaders at the municipal, district and provincial levels in Farah province, Afghanistan.

While it may well be a matter of policy not to identify local people, the effect is nonetheless one of reproducing the sense of homogeneity among those benefiting from US humanitarian (or otherwise) intervention.

Conclusion

In this chapter, I have sought to demonstrate how military image banks can be a rich source of data for analyzing how militaries represent themselves to the public and their own personnel. I have also sought to demonstrate the efficacy of a systemic approach to analyzing images in which claims to *what* is being accomplished discursively is determined by considerations of *how* semiotic resources are actually selected and combined. This is illustrated through a study of images that purport to represent the humanitarian actions of US DoD personnel as military aid workers.

Critical military studies examines military institutions and practices in relation to the broader society to which they belong. Rather than reifying them as fixed and stable entities, military institutions are understood to be normative organizing structures that are both contested and negotiated. Accordingly, the processes by which these institutions, their practices and their subjectivities come to be constituted and normalized are scrutinized and challenged through critical military research. Of particular interest are the ways in which militaries and their actions are made to be visible to the broader population. How militaries and their actions are rendered seeable is understood to be the product of socially constructed ways of seeing rather than a neutral or natural process. What is visible is never simply a matter of what is available to be seen but rather the production of limits of visibility: what can be seen, who can do the seeing, and the perspectives they substantiate. This, I argue, is best accomplished through detailed examination, by looking carefully, and documenting how images render particular ways of seeing by drawing resourcefully upon socially defined visual elements and their meaning potentials. By critically reflecting upon the options that were selected in comparison to others that were available, we can better interrogate images that afford one way of seeing the military and its actions over another.

Studying how these online image repositories are curated and the content of they make available necessitates a systematic methodology that guides the assembly of a corpus of images and detailed description in support of analysis. In keeping with Rose's criteria, the method I have presented entails the careful examination of images and analyzing them as communicative events. Researching banked military images requires collecting and studying them comparatively, taking into account the kinds of inclusions and exclusions the images afford as well as the underlying practices of producing and consuming those images. Finally, a social semiotic approach necessitates that visual communication be understood as being constituted by practice as much as discourse; as analysts, we need to contextualize our readings of military imagery in relation the practices and knowledges that produced them.

Note

1   Since first drafting this chapter the US DoD has replaced DefenseImagery.mil with the Defense Imagery Management Operations Center and outsourced its image archive to T3Media, an image bank and archiving service. The new site is only available to registered military personnel and promises future civilian image search capabilities. Nonmilitary personnel are redirected to Defense Video & Imagery Distribution System and the main US DoD website. There have been no changes to the official DoD Flickr site and the changes in access most likely due to pricing arrangements with the new hosting service.

References

Barthes, R. (1977) Rhetoric of the Image. In *Image, Music, Text.* New York: Farrar, Straus and Giroux.

Boje, D. M. (2008) Storytelling Organizations. Los Angeles: Sage.

Fairclough, N. and Wodak, R. (1997) Critical Discourse Analysis. In Van Dijk, T.A. (Ed.), *Discourse as Social Interaction (Discourse Studies: A Multidisciplinary Introduction, Volume 2)*. London: Sage, pp. 258–284.

Frosh, P. (2003) *The Image Factory: Consumer Culture, Photography and the Visual Content Industry.* New York: Berg.

Halliday, M.A.K. (1985) *An Introduction to Functional Grammar.* London: E. Arnold.

Hansen, A. and Machin, D. (2008) Visually Branding the Environment: Climate Change as a Marketing Opportunity. *Discourse Studies* 10(6), 777–794.

Kress, G. R. and Van Leeuwen, T. (1996) *Reading Images: The Grammar of Visual Design*. London: Routledge.

Kress, G. R. and Van Leeuwen, T. (2006) *Reading Images: The Grammar of Visual Design* (2nd edition). London: Routledge.

Kuus, M. (2009) Cosmopolitan Militarism? Spaces of NATO Expansion. *Environment and Planning A* 41(3), 545–562.

Lash, S. (1990) *Sociology of Postmodernism.* London: Routledge.

Lutz, C. (2002) Making War at Home in the United States: Militarization and the Current Crisis. *American Anthropologist* 104(3), 723–735.

Machin, D. (2004) Building the World's Visual Language: The Increasing Global Importance of Image Banks in Corporate Media. *Visual Communication* 3(3), 316–336.

Machin, D. (2007) *Introduction to Multimodal Analysis.* New York: Hodder Arnold.

Machin, D. (2013) What Is Multimodal Critical Discourse Studies? *Critical Discourse Studies* 10(4), 347–355.

Machin, D. and Jaworski, A. (2006) Archive Video Footage in News: Creating a Likeness and Index of the Phenomenal World. *Visual Communication* 5(3), 345–366.

Machin, D. and Mayr, A. (2012) *How to Do Critical Discourse Analysis: A Multimodal Introduction*. Los Angeles: Sage.

Machin, D. and van Leeuwen, T. (2005) Computer Games as Political Discourse: The Case of *Black Hawk Down*. *Journal of Language and Politics* 4(1), 119–141.

Manzo, K. (2008) Imaging Humanitarianism: NGO Identity and the Iconography of Childhood. *Antipode* 40(4), 632–657.

Roderick, I. (2009) Bare Life of the Virtuous Shadow Warrior: The Use of Silhouette in Military Training Advertisements. *Continuum: Journal of Media & Cultural Studies* 23(1), 77–91.

Rose, G. (2012) *Visual Methodologies: An Introduction to Researching With Visual Materials* (3rd edition). Thousand Oaks, CA: Sage.

Van Leeuwen, T. (2005) *Introducing Social Semiotics*. New York: Routledge.

Van Leeuwen, T. (2008) *Discourse and Practice New Tools for Critical Discourse Analysis*. Oxford: Oxford University Press.

Woodward, R., Winter, T. and Jenkings, K. N. (2009) Heroic Anxieties: The Figure of the British Soldier in Contemporary Print Media. *Journal of War & Culture Studies* 2(2), 211–223.

25

CRITICAL METHODOLOGIES FOR RESEARCHING MILITARY-THEMED VIDEOGAMES

Daniel Bos

The relationship between the military and videogames has evolved dramatically over the last few decades. Along with their use by militaries for the purposes of training, recruitment and increasingly the rehabilitation of soldiers, military-themed videogames have also come to dominate commercial entertainment landscapes worldwide. As Stahl (2010: 112) suggests, "videogames are increasingly both the medium and the metaphor by which we understand war." This infiltration of virtual warfare into public life has spawned a burgeoning critical academic interest which has sought to trace the political, cultural and social implications of the relationship between videogames and the military. But despite this growing scholarly interest, analysis has predominately focused on the 'videogame-as-text'. This has involved deconstructing and commenting on the relationship between virtual worlds and the military ideologies, values and sensibilities apparent in their visual schemes and playful structures (Leonard, 2004; Power, 2007; Shaw, 2010). What remains unclear, however, is how players themselves consume, engage and interact with the explicitly militarised content of military-themed videogames.

It is the players who are the focus of this chapter, along with the methodological approaches that might be employed to explore their practices as players, their engagements and experiences. More broadly, my aim here is to explore methodologies which allow for an insight into the embodied and experiential aspects of military-themed gaming. The chapter is structured as follows. First, I will outline the significance of military-themed videogames in social science research. Second, I will expand on some of the methodological perspectives already discussed within the literature which advocate a player-based approach. In doing so I provide context for the remainder of the chapter. Finally, I draw on my research which explored the videogame series *Call of Duty: Modern Warfare*, and discuss how different qualitative and experiential methodologies might be used effectively. Overall, the chapter argues for the need to develop a set of methods enabling more detailed understandings of players and their interactions.

Military Videogames and the Absence of Players

Videogames have emerged as a vital medium through which popular cultural militarism is articulated and reinforced. Military-themed videogames such as *Battlefield*, *Medal of Honour* and *Call of Duty* are but a few key titles in a hugely popular cultural field that allows players to experience and virtually partake in historical, contemporary and futuristic military and geopolitical conflicts. The growth of gaming has encouraged academics to critically explore and problematise the relationship between play and virtual war. Studies in this vein have centred largely on an ideological critique of militarised virtual worlds and the ideals they subsequently entrain (Schulzke, 2013). Videogames arguably act to mirror contemporary geopolitical tensions and cultures (Power, 2007; Gagnon, 2010; Salter, 2011), to stereotype people and places (Höglund, 2008; Sisler, 2008; Shaw, 2010) and to curtail deliberation over the legitimacy of and justification for state-based forms of military violence (Leonard, 2004; Stahl, 2010). While this videogame-as-text approach remains an important intervention in broader explorations of the ways that popular ideas of militarism find expression in everyday life, the focus on ideological critique is limiting, and limited methodologically.

Most significantly, as Schulzke (2013) has argued, critical attention to the ideological aspects of military-themed videogames has been overly presumptive, particularly in relation to their alleged societal effects. For example, critics have been quick to problematize videogames such as *America's Army* because of its origins as an American military recruiting and public relations tool. But for Schulzke (2013: 72), merely demonstrating this sort of "connection does not tell us what ideological message the games promote or [how] military games are actually experienced." These 'ideological' critiques are, in some ways, unsatisfactory in that they overlook the practices and experiences of the millions of people who come to interact and engage with videogames and who, in turn, generate the meaning of the game itself. This is not to suggest that such critiques are redundant – they remain important in revealing forms of popular militarism. The point remains, however, that by turning our focus to the players themselves, we can start to understand in more detail how, exactly, militarism enters into and is reproduced as part of the everyday.

To date academic research on gaming has tended both explicitly and implicitly to present players as passive consumers of militarised content. Generally, this work conceptualises audiences by way of a 'hypodermic needle' model, where omnipotent media industries 'inject' hegemonic ideologies into a receptive and submissive public (Ruddock, 2000). Playing military videogames, it is argued, "contribute[s] to an acceptance of the militarization of society" (Leonard, 2004: 1). Such a linear imaginary of player engagements has been criticised by cultural and media theorists alike, who instead emphasise the role of the audience and their capacity to generate meaning in their interactions with cultural texts (Hall, 1973; Fiske, 1989). Taking this forward, Thomson (2008: 21) argues that in respect of military-themed games, the "interaction between game and player" can be understood through "processes of encoding and decoding, as well as resistance and rejection." Players are capable of sophisticated and multiple readings which can deviate completely from the intended encoded meaning. Furthermore, the meaning derived from this interaction is also contingent on a range of factors including an individual's investments, their social and cultural background, and the context (i.e. the immediate space) in which the videogame is being played.

In short, studying the players of military-themed games offers greater nuance and insight into the ways in which both games and militarism might be understood. Player-based approaches offer opportunities to further understand the significance of videogames in generating and circulating particular imaginations of the military and their activities, and offer the potential for grounded accounts of how militarism is understood and engaged with in the everyday.

A Player-Centred Approach to Understanding War Play

Nascent exploration of players in the wider videogame literature, according to Egenfeldt-Nielsen et al. (2013), can be divided into two perspectives: that of the 'active user' and that of an 'active media'. Active user perspectives focus on what players actually do with videogames, adopting qualitative methodological approaches to explore the individual's experience. An active media perspective, on the other hand, is rooted in a media-effects tradition, which examines the role of videogames in influencing a mostly passive recipient. Such studies tend to be informed by the theories of behaviourism and social psychology, and often use quantitative methods. There are a number of important differences between these two approaches, both of which shape how we come to understand and define the relationship between players and military-themed videogames.

Within active media approaches there remains a critical interest in academic and media commentaries on the supposed effects of military-themed videogames on individual behaviour. Commentaries have positioned such games amid debates around the effects of videogames on violent and psychological individual behaviour (e.g. Greitemeyer and Mügge, 2014). This has encouraged research employing quantitative and scientific approaches towards data collection in order to confirm or falsify hypotheses.

A good example of this approach is Festl et al.'s (2013) study of German gamers. This research assessed the extent to which individual engagements with videogames might be associated with a greater development of militaristic attitudes. In total 4,500 gamers were contacted via telephone survey and questioned about their attitudes to the military vis-à-vis their engagement with videogames. These were measured alongside various social demographic data, and together went to define what Festl et al. (2013) called the New Militarism Short Scale (NMSS). The scale reflected three thematic lines of inquiry: (1) soldier admiration, (2) army necessity and (3) terrorist threat. Individuals were asked to respond to six statements relating to these themes using a Likert-type scale. Overall, the research led to the conclusion that the type and frequency of play, and whether or not the individual played military first-person shooter videogames, did not appear to influence and shape militaristic attitudes. Instead, such attitudes were suggested to be attributable to age (with older persons being more militaristic), to lower educational attainment and to incidences of authority-orientated aggressive personalities.

While quantitative approaches are suitable for collecting factual aspects of people's engagements and purport to provide large representative samples, the methods and conclusions provided can be considered limiting. These studies can often overlook the "micro-reality of confusingly kaleidoscopic everyday experiences" (Schrøder et al., 2003: 31) and the lived-in realities of virtual war. The conditions under which this sort of research often takes place can also be readily scrutinised, as they often overlook the contexts in which videogames are played (see Ferguson, 2007 and Egenfeldt-Nielsen et al., 2013 for more detailed discussion). Summarily, the finer details of the multifaceted and contingent ways in which military videogames are consumed, understood and internalised, and how this relates to attitudes and understandings of the military, is masked by/in active media analyses.

Qualitative methods and an active user perspective offer alternative ways of gaining accounts of players' engagements, and are increasingly used to explore audiences' reception of popular and visual cultures (Schrøder et al., 2003; Rose, 2012). Moving beyond a purely academic reading and privileging everyday audience engagements has the potential to reveal "how players connect their war-themed video game experiences with their real-life understandings of war and politics" (Penney, 2010: 194). Active user approaches do not, however, suggest an unequivocal relationship between play and militarised attitudes. Huntemann's (2010: 232) study, for instance, points to complexity, and concludes that "while players clearly do not wholly accept the ideology about militarism embedded in these games, they do not wholly reject it either." Qualitative approaches in general enable individuals to respond on their own terms and to clarify and expand on particular practices and thoughts (Bertrand and Hughes, 2005: 74–82). Thus, talking to and opening up a dialogue with players enables them to define experiences and understandings on their own terms. However, the *practice* of playing virtual war has been consistently overlooked in both active media and active audience approaches. Understanding these sorts of practical experiences requires careful methodological consideration. In the next section I draw on the methods employed in my own research to provide an insight into how we might understand what it is to play war.

Researching Players in Practice

My own research has set out to understand the ways military and geopolitical sensibilities are produced, consumed and represented in commercial military-themed videogames. One key aspect of the research explores how the consumption, interpretation and experience of gameplay is entangled with dominant geopolitical and militaristic visions of the world. Here, I attempt to draw connections between hugely successful commercial enterprises in popular culture (specifically the *Call of Duty* franchise), mass popularity and engagement, and individual and small-group understandings of the military. *Call of Duty* is a hugely popular videogame franchise which has commanded $11 billion in sales since the release of the first in the series in 2002; to date 175 million copies of its games have been sold worldwide (Makuch, 2015). Thus, in focusing my research on players of the *Call of Duty* games, I anticipated being able to gain interest from a wide range of potential participants.

I employed a number of methods to recruit participants for the study. In the early stages of the research, I attempted to use online gaming forums and official *Call of Duty* fan forums to generate participants. However, the response rates on these forums were low. Initially, I also engaged with participants via email to negate issues of location, and considered the possibility of using the Voice over Internet Protocol (VoIP)[1] system used on the multiplayer option of *Call of Duty*. However, this often caused difficulties in terms of arranging appropriate times to 'talk', and furthermore, email exchanges were often met with partial responses from participants. Therefore, I decided to arrange face-to-face interviews. I undertook a poster and leafleting campaign focusing on the local videogame stores, videogame centre and the universities within the city. In total, I completed thirty-two interviews with people between the ages of nineteen and forty in phases across 2010–2013. The vast majority of participants were male, with only one female getting in touch for an interview – a situation illustrative of the gendered nature of contemporary military-themed gaming. While participants largely responded through the poster campaign, others were recruited after discussions with family and friends. Much like the recruiting process, then, the methodological elements of my research required frequent adaptation. In the following section I will provide a critical reflection on these methods and their utility for researching military videogames. I begin with the face-to-face interviews before discussing video-based ethnographic work.

Face-to-Face Interviews

Through the recruitment campaign I arranged to meet participants face-to-face. The interview questions themselves were designed in a way as to advance a wider understanding of consumptive behaviour, alongside player reflections on militaristic and geopolitical content. Standard interview procedures were followed.[2]

The interviews were designed to be semistructured and were orchestrated around three broad themes of questioning. First, players' background in videogames, their playing habits and their engagement with the *Call of Duty* series were discussed. These questions eased participants into the interview process and worked to build rapport. Second, I began to ask after players' attitudes towards geopolitical and militaristic content. Finally, I attempted to engage with the experiences of playing war. I sought to acknowledge the positive and negative experiences of play, the immersive qualities of the games, and the technologies, practices and in-game moments which amplified certain affective and emotional states.

However, attempting to solicit responses concerning 'experiences' often proved difficult. Many players found it difficult to clearly articulate their experiences of playing virtual war away from the game (i.e. where players were not near their games consoles). Indeed, when asked to reflect on experiences of gameplay players were hesitant and found it difficult to recall their experiences. As a result of these issues, I decided to adopt another approach – the gaming interview – in order to overcome some of the ambiguities of recounting playing virtual war.

The Gaming Interview: Talking and Playing

The gaming interview was first used by Schott and Horrell (2000) as way to explore how female gamers attribute meaning to everyday experiences of play. This technique involved the researchers interviewing female players in their homes. Schott and Horrell (2000: 40) claim this approach

> provided direct access to the girl gamers' playing style and habits, generated new questions and permitted the girl gamer to express their views on gaming whilst engaging directly with the technology.

The gaming interview was performed in the location where the 'girl gamer' played, in this case the domestic setting. The interview process itself was seen more as purposeful discussion, and accordingly meant "a more 'play like' atmosphere and generate[d] questions about female playing experiences as they occurred" (Schott and Horrell, 2000: 40). Adopting this technique for my research offered a means of situating questions and responses in the context of the playing of war, and of questioning players as they were immersed in militarised virtual environments. Participants were given the freedom to select a game to play from the *Call of Duty* series, and while they played I observed and asked interview-style questions as I would have done in a face-to-face interview.

These interviews were conducted on the university campus, as this was the most viable location for both participants and researcher. As such, a space was booked, usually a seminar room, in which to set up a videogame console. Despite its methodological potential, the gaming interview did come with a range of practical limitations. Some participants, for example, declined outright to play the game at all. For example, one participant noted that he used a PC at home to play *Call of Duty*, and so wasn't able to play on the console which had been set up for the purposes of the research. For certain participants it was also "difficult to talk and play at the same time" (Alexander, nineteen years old). During the interview, which usually lasted between forty-five and sixty minutes, participants often stopped playing in order to engage more in the discussion. While the gaming interview certainly provided a means for players to discuss the game content and mechanics in a more direct manner, to some extent it occluded key details, particularly those related to the immersive qualities of videogames. It also, because the interviews took place on a university campus, stripped the experience of playing war from the context in which it is situated (usually the home).

Video Ethnography

In taking these methodological issues forward, and in wishing to get to grips with the experiences of playing war in situ, I devised a video ethnographic approach. Within the social sciences use of video ethnography has become increasingly prevalent, and has developed in tandem with interest in the multiple relationalities, experiences and practices that are constitutive of everyday life (Pink, 2013). Particularly for researchers interested in videogames, the video camera offers a way of capturing the complex encounters between screens, bodies, technologies and environments (Giddings, 2009; Ash, 2010; Thornham, 2011).

There were a number of practical benefits of employing video cameras in my own videogame research. It allowed me, for example, to capture and record the embodied and verbal exchanges that occurred during gameplay. In doing so the camera provided a means of capturing the fleeting and transitory details of playing virtual war, details that would have been elusive and difficult to capture in discussion, or with pen and notepad. Recorded footage could be replayed for the purposes of detailed transcription and analysis of body language and other nonverbal cues. In the case of this research, the final video was analysed using a preliminary categorisation that enables the transcribing of the "basic aspects of the activities and events that have been recorded" (Heath et al., 2010: 64). This included annotations that referred to the time, embodied movements and particular in- and out-of-game situations, and noted the context in which things were said. The data was then coded, enabling the drawing-out of emerging themes. To develop on the utility of a video ethnographic approach to the study of military themed videogames, I want to briefly critically reflect on three issues: place, technology and experiences.

The Place of Virtual War

The use of video ethnography first enables us to capture the setting in which virtual war is played, and the ways spaces of play are prepared and organised. While games often virtually locate players in distant lands (a fact often noted in orientalist critiques), gameplay itself is rooted and occurs in specific places. Insofar as it is often online and networked, it is enabled by a vast assemblage of social, technological and material relations. For Payne (2010), methodologies need to be sensitive to these sorts of contexts. Taking 'ludic war'[3] seriously therefore means arguing that:

> the *where* (i.e., social setting) and *how* (i.e., social relations) must be considered alongside the *what* (i.e., video game text) of gameplay, as well as its connection to the culturally dominate symbolic regimes.
>
> *(Payne 2010: 208, emphasis in the original)*

Payne's ethnography of a local area network (LAN) gaming centre is useful in developing this notion of the place of virtual war.[4] By attending to the social relations and situated context in this case, Payne (2010: 208) remarks how "power hierarchies in fictional, war-torn synthetic worlds [become] reified and replayed in the real world." As such, these observations reveal the ways the ideological and militarised game content might often pervade the social spaces of play.

In the initial stages of my research, pilot studies had been organised and set up in a studio on university campus. This was largely due to accessibility, as previously noted, but also the availability of technology, such as television screen and video cameras (see Figure 25.1).

However, it soon became clear that inviting participants into a prescribed environment detracted from the everydayness of play. Specifically, when undertaking my video ethnography pilot studies, the artificial setup revealed the importance of place/space in the research process and the domestic geographies of media consumption (Adams, 2009). Simply, for many players, their interactions with videogames are largely practiced in the home.

As such my video ethnographic research was redesigned to examine players' interactions in and with the place in which they typically played the game (usually a domestic setting). In this case I turned to friends who were keen players of the *Call of Duty* series. While I do not have the space here to discuss the implications of using friends in research (see Brown, 2003; Taylor, 2011), this approach did offer a number of significant methodological opportunities. For example, rather than being the newcomer to the situation and space, I was able to adopt the position of intimate insider, which allowed for the generation of a much more in-depth understanding of play than perhaps could have been achieved with nonfriend participants. It also meant I could remain in regular contact, and generally, have an increased level of

Figure 25.1   Location of initial video ethnographies: Newcastle University campus.

understanding concerning intended meanings (Taylor, 2011). On the other hand, the slippage between 'friend' and 'researcher' was a constant feature of the research process.

The setup of the ethnography involved placing a video camera in a position that would capture both myself and participant engaged in play. I initially planned to introduce another camera, facing the screen, to capture the gameplay. However due to the layout of the rooms this was unfeasible. Moreover, the idea was to ensure that the video camera(s) would remain unobtrusive to the research process. Further studies might consider ways of recording both the gameplay and player simultaneously.

Playing in the participant's home offered a comfortable and familiar setting as opposed to the 'laboratory setting' of the university. This meant that participants would be engaging in play, using their own controls, consoles, play configurations, and gameplay set-up. The participant was told to choose a game and gameplay mode that they would usually engage with. This encouraged use of the multiplayer option of the games (which were often unavailable in the studio ethnography situations), and allowed me to participate in the gameplay myself.

The footage provided an intimate insight into the ways the room was organised around playing virtual war. Curtains were drawn to avoid glare from the sun on the screen and the player's faces and seating were adjusted accordingly throughout the filming process. Thus, the footage captured play in its context, offering an intimate insight into players' customary practices while facilitating virtual war. Situating play in its context therefore offers the potential for useful insights into how militarism and militarisation emanates from and is constituted in place.

Technology and Playing Virtual War

Alongside the environment, home-based video ethnographies also allowed for a deeper insight into the complex relationship between various human and non-human actors in the playing of virtual war. Various technologies significantly shape and act upon the body during play. The forced feedback technology of the videogame controller which vibrates in relation to the game's content, the first-person perspective that permits a particular field of vision, and the audiovisual arrangements have significant impact on the experiences and senses of the player. The footage collected began to show how these heterogeneous elements came to constitute the playing of war (Taylor, 2009). But after watching and analysing the footage in detail, the fragility of this assemblage also became evident. For instance, the multiplayer mode was contingent on a stable Internet and server connection. On numerous occasions the connection would break or become 'glitchy', halting the gameplay and visibly breaking the players' immersion in the virtual world. Overall the event of gameplay was shown to be contingent and reliant on constantly shifting and unfolding relations between player and screen world.

While the assemblage of technologies involved in playing war can be captured and ana-lysed via filming, it is also important to consider the introduction of the technology of the video camera and its influence on the research process. In this respect, introducing the video camera into a gaming situation can evoke a 'camera consciousness' (Pink, 2013). The camera's presence may, simply, encourage participants to alter, modify, and transform their normal practices and behaviours. The influence that the video camera had on the research process is difficult to ascertain. However, despite efforts to make the video camera a less obvious presence, it was often met with flippant remarks and continual, nervous glances. Therefore, it's clear that we need to think carefully about the use and placement of recording equipment in future ethnographic research into war play, including the imposition of the researcher themselves.

The Experiences of Playing Virtual War

Research into players has often overlooked the emotive and affective ways play acts upon bodies. As Shaw and Warf (2009: 1335) argue, we need to consider virtual worlds as affective worlds which "increasingly '[spill] out' of the screen to affect the player in banal, exciting, or unexpected ways." Such a perspective poses new challenges for research into military-themed videogames, and requires us to consider how militaristic cultures, logics and ideologies reso-nate with players in 'more-than-representational' ways (Dittmer, 2010).

In this instance, turning the video camera onto the players revealed the visceral and affec-tive states stimulated by playing virtual war. Participants would lean forward on the edge of their seats, hunch their shoulders, rest forearms on legs and fix their eyes intently on the screen. Facial contortions and sudden jolts of bodily movement illustrated and accompanied intensive moments of gameplay. Verbal utterances of joy, frustration and anger alluded to the wide range of mental and experiential states we can associate with playing war.

Despite clearly demonstrating the affective relations forged between player and screen world, this footage did not necessarily give an indication or make clear the affective relations between *military content*, player and screen world. This sort of concern is inherent in video ethnography and is raised by Simpson (2011), who used a video ethnography to document the practices of street performance:

> While in [the] video we can see the minute detail of bodily movements and non-verbal communication, alongside verbal communication and other sounds, it arguably provides *little in the way of a sense of the felt aspects in and of these movements* (Paterson, 2009).
>
> *(Simpson 2011: 350, my emphasis)*

In the initial stages of the video recordings I attempted to remain inconspicuous, principally by not intervening and following rather than leading conversations. However as the filming progressed I began to ask questions and to encourage players to reflect on their actions, behaviours and practices within and outside the screen world. This gave more of an insight and connection between with the screen and real world. In this transcript example with Dean and Gary,[5] developing on Dean's professed nervousness of his avatar being killed in the multiplayer mode of the game, I probed further here around his associated emotional and affective states:

Interviewer: . . . and you talk about it Dean, the last kill and you get nervous . . .?

Dean: Yeah that's like the whole *Sonic*[6] thing where you go under water in *Sonic* and you'd hear the bubble trying to get the air that kind of feeling.

Interviewer: When you're playing now is there anything that does that?

Dean: When you've come up to a group of people it happens. . .

Gary: . . . or when you're reloading . . .

Dean: . . . yeah, when you're reloading and you get that panic. It's all about being killed . . .

Gary: See I don't get it as much because I'm expecting to die a lot when it's like this. [however] When I'm sniping and there's someone close range and you can't aim properly . . .

Dean: I'm doing shocking now . . .

Interviewer: Are there certain peaks [you experience] . . .?

Gary: I don't know it depends what game I'm playing . . . in this you're always pretty much on edge with this.

These purposeful discussions began to reveal, therefore, some of the affective relations between aspects of gameplay and player (relations which varied between players). In this instance, through noting embodiments and physical responses and by asking for follow-up explanations, I was able to obtain further insights into how the gameplay resonated with and cultivated particular affective states. Unlike the gaming interview described earlier, my own intervention was kept to a minimum, thus enabling the capturing of the experiences of playing war in an unaltered fashion. Such an approach could also be complemented with presenting the recorded footage back to the player (see Thornham, 2011). What is clear is that when undertaking a video ethnography, the camera alone is unable to shed light on the significance of the affective relations between player and screen world. Instead, by carefully engaging participants in context, we can gain a further understanding of what it is to play and experience virtual war.

Conclusions

This chapter has examined the methodological challenges faced when examining player consumption and experiences of military-themed videogames. Previous literature has been dominated by ideological criticisms of the videogame itself, and has often overlooked the relationship between the player and game, and the ways in which military values, logics and ideologies are experienced by audiences. Despite the marginalisation of players' voices and experiences from the literature in general, approaches like the gaming interview and video ethnographic methods can illuminate the complex relations and contingencies that are constitutive of playing war. Above all, they tell us that it is important not to generalise about these experiences. Qualitative approaches to games provide a significant opportunity for players to verbalise their understandings of the relationship between war, geopolitics and militaries. Moreover, they encourage us to further explore the complexities of player engagements and practices, especially as this is a key space in which knowledges, practices and moralities exist at the interface of civilian and military worlds.

In this chapter I focused on three aspects of my own research in the audiencing of military videogames. Qualitative approaches, such as interviewing, offer detailed insights into player engagement and understanding. However, consideration clearly needs to be taken over the overall design of the interview process. Accordingly, while further detailed ethnographic approaches offer the potential for broader insights in into the playing war in situ, they need to take into account the sociomaterialities of the player's habitus and the affective relations between body and screen world. Military-themed games are important because of what they contain, but emerge as potent cultures of militarisation because of the sociality of play, and become so because of the contexts in which playing war happens. What is clear, however, is that there is further work and methodological innovation required to adequately grasp the role military-themed videogames should play in broader concerns about the militarisation of everyday life.

Notes

1 VoIP allows players to connect and talk while playing via the use of a headset. This is usually found in multiplayer options of videogames.

2 The interview process began with the participants being given a brief introduction to the research and the overall structure of the interview. Ethical procedures were outlined and an indication of how the interview data would be used. All interviews were recorded using a digital recorder and subsequently transcribed.

3 Ludic war: "the activity of playing war or military-themed video games alone or with others" (Payne, 2010: 207).

4 A LAN gaming centre usually takes the form of a small business whereby customers (gamers) are able to play multiplayer online games individually or together by using a computer or console which is connected to an area network.

5 Pseudonyms were given to all participants.

6 Dean is referring to the *Sonic the Hedgehog* videogame series and a gameplay situation where the avatar would be under water and the player tasked to locate air bubbles in order to keep the character alive.

References

Adams, P. C. (2009) *Geographies of Media and Communication: A Critical Introduction*. Oxford: Wiley-Blackwell.

Ash, J. (2010) Teleplastic Technologies: Charting Practices of Orientation and Navigation in Videogaming. *Transactions of the Institute of British Geographers* 35(3), 414–430.

Bertrand, I. and Hughes, P. (2005) *Media Research Methods: Audiences, Institutions, Texts*. Basingstoke: Palgrave Macmillan.

Browne, K. (2003) Negotiations and Fieldworkings: Friendship and Feminist Research. *ACME: An International E-Journal for Critical Geographies* 2(2), 132–146.

Dittmer, J. (2010) *Popular Culture, Geopolitics, and Identity*. Lanham, MD: Rowman & Littlefield.

Egenfeldt-Nielsen, S., Smith, J. H. and Tosca, S. P. (2013) *Understanding Video Games: The Essential Introduction* (2nd edition). New York: Routledge.

Ferguson, C. (2007) The Good, the Bad and the Ugly: A Meta-analytic Review of Positive and Negative Effects of Violent Video Games. *Psychiatric Quarterly* 78(4), 309–316.

Festl, R., Scharkow, M. and Quandt, T. (2013) Militaristic Attitudes and the Use of Digital Games. *Games and Culture* 8(6), 392–407.

Fiske, J. (1989) *Understanding Popular Culture*. London: Unwin Hyman.

Gagnon, F. (2010) 'Invading Your Hearts and Minds': Call of Duty and the (Re)writing of Militarism in U.S. Digital Games and Popular Culture. *European Journal of American Studies* 5(3). Available at: http://ejas.revues.org/8831#toc (Accessed 18 April 2015).

Giddings, S. (2009) Events and Collusions: A Glossary for the Microethnography of Video Game Play. *Games and Culture* 4(2), 144–157.

Greitemeyer, T. and Mügge, D. O. (2014) Video Games Do Affect Social Outcomes: A Meta-Analytic Review of the Effects of Violent and Prosocial Video Game Play. *Personality and Social Psychology Bulletin* 40(5), 578–589.

Hall, S. (1973) Encoding and Decoding in the Television Discourse. Centre for Cultural Studies, University of Birmingham, CCS Stencilled Paper No. 7.

Heath, C., Hindmarsh, J. and Luff, P. (2010) *Video in Qualitative Research: Analysing Social Interaction in Everyday Life*. Los Angeles: Sage.

Höglund, J. (2008) Electronic Empire: Orientalism Revisited in the Military Shooter. *International Journal of Computer Game Research* 8(1). Available at: http://gamestudies.org/0801/articles/hoeglund (Accessed 26 January 2016).

Huntemann, N. B. (2010) Playing With Fear: Catharsis and Resistance in Military-Themed Video Games. In Huntemann, N. B. and Payne, T. M. (Eds.), *Joystick Soldiers: The Politics of Play in Military Video Games*. New York: Routledge, pp. 223–236.

Leonard, D. (2004) Unsettling the Military Entertainment Complex: Video Games and a Pedagogy of Peace. *Studies in Media & Information Literacy Education* 4(4), 1–8.

Makuch, E. (2015) 175 Million Call of Duty Games Sold to Date, Still Fewer Than GTA. *Gamespot Website*. Available at: http://www.gamespot.com/articles/175-million-call-of-duty-games-sold-to-date-still-/1100-6426188/ (Accessed 15 February 2015).

Paterson, M. (2009) Haptic Geographies: Ethnography, Haptic Knowledges and Sensuous Dispositions. *Progress in Human Geography* 33(6), 766–788.

Payne, T. M. (2010) "F*ck You, Noob Tube!" Learning the Art of Ludic LAN War. In Huntemann, N. B. and Payne, T. M. (Eds.), *Joystick Soldiers: The Politics of Play in Military Video Games*. New York: Routledge, pp. 206–222.

Penney, J. (2010) 'No Better Way to "Experience" World War II' Authenticity and Ideology in the Call of Duty and Medal of Honor Player Communities. In Huntemann, N. B. and Payne, T. M. (Eds.), *Joystick Soldiers: The Politics of Play in Military Video Games*. New York: Routledge, pp. 191–205.

Pink, S. (2013) *Doing Visual Ethnography* (3rd edition). London: Sage.

Power, M. (2007) Digitized Virtuosity: Video War Games and Post 9/11 Cyber-Deterrence. *Security Dialogue* 38(2), 271–288.

Rose, G. (2012) *Visual Methodologies: An Introduction to Researching With Visual Materials* (3rd edition). London: Sage.

Ruddock, A. (2000) *Understanding Audiences: Theory and Method*. London: Sage.

Salter, M. B. (2011) The Geographical Imaginations of Video Games: Diplomacy, Civilization, America's Army and Grand Theft Auto IV. *Geopolitics* 16(2), 359–388.

Schott, G. R. and Horrell, K. R. (2000) Girl Gamers and Their Relationship With the Gaming Culture. *Convergence: The International Journal of Research into New Media Technologies* 6(4), 36–53.

Schrøder, K., Murray, C., Drotner, K. and Kline, S. (2003) *Researching Audiences: A Practical Guide to Methods in Media Audience Analysis*. London: Arnold.

Schulzke, M. (2013) Rethinking Military Gaming: America's Army and Its Critics. *Games and Culture* 8(2), 59–76.

Shaw, I.G.R. (2010) Playing War. *Social & Cultural Geography* 11(8), 789–803.

Shaw, I.G.R. and Warf, B. (2009) Worlds of Affect: Virtual Geographies of Video Games. *Environment and Planning A* 41(6), 1332–1343.

Simpson, P. (2011) 'So, as You Can See . . . ': Some Reflections on the Utility of Video Methodologies in the Study of Embodied Practices. *Area* 43(3), 343–352.

Sisler, V. (2008) Digital Arabs: Representation in Video Games. *European Journal of Cultural Studies* 11(2), 203–220.

Stahl, R. (2010) *Militainment, Inc.: War, Media, and Popular Culture*. New York: Routledge.

Taylor, J. (2011) The Intimate Insider: Negotiating the Ethics of Friendship When Doing Insider Research. *Qualitative Research* 11(1), 3–22.

Taylor, T. L. (2009) The Assemblage of Play. *Games and Culture* 4(4), 331–339.

Thomson, M. (2008) Military Computer Games and the New American Militarism: What Computer Games Teach Us About War. Thesis Submitted to the University of Nottingham for the Degree of Doctor of Philosophy.

Thornham, H. (2011) *Ethnographies of the Videogame: Gender, Narrative and Praxis*. Farnham: Ashgate.

26

PHOTO-ELICITATION AND MILITARY RESEARCH

K. Neil Jenkings, Ann Murphy and Rachel Woodward

In this chapter, we explore the utility of photo-elicitation as a visual research method when applied to military topics. As this section of this volume illustrates, there are a number of visual research methods of utility for military research. Here, we focus on photo-elicitation to argue for its utility for researchers interested in military identities and subjectivities. Photo-elicitation uses photographic images in interview situations to explore life-worlds and experiences, including places and landscapes. It has proven utility across the social sciences as both a powerful means by which life-worlds and phenomena can be understood in the present, and also as a method for exploring the construction of memories and interpretations of past events with reference to photographic records. Its utility in military research contexts derives from the inherent capacity of the method itself, but also because of the specificities of military social life and culture which, as we go on to explain, make the exploration of certain military phenomena amenable to exploration using this method. In this chapter we draw out some of those mutually reinforcing factors to discuss how and why the photo-elicitation method becomes valuable when developing a research methodology for studying military phenomena.

It is now generally accepted that over the past two decades there has been a 'visual turn' across the social sciences which, among other things, has pointed to the significance of the visual as a focus and tool of research. While it could be argued that the use of visual data has always had a place in the social sciences, not least for illustrative purposes, it is notable that visual artefacts have themselves become the central data of research analysis. This visual turn has become consolidated and legitimated over the past ten to fifteen years with the development and formalisation of various visual research methods (see Banks, 2001; Rose, 2001; Pink, 2012, 2013). Taking the visual seriously has developed in different ways in different academic disciplines where both the use of visual data as a resource and the conceptual approaches towards the use of visuality and visualisation vary. As befits military-related research, which is itself very cross-disciplinary, we don't dwell here on the precise differences between disciplinary approaches (for further detail, see Margolis and Pauwels, 2011). It is worth noting, however, how the visual turn has produced multiple approaches to the ways in which images, visual media and visual research methodologies have proceeded in the social sciences. As other chapters in this section show, this range has enriched the study of military phenomena. What we want to address here is a specific approach to the use of visual data in military research – the technique of photo-elicitation – in order to illustrate its utility in exploring the identities and subjectivities of military personnel. We start with a brief overview of photo-elicitation in its various forms as a visual methodology in social science. We then go on to explore why it might be particularly useful in military research contexts where access to certain phenomena is often restricted, and we illustrate this drawing on two case studies from our own work on soldiers' identities and on army wives and their understandings of landscape. We draw conclusions about the potential of photo-elicitation as an interview stimulus and a potential occasion for visual data collection.

Photo-Elicitation as a Research Method

Photo-elicitation as a research method is distinct from the use of photographs for illustrative purposes (whether in academic, journalistic or other contexts) in two ways. First, as part of a methodology, the use of photo-elicitation is framed deliberately as a research method, with all the implications associated with methodological rigour in its deployment as, or part of, a data collection strategy. This includes attention to the details of research objectives, questions of methodological design, the utility of photographs in addressing defined research questions and fitting analytic frames, and consideration of the use of visuals in the communication of research findings. Second, as a qualitative methodology, the explicit intention of the use of photo-elicitation is to facilitate theory-building about social phenomena according to the principles of inductive reasoning. The use of visuals in and as part of a research interview, therefore, should be undertaken with awareness of both methodological and conceptual utility.

Early expositions of the method come from anthropologists' attempts to engage with their respondents in cultural contexts very different to those normally inhabited by the researcher (Collier and Collier, 1986; Harper, 1986). A key driver in the use of photo-elicitation is a desire to move beyond the limits of the interview-based method, where the interviewer is reliant on the memory and descriptive verbal powers of the interviewee, and where the power dynamics of the interview encounter can be heavily weighted in favour of the interviewer, with the interviewee as a more passive respondent (Van House, 2006). The utility of photo-elicitation has been noted in particular in interview settings with respondents who may experience more traditional interviews as limiting or even exclusionary (Epstein et al., 2006). Photo-elicitation, however, is not simply the extension of the traditional research interview through the inclusion of photographs. As Pink (2004) notes, and as we explore later, there is a reflexivity inherent in the method, in that both the researcher *and* respondent engage with visual materials; the method blurs the distinction between data collection and data analysis (Jenkings et al., 2008). Above all, the power of this method arises from its ability to facilitate the establishment of communication between two people on the basis of a shared engagement with an image (Harper, 2002; Hurworth et al., 2005). Although we should be cautious about overstating the power of a still photographic image to bridge different worlds of understanding, in the cases we go on to describe, the methodology can facilitate levels of communication about the self-perception of identity by respondents which might otherwise be difficult to achieve via verbal interaction alone.

Photo-elicitation approaches might include using interviewer-generated photographs, where the interviewer provides the images for discussion with the respondent; participant-generated photographs taken specifically for research purposes by respondents; and participant-generated photographs involving respondents' own existing photographs (auto-driven). In this chapter we discuss the two latter types of participant-generated photo-elicitation.

Photo-Elicitation in Military Research

In this section, we describe two rather different uses of photo-elicitation in military research contexts. In both, the method was deployed in order to investigate ideas around militarised identities and understandings, but the central research questions guiding each project, and the target groups of respondents for each were quite different. Underpinning the decision to use photo-elicitation in both projects was an understanding of the potential utility of the method for reaching an understanding of the lived experiences of military and militarised lives beyond the researchers' own specific lived experience. Collaborative exploration by respondent and researcher of the details captured and displayed in the images was used to develop an understanding of these lives and experiences. In both cases, the researchers conducting these photo-elicitation interviews had some personal experiences of the military worlds which they were setting out to explore, so were able to bring this knowledge to bear on the discussions. What was also evident in both cases was that it was felt, during both fieldwork and the process of data analysis, that the method gave respondents a voice in the analysis, and that this facilitated the drawing of conclusions in ways that a straightforward interview (or indeed questionnaire) would not have done. This was felt to be particularly significant for these two groups of respondents in slightly different ways. One group comprised serving and former British Army soldiers and Royal Marines, and the method enabled exploration of ideas about the lived experience of soldiering which the research team, and the respondents themselves, understood to be missing from wider social discourse about military lives and careers. The other group included the spouses (all wives) of military personnel, whose engagements with the militarised landscapes in which they lived were understood by the researcher to have been absent from military sociological research. In this section, we introduce each of the projects in turn, and discuss the photo-elicitation methods used. We then go in on the final section of this chapter to discuss our broader conclusions about the merits of photo-elicitation in military research.

Exploring Soldiers' Identities

This project[1] explored how both military personnel's individual identities and media representations of military personnel were negotiated by personnel working in the British armed forces. Here, we limit our discussion to the research on individuals' military identities (see Woodward et al., 2009 for discussions on print media representations). The method chosen to do this was auto-driven photo-elicitation. This choice was made on the basis of some conversations about military identities with military and ex-military personnel in our teaching and other research activities. In one conversation in particular, an individual previously unable to articulate and communicate his ideas to his satisfaction immediately became more animated, focused and expansive when able to talk with reference to his photographs. This serendipitous experience led to the fine-tuning of the original research methodology to facilitate photo-elicitation as the principle method prior to the commencement of data collection. We raise this here to make a point about how important it can sometimes be, in research terms, to pursue paths which 'feel' right and to have the flexibility with which to adapt to the emerging realities of data collection and 'access' to a phenomenon.

Military culture is an intrinsically visual culture and official photography for administrative, ceremonial and operational reasons is a ubiquitous practice. Alongside this in soldiers' daily working lives, which frequently involve deployments, the taking of personal photographs to capture places and events, to record activities and to register the presence of particular people is commonplace (see Woodward et al., 2010). Additionally, military photographs, both formal and informal, traditionally adorn military family domestic spaces, and we were confident that interviewees would not struggle to find photographs. We put a call out (via local newspapers) for participants willing to be interviewed and able to bring to the interview up to ten photographs (a number determined on the basis of a pilot interview) which spoke in some way to their military life and identity as a soldier. We completed 16 interviews with serving and former British Army soldiers and Royal Marines, structured around discussions of their ten chosen photographs. The interviews were recorded and transcribed in full, and scanned copies taken of each of the photographs discussed.

Our research on soldier identities allowed us to make three contributions pertinent to military sociological debates. The first of these concerns the conceptualisation of identity where the research confirmed for us the necessity of understanding military identities as rooted in doing rather than with reference to essential categories of being. Much of the pre-existing research on military identities (including our own) had taken social categories – such as gender, sexuality, class, ethnicity – and had sought to apply these to military contexts. Our respondents told a different story and drew on a very different conceptualisation, articulating ideas about military identities as rooted in activities and actions, emergent and fluid as processes of continual constitution and performance. Military identities, we concluded, could be understood as being based on ideas about the professional skills and expertise of soldiers; as being strongly established through ideas of fictive kinship within bonded groups (at a range of scales); and as being developed through the presence of the military operative at military events, including both active operations and more mundane or ceremonial public occasions (see Woodward and Jenkings, 2011). Photographs that respondents brought to the interviews were not ones that had been created for the research project, as is the practice in much photo-elicitation, but were preexisting images in the respondents' own collections. These were often taken by the individual themselves, but others were taken by colleagues at the time or by regimental photographers, and a few respondents also included newspaper images and postcards. While the majority were personal photographs there was often some uncertainty as to who had taken the photograph and with whose camera it had been taken. Most of the photographs presented were hard copies, but digitisation and mobile technologies might well now produce onscreen images and even greater uncertainty as to who the original photographer was. In our sample, photographs were kept in albums and scrapbooks, or were stored away, catalogued or uncatalogued, or were presented to us in the frames in which they were displayed in interviewees' homes. The role each image played in individuals' lives varied greatly. Indeed some participants noted photographs of significance which they could not bring themselves to look at, never mind bring to an interview and discuss. The photographs covered a wide range of subjects, from the more ubiquitous passing-out parade to a dirty toilet in a transit camp. Significantly, just as the photographs helped facilitate discussion, they could often only be made sense of through that discussion – an important issue for image-based research. In many of the interviews, although not all, the discussion of the photographs followed an individuals' chronology of service: basic training as a recruit, passing-out, exercises and specialist training (particularly overseas) and deployment in conflict zones. But rather than just a chronology of events, they were instead discussed as representations of achievement and skills development. Thus identity in these photographs was not just about having membership of a particular group, but about membership which was representative of professionalism and its performance. Such performances might have included passing a commando test, an arctic or jungle survival course, patrolling the streets of a conflict zone, or facilitating the mobility of armoured vehicles. The skills, locations and performances varied, but their importance did not. Of course not all the photographs were of formal or working events, and representations of comradeship with our interviewees' peers, and of events of family significance, were also present.

The second debate to which our photo-elicitation work contributes concerns the role of photographs in the memory work undertaken by former military personnel to confirm and consolidate particular ideas about what being a soldier might mean in the present (Woodward et al., 2011). It is significant here that a large proportion of our respondents were former military personnel, and thus came to discuss at interview a military identity which was in the past. Essentially, the interviews provided an opportunity for respondents to reinvoke a past military identity in the present, with photographs confirming a current subjectivity as the product of past experience. Photographs, it seemed, provided a means to 'contain' memories. This was apparent in the most basic and obvious way in that photographs provided a repository for memory, and were a means by which things, events and people could be remembered in the present. In some ways this is akin to the ubiquitous aide-memoire in the military itself. Photographs, however, also contained memories in the sense that they made them safe, setting limits around access to and the use of photographs, including storage, display and viewing practices, and rendering safe memories that might potentially be traumatic or difficult. This process of memory work and containment was clearly work done actively, in the present, making memories liveable and appropriate to an identity of an individual in the present, such that the individual could accommodate a past that might have contained difficult, challenging or traumatic episodes. Significant to this exploration of memory work was a wider debate about the nature of public narratives of war, the ways in which these are constructed, articulated and developed, and the role of individual memory in contributing to that process. This, as Ashplant et al. (2000) note, is a complex process. What the exploration of photographs and memory work did for our research was to provide a clue to the processes through which ideas about armed conflict as personally experienced and remembered might come to circulate beyond the personal and start to inflect wider public narratives about the nature of war.

The third debate to which the use of photo-elicitation in this study contributed concerned the utility of the method in the co-construction of knowledge (Jenkings et al., 2008). The nature of the photo-elicitation interview encounter involved two people simultaneously viewing the same image. A wide range of details would be discussed, starting with the contents of the image itself, and broadening out to include reflection on its origins (who took it), its history of use and display, and its current position in its owner's life. Necessarily, this involved a two-way exchange rather than the more rigid question-and-answer style synonymous with a structured or semistructured interview. What became apparent during these exchanges was an ongoing reflexivity on the part of the respondent *and* the researcher, where together they worked towards a shared understanding of what that image might mean in more abstract terms. In short, analytic understanding was begun by the research team, not as a later phase of analysis of image and interview text long after the interview had finished, but as an inherent and inseparable part of the interview itself involving the respondent. The analysis was co-produced and interactional, drawing on the reflexivity of the researcher and respondent, rather than solely on the postinterview reflections. Photo-elicitation, then, is a methodology that provides for collaborative interaction with the respondent not just in the process of data collection, but also in the process of analysis and theory construction. It is also interesting to consider the influence of the *military* context in which this use of photo-elicitation took place, and we return to this point in our conclusion.

Army Wives and Military Landscapes

The second use of photo-elicitation which we discuss here comes from a project[2] which explored how army wives understand the military landscapes in which they live. The choice of a visual methodology was an obvious one. Landscapes are inherently both visual and spatial, and are both viewed and experienced. While debate may take place about the ways and means by which landscapes' visual and experiential capacities might be explored and understood, their visuality and phenomenology cannot be contested (for further discussion on approaches to landscape, see Wylie, 2007). By taking a landscape approach, it was reasoned that it would be possible in this research to explore both the very great range of military landscapes encountered by army wives, and the diversity of ways in which those landscapes could be engaged with by the women in question (see Woodward, 2004, 2014). The choice of a methodology was also political, in the sense that we were very alert to the invisibility of military wives' own accounts in public (and much academic) discourse about military wives, the assumptions made (not least in defence and related policy) about their attitudes and expectations, and the diversity of ways in which this group experience life as partner to a soldier. In other words, the methodology had to recognise and respect the agency and voice of this often highly marginalised group. The methodology had to take into account a range of issues to do with rigour, while also ensuring that connections could be maintained between, on the one hand, army wives' engagements with the military landscapes in which a great many live their lives, and on the other hand, their ideas about their lives, identities and experiences as civilians. It was felt significant that the wives should not be approached as passive dupes or 'cultural dopes' (Garfinkel, 1967) to the military machine, or as agents only for the reproduction of the soldier on home turf, but as women with agency. In speaking about their identities as wives, our respondents drew on vocabularies which articulated ideas with reference to a spatiality (and awareness of where things occurred) and place ('I do this here'; 'I am like that there'; 'my husband has a different relationship to this place or that place'). They also communicated an understanding of military landscapes as spaces which they had to constantly 'read' in order to negotiate between their own ideas about how they wanted to live, and the demands of the military as an institution. The research, then, looked at this negotiation of identity and at how this was constituted and expressed by these women with reference to the military landscapes in which they lived.

The data collection process comprised three different activities undertaken in sequence. First, respondents were given a compact, disposable 24-exposure camera and asked to take photographs which captured for them something of their experiences of, and engagements with, the landscapes around them, as army wives. In practice, only the three pilot participants used the disposable cameras provided; the rest preferred to use their own cameras. They were given complete freedom to identify and capture the images themselves. One participant chose to include photographs already taken prior to the research. Once the images had been taken, the cameras and SD cards were sent back to the researcher (Ann) and the photographs printed. The logistics of returning photographic images sounds straightforward, but could be more problematic than would initially seem (with items lost in the post and digital files liable to corruption). Second, respondents were asked to record their thoughts and observations about the images they were taking in a diary to be shared with the researcher. Respondents were provided with a simple exercise book to do this, although not all chose to record their thoughts in this way and one did not do the diary entry at all. Third, after reviewing the completed (developed) sets of photographs and the completed research diaries, an interview was conducted with each respondent in her own home, with the interview structured around the images that the respondent had provided.

The images from the nine women interviewed varied enormously, reflecting the respondents' domestic situation, the rank and job of their spouse, their own employment status and pattern of work, the age of children, and the nature of the military landscape in which the respondents lived. A 'punctum' in the Barthesian sense, the defining feature of a photograph, was in many photos quite obvious, but not in all (Barthes, 2000). Following Hall (1986), a picture may be worth a thousand words, but only when subject to analysis. The photographic data is thus inactive, frozen in time and space once captured; the ability of that image to develop into something active happens through the interpretative process and requires an elaboration of the intended meaning. A heuristic was used to crudely categorise the women on the basis of their approaches to their lives as army wives, and the images, explanatory texts and interviews provided the empirical data which permitted the reading of these women in this way. Three of the women constructed narratives of their married lives in reference to the practical strategies they employed to cope with the demands exerted by the army's requirements of their husband, along with the demands placed on them by family. Three women constructed identities which included resistance to, and the possibility of opposition to, the boundaries set for them by the institution of the military. In a sense, these women constituted pockets of resistance to militarism's effects, and commented on the tensions that these generated with their spouses and with those in military authority. The final three articulated their position as steadfast wives, recognising that the demands of their spouses' careers meant, for them, the subsuming of their own individual independence and aspirations. Of the many conclusions about military wives' engagements with landscape which the use of the photo-elicitation method facilitated, we focus here on two which speak directly to the use of the method.

The first of these concerned the way in which the method enabled an exploration of the tensions between wives and the military, broadly defined. Because the majority were stationed in army housing, often either on base or in proximity to a military base, photographs included images of respondents' local environments: base gates, perimeter fences, the regularity of patch housing, or planting and landscaping features. In turn, such images prompted explorations of the emotions played out through these spaces. A particularly significant idea which emerged through consideration of the visual data was a sense of the dynamic between the respondent and military authority. This in turn led to discussions of the playing-out across space of the conflicting demands of family and the military (classically described by Mady Wechsler Segal [1986] as the interchange between two greedy institutions). It also prompted discussions about the negotiations between wives and their husbands about the dynamic around the demands of the military within spaces of the family home, and demands of the family within the military spaces of a husband's employment. There was, for one woman in particular, a dynamic which she tried to capture in her photographs between ideas of safety and danger, reflecting Tivers's (1999) assessments of the spaces of military bases as simultaneous places of stress and safety. Images were captured by participants and used in discussion and diaries to reflect controls over the personal movement experienced by wives as a consequence of demands of their husbands' work. Restrictions on mobility extended to the ability for some women to engage in local labour markets and social networks. Some women discussed the strategies they used to take the images, and many discussed the difficulties of capturing a photograph of something close to home and part of their daily life which was party to militarily decreed photography restrictions – the main gates of an army base are one example. As members of a 'total institution' (Goffman, 1986) in which all actions are potentially accountable, some participants expressed a concern that taking photographs would be 'discussed' within the camp and even brought to the attention of their husbands. However, the fact that their actions were accountable as part of a university project was reported by respondents as enabling: the legitimacy of their activities conferred by their origins in a university research project was felt to be valuable.

The second conclusion we would like to emphasise concerns the ways in which the method allowed the respondents to talk about scales and sites of military activity and military influence. Photographs enabled participants to do this in ways which spoke directly to observations around geopolitics and intimate, everyday scales (see Sylvester, 2010). A particularly stark example is an image of an aircraft pilot's flying suit, hanging in the space of a domestic hallway, surrounded by the paraphernalia of children's outside activities – raincoats and wellington boots. For the respondent, the dust which fell from this garment – dust from Afghanistan – felt like an invasion of the domestic sphere and was a continual reminder of their lack of control of the boundaries between the familial and military. For another, an image of pots containing annual plants, arranged in the space of a backyard, spoke of her attempts to personalise a space – army housing – which was anything but personal and permanent. There was a tension, then, in a number of photographs which spoke to this dynamic between the work of the spouse on military missions, and the spaces of domestic life back home.

In terms of the utility of this method, we note the ways in which this particular use of photo-elicitation seemed to indicate the limits of the method. There were undoubtedly benefits to the method. In dealing with photographic images, this method employs a practice (the taking and sharing of photographs) readily used and understood in daily life. The use of disposable cameras which could be sent to the researcher after completion was seemingly very straightforward. In taking pictures of everyday life as it is lived, the respondents were providing data which, because it was in so many cases quite mundane, represented daily life much better than more exceptional or dramatic images might have done (Harper, 1986). Yet in asking respondents to take photographs, to record their thoughts and ideas in a diary, and to discuss image and text in an interview, we were aware of the significant commitment being made by people with busy lives. The small sample size generated for this project was in part a reflection of the difficulties of respondents in finding the time and personal capacity to engage with the research.

Discussion

In this chapter, we have discussed the use of photo-elicitation methods and their utility in facilitating explorations of identity among military personnel and military wives. In both cases, the use of images was guided by Berger's observation that photographs in themselves do not preserve meaning. Rather, in quoting Susan Sontag's (1978) *On Photography*, Berger states:

> They offer appearances – with all the credibility and gravity we normally lend to appearances – prised away from their meaning. Meaning is the result of understanding functions. 'And functioning takes place in time, and must be explained in time. Only that which narrates can make us understand.' Photographs in themselves do not narrate.
>
> *(Berger, 1980: 51)*

In our use of photographs for photo-elicitation, we have sought to elicit narratives from participants with specific research objectives in mind. Photographs have to be contextualised: images are situated in space and time at the moment of production and reception. Thus with both of our projects, the images themselves were insufficient for our purposes in exploring identities and subjectivities. Rather, the interpretation of the phenomena by respondent through their choosing and/or taking photographs, and the collaboration of respondent and researcher in developing a contextualising narrative, was what led to a coherent understanding of the phenomena.

The primary indicator of whether or not a particular research method is successful is whether it enables the production of data and allows subsequent analysis which enhances existing debates or enables new conclusions to be drawn. Photo-elicitation certainly enabled this in terms of the exploration of military identities. The point we reflect on here is *why* it worked: what was it about the method which enabled the research team to do what they did? We have two suggestions here. First, taking and looking at photographic images (whether our own or those of others) and entering into a dialogue about images with another person are both common practices and part of ordinary social interaction. There is something very normal about talking about pictures, and the skills required to do it are ones which both interviewee (and interviewer) are very likely to have and to be comfortable using. Research interview situations themselves are anything but everyday events, and for most participants this was their first involvement with a research interview. How to 'correctly participate' was not necessarily clear to them and was actually learned as the interview progressed. It is easily forgotten that a research interview – the formal conversation to be had with a stranger (an academic one at that) about something potentially quite intimate, which is to be recorded, transcribed and pored over – can be an intimidating event. However, discussing photographs, a mundane social practice, can mitigate the strangeness of a research interview. Moreover, it can be (potentially) empowering because of this, in that the respondent has greater facility to lead the discussion on their own terms. We think this is one reason why the method worked so well. Furthermore, the use of photographs allows respondents to use other forms of communication other than just the spoken word in communicating their experiences to the researcher. This is especially the case when respondents are not orally orientated but may have greater confidence in their visual skills. For example, one of the participants from the army wives project deliberately took a shot out of focus to convey an idea about her relationship with her husband and his job that was not easy for her to convey verbally. In photo-elicitation, participants have agency with regards to their choices of images for sharing and thus agency in terms of research project input. This agency is something generally held to be beneficial.

A second explanation for the utility of the method relates to visuality and visual cultures within the armed forces. We speculate that there may be things about the military context which made photo-elicitation a particularly appropriate methodological tool for the exploration of identities and subjectivities, including their landscape dimensions. The British armed forces (and quite possibly, military forces more widely) have a very visual culture. Armed conflict itself involves visual ability and visualisation for the pursuit of military objectives; the development of tools, tactics and strategies to render visible that which is obscured by distance or concealment and may thus be invisible, constitutes a core part of the business of armed forces (see MacDonald et al., 2010). Within armed forces themselves, the task of preparing personnel to execute lethal force through continual processes of training are contingent on continuous observation for the purposes of judgement, critique, development and improvement. The observation and visual documentation of such processes is central to those purposes. In their construction, articulation and reproduction of ideas about what they *are*, armed forces engage in processes of representation, both for internal consumption and external display, and again, these are visible and visualised – hence the significance to individuals in their photograph collections of official photographs showing personnel in dress uniforms, and the seeming ubiquity of certain types of photographs showing key rites of passage. That visibility and those practices of visualisation may be subject to censorship. Military authorities do this, as do individuals – our research participants self-censored their images not only for personal reasons or to accord with specific sensibilities around what could and could not be shared, but also to adhere to military security requirements. Yet, beyond this caveat, in using photo-elicitation within military contexts, we were using a phenomenon – the photographic image – which has significant practical and cultural use in that context. In military contexts, it is normal to take and view visual images as part of the day-to-day practice of the work of armed forces. In military contexts, it is also regular practice to share and review visual images with others, as part of the job or as a regular part of social life. We speculate here too on the utility of this regularised practice in facilitating research, particularly in cases where the researcher may have little previous experience of military contexts and may be working with assumptions that benefit from creative input by participants.

Notes

1   Woodward, Rachel, Winter, Trish and Jenkings, K. Neil (2006–2007) *Negotiating Identity and Representation in the Mediated Armed Forces*, ESRC reference RES-000–22–0992.
2   Murphy, Ann (2007–2013) *Military Landscapes: Place, Space and Identity Issues for Army Wives*, PhD candidate, School of Geography, Politics and Sociology, Newcastle University, ESRC reference ES/F021968/1.

References

Ashplant, T. G., Dawnson, G. and Roper, M. (2000) *The Politics of War Memory and Commemoration*. London: Routledge.

Banks, M. (2001) *Visual Methods in Social Research*. London: Sage.

Barthes, R. (2000) *Camera Lucida: Reflections on Photography*. London: Vintage.

Berger, J. (1980) *About Looking*. London: Readers and Writers Publishing Cooperative.

Collier, J. and Collier, M. (1986) *Visual Anthropology: Photography as a Research Method*. Albuquerque: University of New Mexico Press.

Epstein, I., Stevens, B., McKeever, P. and Baruchel, S. (2006) Photo Elicitation Interviews (PEI). *International Journal of Qualitative Methods* 5(3), 1–9.

Garfinkel, H. (1967) *Studies in Ethnomethodology*. Englewood Cliffs, NJ: Prentice-Hall.

Goffman, E. (1986) *Asylums*. Harmondsworth, Middlesex: Penguin.

Hall, E.T. (1986) Foreword. In Collier, J. and Collier, M. (Eds.), *Visual Anthropology: Photography as a Research Method*. Albuquerque: University of New Mexico Press, pp. i–xvii.

Harper, D. (1986) Meaning and Work: A Study in Photo Elicitation. *Current Sociology* 34(3), 24–45.

Harper, D. (2002) Talking About Pictures: A Case for Photo Elicitation. *Visual Studies* 17(1), 13–26.

Hurworth, R., Clark, E., Martin, J. and Thomsen, S. (2005) The Use of Photo-Interviewing: Three Examples from Health Evaluation Research. *Evaluation Journal of Australasia* 4(1 and 2), 52–62.

Jenkings, K.N., Woodward, R. and Winter, T. (2008) The Emergent Production of Analysis in Photo-Elicitation: Pictures of Military Identity. *Forum Qualitative Sozialforschung* 9(3), Article 30.

MacDonald, F., Dodds, K. and Hughes, R. (Eds.) (2010) *Observant States: Geopolitics and Visuality*. London: Routledge.

Margolis, E. and Pauwels, L. (Eds.) (2011) *The Sage Handbook of Visual Research Methods*. London: Sage.

Pink, S. (2004) Visual Methods. In Searle, Clive, Gobo, Giampietro, Gumbrium, Jaber F. and Silverman, David (Eds.), *Qualitative Research Practice*. London: Sage, pp. 391–406.

Pink, S. (Ed.) (2012) *Advances in Visual Methodology*. London: Sage.

Pink, S. (2013) *Doing Visual Ethnography* (3rd edition). London: Sage.

Rose, G. (2001) *Visual Methodologies*. London: Sage.

Segal, M.W. (1986) The Military and the Family as Greedy Institutions. *Armed Forces & Society* 13(1), 9–38.

Sontag, S. (1978) *On Photography*. London: Allen Lane.

Sylvester, C. (2010) *Experiencing War.* London: Routledge.

Tivers, J. (1999) The Home of the British Army: The Iconic Construction of Military Defence Landscapes. *Landscape Research* 24(3), 303–319.

Van House, N. (2006) Interview Viz: Visualisation-Assisted Photo Elicitation. In Olson, Garry M. and Jeffries, Robin (Eds.), *Extended Abstracts, Proceedings of the 2006 Conference on Human Factors in Computing Systems.* CHI 2006, Quebec, Canada, 22–27 April. New York: ACM Press, pp. 1463–1468.

Woodward, R. (2004) *Military Geographies.* Oxford: Blackwell.

Woodward, R. (2014) Military Landscapes: Agendas and Approaches for Future Research. *Progress in Human Geography* 38(1), 40–61.

Woodward, R. and Jenkings, K. N. (2011) Military Identities in the Situated Accounts of British Military Personnel. *Sociology* 45(2), 252–268.

Woodward, R., Jenkings, K. N. and Winter, T. (2011) Negotiating Military Identities: British Soldiers, Memory and the Use of Personal Photographs. In Hall, K. and Jones, K. (Eds.), *Constructions of Conflict: Transmitting Memories of the Past in European Historiography, Culture and Media.* Bern: Peter Lang, pp. 53–71.

Woodward, R., Winter, T. and Jenkings, K. N. (2009) Heroic Anxieties: The Figure of the British Solider in Contemporary Print Media. *Journal of War and Culture Studies* 2(2), 211–223.

Woodward, R., Winter, T. and Jenkings, N. (2010) 'I used to keep a camera in my top left-hand pocket': British Soldiers, Their Photographs and the Performance of Geopolitical Power. In MacDonald, F., Dodds, K. and Hughes, R. (Eds.), *Observant States: Geopolitics and Visuality.* London: Routledge, pp. 143–166.

Wylie, J. (2007) *Landscape.* London: Routledge.

27

VISUALISING THE INVISIBLE

Artistic Methods Toward Military Airspaces

Matthew Flintham

By England's standards, this place is about as remote as it gets, with salt marshes, creeks and tributaries stretching out before me toward the horizon and eventually, the sea. I am here to take photographs and sound recordings of something that doesn't exist: the restricted block of sky known as EGD207, a giant radial wedge of airspace that stretches up to 23,000 feet and spans most of the square-jawed bay above East Anglia called The Wash. Alternatively known by the British military as RAF Holbeach, the area is one of a number of aerial bombing ranges in the UK used for target practice by the Royal Air Force and fast jets from other NATO alliance countries. Except today, standing on the raised sea wall bank, there is no obvious evidence of military activity, only the cries of waders and seabirds and a churning weather system that plays havoc with camera exposure levels. Roving arcs of light pierce the cloud layers and interrogate the young crops below, giving the landscape a theatricality that would not be out of place on the midwestern plains of Kansas. I pass a solitary walker who warns me that there are men with guns on the sea wall about a half a mile or so in the distance. All I can see is two trucks parked askew on the high banks near a range observation tower, so I shoulder my tripod (which I darkly speculate could be read by a distant observer as a shoulder-mounted weapon) and head over there to take a look. Two giant Ford Super Duty trucks are being unloading by a group of furtive men in nondenominational camouflage and wrap-around shades. I see no weapons or badges but plenty of military kit including radios, flares, utility vests and maps in ziplock bags. I say 'Hi.' They say 'Hi.' Clearly, they are US soldiers tooling up for something, and I ask them if they are on exercise, which they seem to think is the funniest thing they've heard all day. 'Yeah, sorta, . . .,' they laugh. Weirdly, they are happy to pose for a couple of photographs while packing their utility vests and backpacks. 'What's going on?' I ask. 'Plane's comin'. . .' – and sure enough it does.

As I leave, watching the soldiers scurry over the brow of the sea wall and out of sight, a low, thunderous force rumbles under the cloud layer across the bay. A US Air Force (USAF) F-15 Eagle loops around the Wash and carries out a series of low passes over the bombing

357

range. The noise is excruciating and my audio recorder is peaking wildly, but I manage to capture a couple of minutes of usable sound before I have to leave. On the three-hour drive back to London, I consider the images and sounds gathered at the Wash, and how they might fit into a body of artwork I was planning for a 12-month artist's residency at Newcastle University. However, the events of the day, while mundane in so many ways, had seemed like a fracture in reality: I had witnessed a well-choreographed simulation of battle, a two-stage ground and air assault on a landscape that neither knew nor cared. Having briefly entered EGD207, I had also breached a zone of surreality.

Military Airspaces

This chapter details the process of making artworks that relate to military airspaces such as the one just described. I will try to be as frank and critical as necessary when describing methods of artistic practice in relation to the study of military geographies, but the reader is asked to temporarily defer any antipathy when I inevitably stray into the uncertain hinterland between the two. In fact, the very purpose of this chapter is to visit that hinterland, where it is possible for arts practices to engage, interpret, mimic and even subvert the spatialities and language of military activity.

At the Wash in February 2014, I was gathering material and ideas for an artist residency I had recently begun in the School of Geography, Politics and Sociology (GPS) at Newcastle University. The residency was funded by the Leverhulme Trust to collaborate with geographer Dr Alison Williams toward the making of a body of work relating to the visualization of militarised airspaces in the UK. Williams and I were keen to stress the importance of revealing and interpreting the paradoxical and increasingly politicised realm of militarised airspace in the UK. Independently, we had both been investigating, through academic and artistic means respectively, the enormous volumes of Restricted, Special Use, Low Flying, Air Tactical and Danger Area airspaces in use every day across the sovereign British territory, all of which have precisely delineated coordinates and heights, but which can also change form and/or collapse at any specified time (Williams, 2012). These spaces are managed and coordinated independently from civilian air routes but fit seamlessly around, under and over them like a three-dimensional jigsaw or a giant spatial mechanism with moving and collapsible parts. We both concluded that these airspaces are a surprising and sometimes problematic appropriation of the sky by the military and a manifestation of the British military-industrial complex that is woefully underresearched.

Airspaces are by their very nature invisible: we only see the mercurial atmospheric weather forms that pass through them and the air vehicles that depend on them so directly. Williams and I wished to reveal the very structures that make mass air travel possible, but more specifically we wanted to visualise and engage with those airspaces that resist civilian and commercial traffic: vast delineated volumes of space containing military training exercises, artillery testing, small-arms fire, static explosions, drone operations, air tactical training, airborne radar surveillance and any number of everyday lethal activities associated with martial training and defence. The scale and proliferation of these spaces seems to be a somewhat luxurious feature of a sovereign airspace that is otherwise intensively used by civilian and commercial air traffic[1] – an invisible expression of Britain's self-possessed, if subtly conspicuous, militarism.

Military land use and ownership accounts for about 1.5% of the total land mass of the UK (Flintham, 2014). However, despite the public availability of complex aviation maps detailing militarised airspaces across the UK, military use of the British skies is an almost unquantifiable reality, with military traffic constantly moving between huge restricted airspaces, military air traffic zones (MATZ) over airfields and landing strips, and designated low-flying areas. When we take both land and air use into account, there is a sense that much of this militarization has happened broadly in the period since the end of the Second World War and largely under the radar of public scrutiny. Civil and military airspaces are structured according to a set of esoteric military principles established during the Cold War, when the UK was secretly sub-divided into manageable defensible units and when the United States was endlessly rewriting its suicide pact (mutually assured destruction) with the Soviet Union. If the Second World War was the era when the skies of Britain became divided into sectors and when tiny aircraft were gingerly pushed across military planning tables, the Cold War was the period in which airspace became systematised as a technology for delineating vertical territorial boundaries and for managing and controlling the flows of military and commercial flights across nation-states and territorial waters.[2]

Today, both military and civil airspaces have reached a level of refinement and complexity in which thousands of aircraft can fly safely cross the skies, but within which the constant preparation for 'warfare is woven into the filigree of peace' (Hanssen, 2000: 102). I have writ-ten elsewhere that in order to measure and track the increasing militarisation of civil life and ensure martial accountability, it is essential to distinguish between what *is* military and what is *not* (Flintham, 2010). Moreover, there is a healthy stack of writing on the subject of how urban and rural space is created and mediated by hierarchies of state and commercial actors, and how the enclosure of public and 'common' space is a recurring strategy in land use and urban design (Graham, 2010; Minton, 2012). If studies of airspace (and particularly the militarisation of airspace) are to similarly take into account factors of power, control and subtraction from the commons, it would seem necessary to find ways of illustrating how these spaces work. As this chapter will describe, a useful research strategy in this field is to employ art and visualisation methods to illustrate, expose or reveal this complex, aerial architecture for critical analysis.

Revealing the Invisible and the Virtual

The more I understood the immaterial, invisible and virtual dimensions of militarism, the less my own photographic practice seemed to effectively describe the realities and paradoxes of contemporary training and defence. The relics and detritus of past wars may still litter the British landscape and provide alluring and iconic images, and may indeed contribute valuable evidence for historical and cultural analyses of conflict, but they tell us little of the emerging spaces of militarism and its current trajectory of rationalisation, automation, network-centric and asymmetrical warfare.[3] There are few examples of where photography alone is able to communicate the invisible and virtual dimensions of contemporary conflict and militarism. Subsequently, many artists are stretching contemporary art practices to interrogate these new realms by combining media, adopting performative, site-specific and clandestine investi-gative strategies, evoking bureaucratic and administrative protocols to mimic military proce-dures, and deploying military technologies to better understand surveillance and automated, machinic vision.[4] Indeed, many artists have been galvanised to act in response to abuses of powers in the wake of September 11, 2001, or by the more recent revelations of mass surveillance by the US National Security Agency (NSA) and the UK Government Communication Headquarters (GCHQ), or by the increasing use of drone attacks in Afghanistan, Pakistan, Somalia and elsewhere.

My own pursuit of the hidden and virtual forms of aerial militarisation began as a desire to uncover the mechanisms by which civilian life becomes incrementally militarised. However, being a visual artist I also followed a compulsion to engage with these complex spatial structures on a formal and aesthetic level – as vast, invisible edifices (some the size of whole counties) with endless potential as 'found objects' for cultural and artistic appropriation. Their forms are defined by the principles of utility, but also by the topography and contours of the land, and the invisible boundaries of sovereign territory. As such, they are akin to a purely functional architecture, the everyday infrastructure of engineering or perhaps even the most puritanical minimalism of high modernity, and yet they also raised unique conceptual paradoxes. How, for instance, is it possible for airspaces to exist in the minds of pilots and air traffic control as vital aerial structures for managing air travel in crowded skies, and yet have no physical or material presence whatsoever? It might be tempting to think of them as virtual environments, occupying an entirely fictional, immersive realm like the online virtual world, *Second Life*. They are, however, more akin to a mediated, 'coded' or augmented reality, one in which technology generates pathways and spaces which are then imbedded into the users' sensory experience of the 'real world' (Dodge and Kitchin, 2004). In this sense, airspace coordinates and geometries are lived and experienced, and share the same material value as 'real' spaces. Their existence may require a small leap of faith or even a suspension of disbelief by users, yet their risk value as real, functioning spaces is rarely called into question. But like religious faith, their existence and dogma is learned and their rules are observed.

By contrast, the range, intensity and exclusivity of activities within militarised airspaces add a whole level of complexity to the paradox described earlier, one which begs serious analysis and interpretation. Why, for instance, does Britain need over seventy militarised airspaces in its sovereign skies, and what activities connect them to ongoing conflicts around the globe? What can we learn from such invisible militarisation about the increasing systematisation, stratification and segregation of social space? It was in the spirit of these questions that I began the residency at Newcastle University, with the intention to both describe and interpret military airspaces as spatial and cultural phenomena that bridge the seemingly disparate geographies of peace and conflict.

Work in Progress 1: Groundwork

It was clear from the outset that the majority of the work made during the residency would be generated from two principle aerospace documents. The first, known as *ENR5: Navigation Warnings*,[5] is a spreadsheet which lists the names, exact coordinates, altitude and function of every single militarised and special use airspace in the UK. This is one of the few documents in the public domain from which it is possible to plot geographically accurate airspace footprints in geographic information system (GIS) programmes such as Google Earth, Google Maps, GRASS, and ArcGIS. The other document, *ENR 6-5-1-1: Chart of the United Kingdom*

Airspace Restrictions and Hazardous Areas,[6] is a high-resolution PDF published online by the National Air Traffic Services (NATS), which is a map of the entire UK with every restricted airspace colour-coded and plotted. It is probably the single most effective document describing the scale and proliferation of segregated airspace, and implies that that the skies of the UK are a highly regulated spatial mechanism, with volumes and cells constantly opening and closing at predetermined intervals. It is a complex, static chart, but when read with the lengthy appended notes also implies so much movement in space and time.

I began the residency with a single question: is it possible to translate these seemingly inauspicious items of air traffic ephemera into engaging artworks, to use them as urtexts for the development of dynamic, three-dimensional objects and images? I was certain that the documents could easily be decoded and rendered into other media as a form of visualisation. However, the idea of 'applying' a fine art methodology to this subject is somewhat problematic since fine art draws (directly or indirectly) on a canon of work and aesthetic theory stretching back centuries and/or enters a certain dialogue with other work being made in the contemporary arena. Fortunately, there is a rich historical precedent set by certain artists of the early modernist era who engaged forcefully with the new dynamic spaces and flows of industrialization, and the brutal mechanization of twentieth-century warfare. Many such artists felt compelled to develop abstract and conceptual techniques to describe the fusion of ideology and technology, and the role of the human body accelerated in space.[7]

The first task was to identify as many militarised airspaces in the UK as possible, so in the spirit of abstraction and appropriation, I imported the *ENR 6–5–1–1* airspace chart into a photo editing programme, isolated and extracted all the relevant elements and created a separate high-resolution file from which I could pull out individual airspace elements when I needed them.[8] By removing the individual elements from the master chart and then ranking them by size in the new document, I had created a kind of cartographic typology of military airspace, something which must go down as one the more obscure acts of visual research. This repository of airspaces will hopefully serve as a resource for making future two-dimensional graphic works.

Work in Progress 2: Graphic Works

My research and arts practices are predominantly concerned with studying architectural and landscape subjects as a means to elicit the tacit, often invisible power relations between people and places. This process often involves exploring different forms of cartographic representation or extrapolating visual material from very basic geographic data. With these methods in mind, I began to think about ways of fusing airspace coordinate data with images of the places they actually refer to, actively seeking out places in the UK with unusual militarised airspace and taking photographs of the surrounding landscapes. The following photographic journey around the UK was an attempt to capture the collision of the military and pastoral.

At the RAF Holbeach bombing range, described earlier in the chapter, the landscape is flat, featureless and unassuming (despite a prowling unit of US combat troops). Yet it also exudes a certain peril, at once latent in the endless tributaries and mudflats and imminent in the destructive hardware circling overhead. There is also an invisible volumetric structure hanging over the entire region, defining military occupancy and legitimising a tactical and strategic vision of the landscape.

Figure 27.1   The Wash: high angle dives and loft/toss bombing attacks.

Source: Flintham (2014), colour c-print.

Later, at a visit to the perimeter of Stanford Training Area (STANTA) near Thetford Forest, I encountered the very photogenic Devil's Punchbowl, a large circular depression in the ground reminiscent of a meteor impact crater. While it is almost certainly one of many ancient dolines or chalk 'swallow holes' in the area, there are numerous local legends attached to the site, including the almost obligatory extraterrestrial object falling from the sky. The correlation between virtual military airspaces and supernatural or unknown aerial events may at first seem tenuous, but opening a visual channel between them will, hopefully, challenge certain assumptions about the kind of invisible things we choose to believe in. On another trip, I attempted to commune with the complex airspace structure at MOD Shoeburyness (otherwise known as Foulness Island), but having been on the island a number of times before (as a PhD researcher), I knew that photography would not be allowed. As a weapons testing and demilitarisation facility, MOD Shoeburyness is actually managed by QinetiQ, a private sector defence and security organisation that very much values its privacy. Undeterred and still hoping for a good photo opportunity, I followed the perimeter fence of the MOD's mainland facility at Pig's Bay on the edge of Southend-on-Sea, which I knew to be within the island's overarching airspace structure. Instead, I encountered an extraordinary scene: over 100 black Christians of all ages in white robes were engaged in a mass baptism at the sea shore. The event was so unexpected and so at odds with my intension to photograph the nearby facility that it took me some time to actually believe what I was seeing. I simply sat on the beach watching the priest, his deep red robes flowing in the sea, administer the sacrament to one devotee after another. Were they drawn to this place for a particular reason, and were they aware of the military significance of the area? I was not inclined to photograph the scene, but somehow the event added an unforeseen complexity to my understanding of the plurality of militarised environments. It seemed astonishing that an event of such profound spiritual significance for those taking part could take place on the very edge of this secure weapons testing facility, where artillery shells arc over the estuary nearby, static explosions sound out over the bay and beyond, and groups of scientists are constantly fine-tuning the efficiency of Britain's deadliest ordnance. These disparate worlds may occupy adjacent and overlapping spaces, but both are part of the intricate matrices of place.

The heathland and quarry lakes to the east of the twin US bases of RAF Lakenheath and Mildenhall resound with tumultuous birdlife: Canada geese and swans splash down at leisurely intervals, and nebulous murmurations of starlings unfold and explode across the sky following the mysterious logic of biological 'phase transitions' (Mora and Bialek, 2011). In autumn, aerial life here is vibrant and often cacophonous, as is the descending whine of military turbofans and props as they power down for landing at the twin bases. Lockheed C-130 Transporters and Boeing Stratotankers are an almost mundane site in the skies as they spiral in and out of the MATZ that segregates most of the lower airspace from unauthorised aircraft. Again, the collision of the military and the pastoral is experienced as a weird congruence in the landscape, an uncanny accord.

Back in the studio, I tested making isometric, graphic representations of the airspaces relevant to each visited place, and transposed them over photographs of the landscapes that they permeate. Here, I hoped to reveal the invisible boundaries and military structures at work in the British countryside. For instance, by inferring that seemingly bucolic, idyllic places can *also* be highly regulated and often dangerous, militarised environments, I hoped to draw attention to the British landscape as a place augmented by virtual, machinic and electromagnetic technologies, and enacted as a simulation of conflict zones elsewhere around the globe. Landscape is always the unwitting recruit in any process of militarisation, and is often perceived in objectivised and strategic terms – an alternative, parallel reality of impending violence and potential destruction.

The process of making images that reflect this plurality in the landscape poses significant challenges. It quickly became clear that the airspace renderings I had been designing acquired an unexpected symbolic quality, as if the landscape images had been stamped with an esoteric, cartographic pictogram. For the moment this seems like an intriguing substitute for a phenomena that is essentially unrepresentable.

I have been on many image-gathering trips to military facilities around the UK and abroad, sometimes by invitation and other times employing more unilateral and surreptitious methods to observe or gain access, but it is always the unexpected events that leave the biggest impression. They are also the events that best serve to challenge our assumptions about power and where it resides in the landscape, and question the apparent exclusivity of military occupation and land use. Collecting such images and combining them with military spatial information was a way for me to confront and examine the systematisation of space and its effect on the intimacy and specificity of place. Furthermore, the value of visiting military landscapes lies in experiencing the peculiarity and multiplicity of such places, a notion which itself challenges those dominant or accepted narratives that so often colour our understanding of land use and ownership.

Work in Progress 3: Objects

Next, I began a process of three-dimensional cartography, building models of military air-spaces set on to Ordnance Survey maps (1:25,000 scale) laminated onto table-sized sur-faces (which resemble the planning tables used during military briefings to describe tactical and strategic operations).[9] The first attempt was a wireframe model of the airspace above STANTA in Norfolk, chosen because of its significant role in the complex of military spaces across East Anglia. This was also a good opportunity to test the viability of certain materials and working methods. It seemed important to try and envisage the airspace frame as a kind of vertical extension of the map, following the map's stylistic conventions (colours, thickness of lines, etc.) as closely as possible. I sourced a batch of 1.5mm lengths of silver steel (more rigid and self-supporting than even carbon fibre) which I then covered with red heat-shrink sleeving (red to match the colour of the Danger Areas on the NATS airspace chart). Using the coordinates on the *ENR5: Navigation Warnings* spreadsheet, it was then possible to plot the footprint of the airspace on the planning table map, and then mathematically determine the altitude of the volume in relation to the scale of the map.

For the next assemblage, I chose the Otterburn Army Training Area which covers nearly 250 square kilometres of Northumberland and is used for artillery firing and multiple launch rocket systems, small-arms training, helicopter and fast jet support among many other activities – all contained within the Danger Area airspaces known as EGD512 and EGD512A.

Figure 27.2   The Martial Heavens.

Source: Exhibition by Matthew Flintham at Ex Libris Gallery, Newcastle University, January 2015.

The model describes the airspaces in relation to the scale of the map and gives a sense of the vast scale in which military training operates. Moreover, the delicate and time-consuming process of assembling the wire frame model, with multiple armatures and supports to hold the gluing pieces together, proved to be a fascinating process, one which suggested other ways of representing objects or events on a geographical scale. The armatures suggest a kind of precarious architecture on the map surface, and extraneous bits of wire implying vectors and trajectory paths erupting over the Cheviot Hills. It now seemed possible to describe the spatial tumult characterised by military training areas where missiles, static explosions and aircraft bombard the landscape in a simulation of warfare elsewhere. For the moment, it is necessary to limit the scope of the models to Danger Area airspaces alone, but there is certainly potential to develop these structures in more expansive or abstract terms.

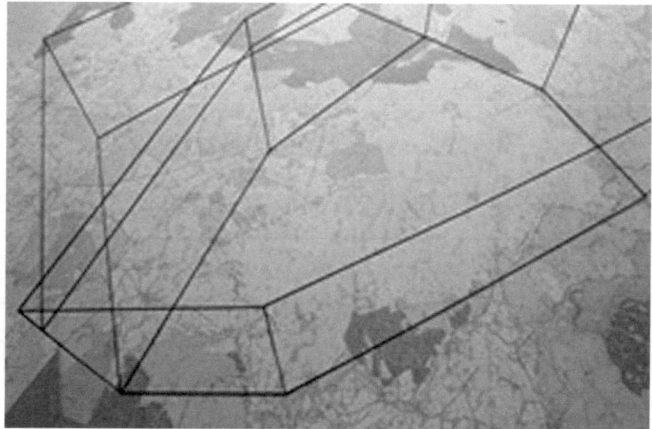

Figure 27.3   Otterburn airspace.

Source: Details, Matthew Flintham, mixed media, variable dimensions.

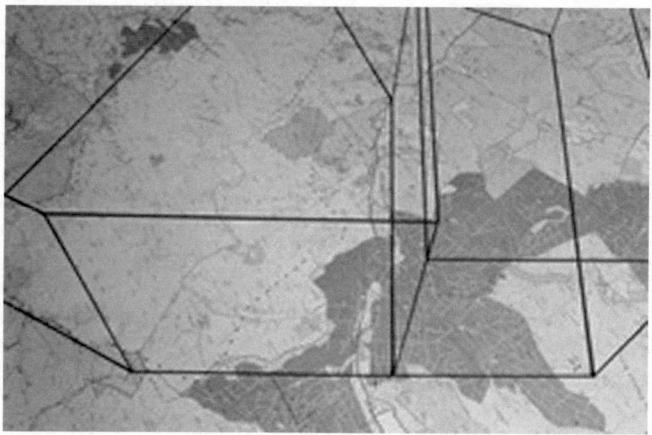

Figure 27.4   Otterburn airspace.

Source: Details, Matthew Flintham, mixed media, variable dimensions.

A third wireframe assemblage describes Danger Area EGD203 over the Sennybridge Training Area (SENTA) to the north of the Brecon Beacons in Wales, which extends up to an upper limit of 23,000 feet over 37,000 acres of forests and rugged hills used for infantry training and artillery firing. The three completed wire frame assemblages proved to be a challenging subject to photograph, with optical lens distortions compromising the form's unusual asymmetrical geometries. However, the most useful technique was to shoot from above, using a small lens aperture, which had the unusual effect of compressing the three-dimensional wire frame into the map itself, making the model seem like a spatial illusion and somehow part of the pictographic geography of Otterburn and the Cheviot Hills.

The process of making these airspace assemblages was not only an attempt to describe an invisible aspect of military spatial production, but also a way of trying to think about spaces that are beyond the scale and physiology of human vision, and somehow capture or freeze the emerging geographies of military globalization.

Work in Progress 4: *The Martial Heavens*

Conveying movement through and within structured airspaces was something I was keen to combine with a critique of military proliferation. Consequently, I took a camera on a number of commercial flights and shot high-definition digital footage of the skies over Britain and the English Channel. Later, I assembled this cloudscape footage as a short film, inserting all seventy-six militarised airspaces from the chart repository described earlier. The red

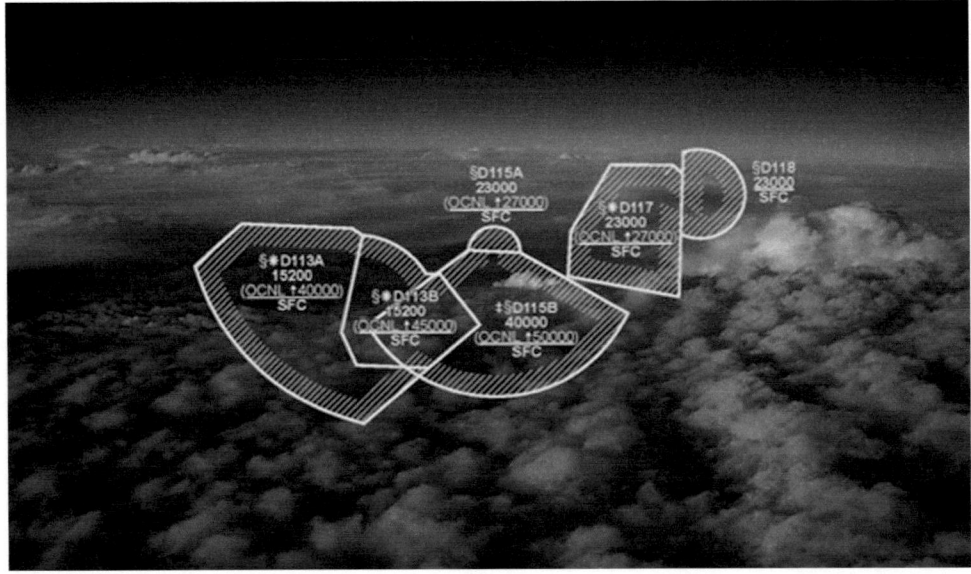

Figure 27.5   Frame from *The Martial Heavens*.

Source: Flintham (2014), digital video.

charts appear transposed over the black-and-white skies in a linear sequence from smallest to largest over an eight-minute period. The effect slightly resembles an advanced head-up display (HUD), a device which projects navigation images onto the aircraft cockpit glass during flight, but it is also a very direct way of presenting both the complexity of military airspace design and their sheer proliferation across the UK. The use of black-and-white footage of clouds implicitly evokes aerial films from the Second World War and before, but also an otherworldly dimension beyond the terrestrial plane. The slow, sequential presentation of the airspace charts over the cloudscapes also suggests a form of mechanical and visual automation, one that alludes to the increasing use of drones and semi-autonomous aerial vehicles.

Speculations

This chapter described a body of work made during an artist residency in the School of Geography, Politics and Sociology at Newcastle University. The photographic works, airspace planning tables and film described were made during a 12-month period and relate to the hidden geographies of military aviation. However, the residency was also a process of researching, rejecting and reevaluating work and working practices. As an artist and academic researcher, there is a tendency to separate both my practices (or to forget that they are mutually constitutive), so using aesthetic processes to think creatively about critical military research was a way of invigorating, bypassing or even short-circuiting certain methods that had become entrenched in my own practices. This process has also reaffirmed a belief that the hybrid model of practice, the artist-as-researcher (or vice versa), is one in which theory, speculation and conviction grow out of doing and making.

The discipline of making a body of work that relates to a specific area of military activity has required learning a set of new practical skills and developing an investigative research methodology. It has also opened up a whole set of questions relating to the representation of complex, immaterial subjects which will almost certainly stimulate new work in the future. Such questions are shared by Toscano and Kinkle in *Cartographies of the Absolute* (Toscano and Kinkle, 2015), which details the seemingly immense task of representing and revealing the totalizing and often invisible structures of late capitalism. The same task could equally apply to tracking, measuring and exposing the logistics of military globalization and the increasingly dematerialized technologies of state power. However, as we have seen, an increasing number of visual artists, designers and filmmakers are developing radical and counter cartographies to expose increasingly complex systems of control, surveillance and conflict. It seems clear that the greater the complexity of the subject, or the more we are coerced into powerlessness, the more critical researchers will need to harness the aesthetic and the creative to map the dispersions and flows of power.

Making art that is exploratory in nature, and which at times has no obvious or immediate value, feels like the antithesis of military order and strictures. Furthermore, the methodology for making art often seems distinctly alien to the military subject, a pointless endeavour that fails to realistically appreciate the art of war, personal sacrifice or the function of national defence within the maelstrom of global politics. However, the real value of art as military critique is not simply about fighting guns with roses, but to offer a range of strategies that reveal the invisible technologies of power, challenge questionable policies or simply re-present a seemingly hopeless situation in a new light.

Notes

1   On any given day the National Air Traffic Services (NATS) handles approximately 7,000 flights in UK airspace. See NATS website: http://www.nats.aero/question/the-number-of-flights-in-uk-airspace-continues-to-grow---how-can-you-guarantee-safety/ (Accessed 1 July 2014).

2   The Chicago Convention of International Aviation was signed in 1944 and went into effect in 1947, to facilitate the regulation and coordination of international air travel. See the International Civil Aviation Organization website for more details, http://www.icao.int/publications/pages/doc7300.aspx (Accessed 25 January 2016).

3   Network-centric warfare is an emerging military doctrine adopted by most advanced military powers to describe the forms of training and engagement that rely on decentralized but connected networks of 'sensors' across the battle space. These sensors might include satellites, drones and weapons systems but also vehicles and soldiers on the ground, gathering a broad spectrum of data and feeding it back to command.

4   For example, Harun Farocki's *Eye/Machine III* (2003) utilizes operational images of cruise missile deployment, while John Gerrard's *Exercise (Djibouti)*, 2012, draws on the choreography of military training exercises to build complex computer-generated moving images. Charles Stankievech's body of work (including the recent *The Soniferous Æther of the Land Beyond the Land Beyond*, 2013) deals at once with the minutiae of electromagnetic signals intelligence and the spectacle of military architecture in remote landscapes, while Steve Rowell's collaborative installation with SIMPARCH, *Gloom & Doom I*, *II*, (2006), is a sonic boom simulator which recreates the effects of living beneath military-controlled supersonic airspace. Recent exhibitions such as *Forensis* at Haus der Kulturen der Welt, Berlin (2014) have also highlighted a spectrum of visualization and artistic methods employed by architects, artists and creative technologists to uncover abuses by the security agencies, military and paramilitary organisations.

5   *ENR5: Navigation Warnings*, available from NATS. See http://www.ead.eurocontrol.int/eadbasic/pamslight-1D336E977510C275B31DA3449E1D5548/7FE5QZZF3FXUS/EN/AIP/ENR/EG_ENR_5_1_en_2015–02–05.pdf (Accessed 11 February 2015).

6   *ENR 6–5–1–1: Chart of the United Kingdom Airspace Restrictions and Hazardous Areas*, available from NATS. See http://www.ead.eurocontrol.int/eadbasic/pamslight-1D336E977510C275B-31DA3449E1D5548/7FE5QZZF3FXUS/EN/Charts/ENR/AIRAC/EG_ENR_6_5_1_1_en_2015–02–05.pdf (Accessed 11 February 2015).

7   One only has to look at Soviet avant-garde artists of the early twentieth century (particularly the work of El Lissitsky, Naum Gabo, and Kazimir Malevich among others), the Italian Futurists (such as Benedetta Cappa, Luigi Russolo, Umberto Boccioni, etc.), or Bauhaus artists such as Laszlo Moholy-Nagy, to see how artists have creatively interpreted the dynamism of industrialisation, the mechanisation of warfare, and attempted to capture movement and duration in space. Indeed, much of this work seems prescient in the way that invisible forces in space are sometimes rendered as volumetric, abstract or geometric structures. Such imaginary or hypothetical structures resemble nothing less than the invisible architecture of controlled airspace or other invisible parameters that mediate our movement in the world.

8   Permission sought and granted to use NATS chart, *ENR 6–5–1–1*.

9   Permission sought and granted to use Ordnance Survey maps in artworks made during the residency.

References

Dodge, M. and Kitchin, R. (2004) Flying Through Code/Space: The Real Virtuality of Air Travel. *Environment and Planning A* 39, 195–211.

Graham, Stephen (2010) *Cities Under Siege: The New Military Urbanism*. London: Verso.

Flintham, Matthew (2010) *Parallel Landscapes.* Unpublished PhD thesis, p. 222.

Flintham, Matthew (2014) The Military Spatial Complex: Interpreting the Emerging Spaces of British Militarism. In Deriu, Davide, Kamvasinou, Krystallia and Shrinkle, Eugénie (Eds.), *Emerging Landscapes: Between Production and Representation*. London: Ashgate, pp. 55–65.

Hanssen, Beatrice (2000) *Critique of Violence: Between Poststructuralism and Critical Theory*. London: Routledge.

Minton, Anna (2012) *Ground Control: Fear and Happiness in the Twenty-First-Century City*. London: Penguin.

Mora, Thierry and Bialek, William (2011) Are Biological Systems Poised at Criticality? *Journal of Statistical Physics* 144(2), 268–302.

National Air Traffic Service (NATS) airspace chart ENR 6–5–1–1 (25 July 2013). Available at: http://www.nats.aero/question/the-number-of-flights-in-uk-airspace-continues-to-grow---how-can-you-guarantee-safety/ (Accessed 1 July 2014).

Toscano, Alberto and Kinkle, Jeff (2015) *Cartographies of the Absolute*. Alresford, Hants: Zero Books.

Williams, A. J. (2012) Reconceptualising Spaces of the Air: Performing the Multiple Spatialities of UK Military Airspaces. *Transactions of the Institute of British Geographers* 36(2), 253–267.

28

TAKING LEAVE

Art and Closure

Gair Dunlop

Taking leave: a process of departure, an invocation of the right to leisure, and a sense of the end of a phase of life.

All of these factors have some purchase in the analysis of UK military sites that have become run-down, redundant, and abandoned. The process of abandonment can stir deep emotional responses in personnel, surrounding residents (most recently at RAF Leuchars, in the process of drawdown and transfer to the army) and in elements of the wider public. Site closures are occasions when the mobilisation of memory becomes a shared process, engaging civilian and military populations. The role of film and photography in commemorative events in military culture is well understood institutionally. Military unit photographers accompany royal visits, passing-out parades, disbandments and a wide variety of formal occasions. Media training, press officers and an active public relations organisation produce many contexts for image-making on bases. Displays, parades, speeches and dances are held; these events have a public face and are widely reported, drawing on local media and publics to share the specificity of the site and its histories. Unusually, in the case of the drawdown and closure of RAF Coltishall in North Norfolk, three contemporary digital artists were invited in to document the site, the personnel and behaviours around withdrawal from one of the RAF's most iconic sites.

Planned in 1936 with construction beginning in 1938, Coltishall was originally intended as a bomber station, with a typical 'Expansion Period' design based around three large hangars in a curve at the edge of concrete apron, leading onto a grass airfield. Elements of the design were planned by Edward Lutyens, with a spacious and elegant consideration of accommodation and infrastructure. After heavy RAF losses over France in the early period of World War II, Coltishall was redesignated as a fighter defence station in June 1940, and continued as a fighter interception station throughout the Cold War period, flying Javelins and Lightnings. Its final aircraft, the reconnaissance and ground attack Jaguar, entered service in 1974 and continued until the last aircraft left for RAF Coningsby in Lincolnshire on 3 April 2006.[1] Generations of aircraft and personnel associated this airfield with the historical high point of

Figure 28.1   Stripped control tower, RAF Coltishall, June 2008. After the removal of the radar and flight control equipment, the sole occasional occupant of the control tower was a security guard.

Source: Photo by the author.

the RAF, and its relatively unmodified condition made it easy to imaginatively connect with the high days of wartime drama. This unmodified condition also meant that the station was effectively doomed once the Jaguar became obsolete. The communications, electronics and work spaces could not be rescaled for more modern aircraft. Accordingly, attempts were made to find new uses for the airfield. A freight hub adjunct to Norwich Airport was floated, but road links were poor. A rather fanciful eco-settlement was proposed by developers Barton Willmore,[2] and initial interest was shown by long-established Norwich printers Jarrold and Sons in relocating. The first chill winds of recession meant that these plans came to nothing. The sole reuse of the enclosed site to date is the renovation and repurposing of the enlisted men's quarters as H.M. Prison Bure (operational from 2009), with its attendant security features. The housing stock, outside the perimeter fence, has been renamed Badersfield after the notable World War II pilot, and is mainly private family housing with some renting. Most recent developments are as a consequence of the site purchase by Norfolk District Council, who plan to lease it as a solar energy generation site while preserving its architectural and historical features.[3]

Artists Strategies in Military Environments

Methodologies evolving between contemporary visual artists and cultural geographers are bringing new perspectives to bear on landscape studies and lived experience. Curator Nato Thompson has coined the phrase 'experimental geography' to describe this emerging field, seen as "a new lens to interpret a growing body of culturally inspired work that deals with human interaction with the land."[4] Jane and Louise Wilson's *Gamma* film project,[5] for example, looks at the former Greenham Common cruise missile storage facility as an iconic site, one where irreconcilable beliefs around safety, security and the future clashed and still carry resonance. Artists and artists' groups such as the Centre for Land Use Interpretation,[6] Trevor Paglen,[7] and the Wilsons share a practice of art production which explores 'spectacles in space' and enable us to reflect on the sides of our social organism which lie hidden in plain sight.

Strategies developed by the Artist Placement Group,[8] where artists became involved in industry and in state organisations, are key precedents in my own practice. Invitations, not commissions, were the core of its procedures. Industry partners were not to expect specific outcomes or illustrative addenda to their productions. The artist's presence in and engagement with the organisation was as an 'incidental person' who, Latham argues, "may be able, given access to matters of public interest ranging from the national economic, through the environmental and departments of the administration to the ethical in social orientation, to 'put forward answers to questions we have not yet asked'."[9] The artist in this context is someone whose presence is authorized but not bound by a particular role in management or utilitarian daily structure, who can act as a catalyst for discussion and reflection on the histories and outcomes of the military site, as seen from its surviving facilities and the surrounding communities.

In the case of the RAF Coltishall project, the freedom to wander between different operational facilities at will was matched by the ability to enter officer's mess, sergeant's mess and other ranks facilities. (After some deliberation, it was decided by the base commander that the artists should however dine in the sergeant's mess). The impending redundancy of the aircraft flown from Coltishall (SEPECAT Jaguar) itself added to the scope for freedom of movement. As it was an aircraft in active service since 1971, there were no associated restricted areas or secret processes to be negotiated. On the other hand, this also meant that it was problematic to extrapolate findings and elements of filmed behaviours to more contemporary airfields and airframe support communities. Informal cross-referencing of observations between obsolete and current practice became possible with visits to RAF Marham during the four-year course of engagement with Coltishall. RAF culture seemed consistent; levels of secrecy and access, however, were more tightly constrained.

Three artists (myself, Louise K. Wilson and Angus Boulton) were invited by English Heritage onto the RAF Coltishall base in North Norfolk over a period of four years, from the early stages of its closure programme until after the gates were locked and future uses had been partly decided. All three artists involved in the Coltishall project had a personal relationship to militarised space, whether through family history or campaign activism. This personal interest was re-stimulated in different ways during the initial period of site access. Visual, sonic and digital arts practices can be seen as magpie methodologies, hybrid forms of knowledge and information gathering. All three artists involved in the project had to approach the personnel and the environment through the prism of what was already understood about the meaning and economies of the image on RAF sites, and then try to push that understanding and the concomitant results a bit further. Coltishall, as the last functioning fighter interception airfield from the Battle of Britain era, carried enormous resonance as a birthplace of RAF traditions.

My participation was down to coincidence; the interest on the part of English Heritage stemmed from a site-specific artwork I had made in the area the year previously. "Vulcan: sublime, melancholic," was a full-size line drawing of a Vulcan nuclear bomber, etched by light exclusion onto the lawn of Bolwick Hall, near Aylsham[10] (Figure 28.2). Alluding to both the tradition of regimental markings on rural hillsides and the miniature siege works

Figure 28.2 "Vulcan: sublime, melancholic," viewed from the air. Piloted by Wing Commander Willie Cruickshank, RAF Coltishall, June 2004.

Source: Photo by the author.

constructed by the obsessive Uncle Toby in Sterne's *Tristram Shandy*, the work also made reference to the linkages between rural estate and military airfield and the occluded histories of the Cold War V force, spread throughout the Eastern Counties of England. An invitation to document this artwork from the air, extended by Wing Commander Willie Cruickshank at the nearby Coltishall base, was my first encounter with the pilots; they had noticed and identified the drawing, and were intrigued and curious as to its meaning. Their identification of the work with the regimental carving tradition opened up a mutual ground for discussion on meaning of the work, and care to cultivate multiple interpretation was key to the later work.

Discussion between the three artists began to focus on ways in which forces acting on the site and on the personnel experiencing a range of emotions (loss, nostalgia, uncertainty) were articulated through formal ceremony, small individual mark-making, and major structural changes. An important part of the work was getting a sense of the 'technological imaginary' which motivated the personnel, and ways in which official imagery, a sense of nature and of nation fed this sense of place for the institution and its participants. As a result of this process, I was able to develop three artworks from the engagement with the airbase.

I will suggest that a 'toolkit' based around four conceptualisations became a useful framework of practice for the project, helping focus attention on changes and continuities as the site evolved and the relation of the RAF to its past and its future in both local and strategic terms changed. First, *Sign Into Abstraction* referred to the changing qualities of signs and functional objects, which become enigmatic over time as personnel familiar with their functionality receded. This process became key to my main photographic production on-site. Second, a *green world/closed world dualism* denoted the combination of tranquillity and alertness embodied by the rural fighter interception network, which was particularly intense at Coltishall due to its history. Pastoral, technological, networked and entropic tendencies all combined, forming a very particular genus loci. These elements are also useful in consideration of landscapes of training, as for instance in the work of Patrick Wright.[11] This conceptual key informed my observational video work "Dispersals." Third, *GeoMirroring as Geography* referred to the disposition of individual sites in a Cold War system where each side reflects the other, sometimes in surprisingly exact ways. Former Russian airfields in East Germany echo the architectures and design of NATO facilities in an extraordinary super-symmetry. This became a key focus of Angus Boulton's work on-site. Finally, *Mirroring as Simulation* referred to the virtual airfield and missions conducted in the Jaguar Flight simulator, operated as a privatized concern by Thales Defence contractors and staffed by former RAF personnel. This became the cornerstone of my short twinscreen film exploring the paradoxical nostalgias available in a virtual Coltishall "Simulator/Realtime."

Is the 'imaginary life' of airfields, research centres, etc. – as embodied in film and popular culture – a help or a hindrance in exploring such places? My argument is that they are unavoidable, and a consideration of the 'aura' of such sites is an important part of understanding their enduring effects. Second World War drama is a shared cultural experience, almost over-familiar. By contrast, the Cold War as lived experience is more distant. The end of a runway in itself is a banal place; it is only with the cultural weight of cinematic apocalypse that they become hugely charged places in a war that exists primarily in the imagination (Figure 28.3). In what follows, I discuss in more detail each of the four 'toolkit' conceptualisations and in doing so imply a range of methodological considerations inherent to critical military research.

Figure 28.3   Runway 22 end, May 2006.

Source: Photo by the author.

Documenting Drawdown

Sign Into Abstraction

It might be thought that a military airfield would be drab and colourless. On the contrary, such militarised landscapes are modernist environments par excellence. Typically they consist of an interlocking series of utilitarian structures, where highly codified behavioural cues prevail. Large and small coloured shapes in paint – both on the walls and on the ground – serve to delineate zones of activity, indicate access and emergency exit points, and modulate behaviour. The imagination is immediately engaged through an intense curiosity and effort to match our received ideas with the reality around us. These highly practical signs – or icons, in the structuralist worldview – were immediately understood and acted upon by all trained

Figure 28.4 "Return to ESA," a spent cartridge safe on the firing range, RAF Coltishall.

Source: Photo by the author.

personnel. As RAF Coltishall closed and equipment was removed, the signs and indicators remained. With fewer people who understood their meaning, the images drifted from purposeful signification into bold abstraction. Airfield arrestor markers, for example, became bold pop art placards. Photographed in series, the range of marks, objects and surfaces serve to remind us of how much of the daily environment we take for granted. A shift into redundancy brings the surfaces forward in a new light. The first resulting artwork from my Coltishall experiences was a sequence of medium-format photographs, working to explore the space as a liminal zone where meaning and function were drifting away, while natural forces of decay and erosion were creeping forward. The cultural memory is informed by the work and writings of war artists such as Paul Nash (*Aerial Creatures*, 1944),[12] and by the literary insights of writers such as Rex Warner (*The Aerodrome*, 1941).[13]

Figure 28.5   Arrestor marker, runway 04, June 2008.

Source: Photo by the author.

Green World/Closed World

Military landscapes in the UK, annexed from a rural hinterland, embody elements of idyll and dystopia. The sense of idyll and the sense of high-tech alertness need to balance in order to make such places function. An interesting analogy to these states of mind was elaborated by Canadian cultural critic Northrop Frye, with his concepts of 'green world' and 'closed world' dramas.[14] The green world represents a space outside normal time and social rules where the rigidities of order can be overcome, subverted or at least ameliorated. As a structuring metaphor, these concepts have been deployed by cultural theorist and historian of technology Paul N. Edwards. In relation to computing and structural metaphors of Cold War confrontation, the closed world is one of Manichean struggle, all or nothing strategies, and perpetual mobilization.

Figure 28.6 Briefing room, RAF Coltishall.

Source: Photo by the author.

> A 'closed world' is a radically bounded scene of conflict, an inescapably self referential space where every thought, word, and action is ultimately directed back toward a central struggle. It is a world radically divided against itself. Turned inexorably inward, without frontiers or escape, a closed world threatens to annihilate itself, to implode.[15]

Meanwhile:

> The green world is an unbounded natural setting such as a forest, meadow, or glade. Action moves in an uninhibited flow between natural, urban, and other locations and centers around magical, natural forces – mystical powers, animals, or natural cataclysms. . . . The green world is indeed an 'open' space where the limits of law and rationality are surpassed.[16]

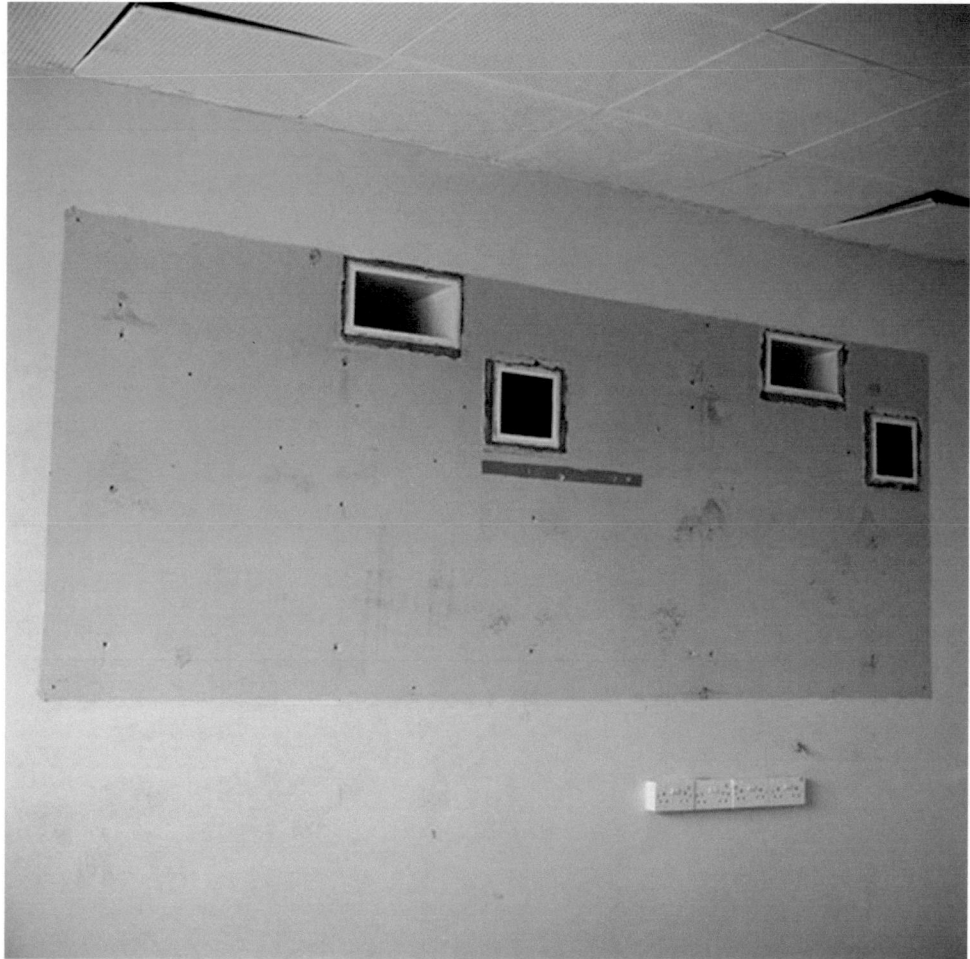

Figure 28.7 Briefing room projection windows.

Source: Photo by the author.

These two mindsets are usually seen as opposites, but they coexist in military and state controlled land such as test sites, experimental facilities, and training grounds where the 'normal world' is held at a distance. The idyll for RAF personnel can be seen as an intimate relation to the past, but also as a more agreeable companion emotion to the boredom of waiting, in the context of the infinitely postponed decisive moment of Cold War–era conflict. Reverie, alertness, instant response and a deep sense of historical continuity formed a constellation of feeling and allegiance at RAF Coltishall.

The green world and the closed world function as two poles of dramaturgical theory; the mental arenas which we keep in our heads as part of our 'technological imaginary'. The Cold War combined stasis with instantaneity, the idea of the generic England being protected becoming entwined with the retrogressive and rural.

Figure 28.8   Ivy invading dispersals blast wall.

Source: Photo by the author.

The presence of a vast workforce is receding from everyday life in the UK, and the ways in which departure was marked brought to mind some of the earliest uses of film and photographic media. The long line of personnel leaving a practice parade was strongly evocative of very early film footage of staff leaving factories and shipyards. During the course of the slow closure, visits were made at key times, and filmic notice taken of the preparations for the final closure rituals. The seeming informality of pilot discussions on formation flying contrasted strongly with the relentless drilling by parade sergeants for the other ranks. The humour of these events, not often noted, was foregrounded.

Figure 28.9   Rehearsals for closure parade, RAF Coltishall, summer 2006; still from "Dispersals."

Figure 28.10   Last unit photograph, engine workshop, RAF Coltishall, summer 2006; still from "Dispersals."

GeoMirroring as Geography

Mirroring has been a guiding principle in the on-site work of Angus Boulton and myself. Boulton is concerned with the historical and geographic mirroring involved in the balance of terror. Previous projects have included *Cood Bay Forst Zinna*,[17] a video exploration of an abandoned site, the raison d'être of which is unclear. The viewer is disorientated: is this a sports camp? an abandoned holiday centre? Traverses through birch woods, drained swimming pools and sports grounds are suddenly disrupted by a discarded tank track or guardhouse. Rich layers of association are gradually built up, until we see the hermetic camp for what it really is: a frozen, self-contained world containing the evidence of its own erasure from geopolitics and memory. The viewer eventually realizes that this camp is an abandoned Red Army enclave in the Brandenburg Forest, abandoned at the end of the Cold War.

At Coltishall, Boulton has been able to make video which echoes and mirrors material from the former East Germany and Poland. Uneasy questions arise about equivalence, morality and the mutual interdependency of two supposedly antagonistic military blocks.[18]

Mirroring as Simulation

Piecemeal privatisations have left many seemingly military structures and resources under the control of corporate entities. In this case, the prime example of relevance to the project outcome was the privatization of the Jaguar simulator facility, controlled by Thales, a French-based multinational. Most of its personnel were ex-UK military: head of unit Clive Crouch, for example, was a former Vulcan pilot. This aspect of the airbase offered the opportunity for the third of my works: "Simulator/Realtime." The final decision to ground the last two Jaguar squadrons (6 and 41) had not yet been taken. The frail, leaky aircraft were transferred north. Pilots accordingly drove back to Coltishall from RAF Coningsby in Lincolnshire for simulator training. World War III, Middle East invasion, and Balkans crises continued to take place on a regular basis in a nondescript industrial building on the Coltishall site. It had been decided that it was uneconomic to move the facility, with its screens, projectors, computer servers and control room.

Unhappy with the move, and conscious of the imminent withdrawal of their aircraft from service, pilots still insisted on 'flying' from Coltishall. Dressed in full nuclear/chemical suits, they sweated their way through engine failures, missile attack, refuelling scenarios and attack runs. Carefree about airfield safety, pilots would career across the 'grass', squeeze through impossible gaps between buildings, and show a general disregard for normal rules. Outside, as the base completed its closure, structures were uprooted, signs taken down, and more

Figure 28.11   "Simulator/Realtime"; digital video still, nine-minute artwork by the author.[19]

buildings were sealed. The virtual Coltishall of the simulator became increasingly more 'functional' and homely than the real one. The electronic half-dome presented the pilots with an opportunity to hold on to their lost home for a little longer.

My response to this paradoxical nostalgia became a resimulation of the pilot's carefree progress: following in a car to as great an extent as possible the freedom of the simulator in journeying round the base. The two sequences were then time-mapped together and the

Figure 28.12   "Simulator/Realtime"; digital video still, nine-minute artwork by the author.

Figure 28.13   Wall graffiti, RAF Coltishall.

Source: Photo by the author.

Figure 28.14   Wall painting, 6 Sqdn Hangar RAF Coltishall.

Source: Photo by the author.

relative sparseness of the real-world airbase became highly visible. The pilot's nostalgia made sense. The engineers, guards and ground personnel are left with more prosaic forms of memory and mark-leaving. The wall art and graffiti became all that remains on site of a multibillion-pound aircraft programme, marked by photographic recording.

Conclusion: Edge Effects

As the final closedown came nearer, base personnel became increasingly focused on the closure parade and flypast. It was clear that this display was to function as an internal marker; the public were invited, but the meaning and emotional resonance of the departure was to be marked for the RAF itself. Video observation and recording of behaviour on-site became inflected by an awareness of this significance. On the day of closure, blessings and speeches paused momentarily. Eyes scanned the horizon. Dots appeared, and suddenly a wave of noise and vibration swept the field. A formation of low-flying aircraft had gone, almost before being fully seen. Cameras and binoculars wobbled uncertainly. The skyful of roaring silver metal was too big to register. Within the ceremonial context, punctuation of parade formality by an explosive irruption of noise and awe fulfil a key role. Preceded by Hawker Hurricane and escorted by a Typhoon Eurofighter, a 'Diamond nine' formation of Jaguars hurtled over the parade ground and into history. The past and the future – represented by Second World War aircraft and the 'new' Eurofighter – bookended their farewell.

The methodologies and techniques which were developed in response to the RAF Coltishall project have further application in other civil and military contexts. We can approach research questions through a convergence of approaches derived from cultural geography, contemporary arts practice, science studies and history. The four polarities which contributed to analysis (sign/abstraction, green world/closed world, mirroring geographically and mirroring in simulation) offer ways to order and construct a response to closed sites of many kinds. In particular, a film production methodology brings together the cultural-visual elements of the technological-military imaginary with their consequences as they play out over sites and communities. Questions, tacit knowledge and the otherwise inexpressible can be externalized and given a tangible form. It is easily graspable by both management granting access and by workforce, retired personnel, witnesses and residents responding to the immediacy of the visual in prompting memory. It enables multiple input, and a flexibility of output format gives a sense of the impossibility of one view being sovereign. Visual research is increasingly being recognized in geography and within the history of science and technology (Gallison, 2015)[20] and offers a truly interdisciplinary epistemology for knowledge exchange. Drawbacks include the artificiality of assumption that the changes and moments recorded have shared significance beyond the production team.

However, the testing of ways to question and extend understanding of place mean engagement with lived experience. Landscape is at once cause and effect, process and product, material and cultural. The artist as an 'incidental person' – someone whose presence is authorized but not part of the management or utilitarian daily structure – acts as a catalyst for discussion and reflection on the histories and outcomes of the military establishment as seen from its surviving facilities and the surrounding communities.

Boundaries and edges between civil and military in institutions such as the UKAEA (UK Atomic Energy Authority) are often porous, blurred and vague. If we accept that military strategic requirements for defence industries dispersal in postwar planning are a major factor in national infrastructure, then a vast area of research and visual investigation is open to us, of continuity between a postwar modernity and contemporary anxieties of despoliation, blight, and risk. The idea of the 'military-pastoral complex' (Flintham, 2012)[21] can be extended and interrogated to tease out strands. In addition, a visual sensibility can tease out revealing aspects of the imagery generated by and around technologies in the postwar period, and recombine them using digital tools to make a reflexive experience for the viewer. Visual arts techniques are ideal tools for investigation in such issues, given that these sites already constitute a collage; of technologies, hierarchies, objects and contested memory.

Notes

1 Station historical references from Jennings, M.D. (2007) *Royal Air Force Coltishall: A Station History.* Cowbit, Lincolnshire: Old Forge Press.

2 Batson, R. (2007) *Eco-Community Vision for RAF Coltishall.* Eastern Daily Press, 11 September, http://www.edp24.co.uk/news/eco_community_vision_for_raf_coltishall_1_698951 (Accessed 16 February 2015).

3 'RAF Coltishall site £50m solar farm plan approved', *BBC News*, 19 December 2014, http://www.bbc.co.uk/news/uk-england-norfolk-30546048 (Accessed 14 February 2015).

4 Thompson, N. (2008) *Experimental Geography.* New York: Independent Curators International.

5 Wilson, J. and Wilson, L. (1999) *Gamma.* Four projection installation, http://www.tate.org.uk/art/artworks/wilson-gamma-t07698/text-summary (Accessed 6 April 2015).

6 Holte, M.N. (2006) 'The Administrative Sublime, or the Center for Land Use Interpretation', *Afterall* (13). See also http://www.clui.org (Accessed 15 February 2015).

7 Paglen, T. (2012) 'Geographies of Seeing', *Photoworks: Agents of Change* (19) Autumn/Winter Brighton Photoworks Annual.

8 Hudek, A. (2010) *The Incidental Person*, http://apexart.org/exhibitions/hudek.htm (Accessed 6 April 2015).

9 Latham, J. *The Artist as Incidental Person: New Role Vis-à-Vis Government.* John Latham Archive (JLA) 13/4231, Folder 335. Quoted in Hudek, 2010.

10 Dunlop G. "Vulcan" (2004). Drawing 111 ft by 93 ft, Bolwick Hall nr. Aylsham, available at https://vimeo.com/13819937 (Accessed 6 April 2015).

11 Wright, P. (1995) *The Village That Died for England.* London: Jonathan Cape.

12 Nash, P. (1996) *Aerial Creatures.* London: Imperial War Museum and Lund Humphries. The 1944 essay of the same name is available at https://www.flickr.com/photos/gair_dunlop/sets/72157626583252950/ (Accessed 6 April 2015).

13 Warner, W. (1941) *The Aerodrome: A Love Story.* London: John Lane and Bodley Head.

14 Frye, N. (1957) *Anatomy of Criticism: Four Essays.* Princeton, NJ: Princeton University Press.

15 Edwards, P.N. (1996) *The Closed World: Computers and the Politics of Discourse in Cold War America.* Cambridge, MA: MIT Press, p. 12.

16 Ibid., p. 13.

17 Boulton, A. (2007) *Restricted Areas.* Manchester: MIRIAD; Manchester Metropolitan University and The Wapping Project.

18 Information is available at http://www.angusboulton.net/0826–2/ (Accessed 6 April 2015).

19 Available at https://vimeo.com/13819865 (Accessed 6 April 2015).

20 Galison, P. (2015) 'Visual STS', in Carusi, A., Hoel, A.S., Webmoor, T. and Woolgar, S. (Eds.) *Visualisation in the Age of Computerisation.* London: Routledge, pp. 197–225.

21 Flintham, M. (2012) The Military Pastoral Complex: Contemporary Representations of Militarism. *Tate Papers*, p. 17, http://www.tate.org.uk/research/publications/tate-papers/military-pastoral-complex-contemporary-representations-militarism (Accessed 6 March 2015).

29

OVERT RESEARCH

Fieldwork and Transparency

Neal White and Steve Rowell

The Office of Experiments (http://www.o-o-e.org) has been at the forefront of critical investigations of the topography of a postindustrial knowledge society – in particular, documenting and recording, analyzing and engaging with its physical sites and enclosures. In this contribution, we will illustrate one of a series of incursions into the techno-scientific and military-industrial complex that forms part of this landscape, one that yields and conceals a highly complicated network consisting of state and private, scientific and military-funded layers. In our attempts to find appropriate research methods that would allow us to map and document this complex, including those using its own surveillance technologies, we have come to understand through a critical sensibility the nature and scale of a military-industrial and military-scientific infrastructure. The slab concrete labs, anonymous glass towers and distant testing ranges punctuate our contemporary landscape and are familiar to us all, yet on close inspection they act as sites of refusal, inaccessible to anything other than visual record, signposted as 'out of bounds', 'forbidden, no trespassing', or classified as legal space as part of the UK Official Secrets Act. The passer-by is encouraged in their myopic distracted activity, to just keep passing.

Accompanying this short text, then, are visual fragments – documentary evidence of a research project whose focus was initially based on the naively simple premise of identifying extraordinary scientific sites that might be, for a variety of reasons, inaccessible to the public. Our motivation was to openly document the huge science infrastructure that the UK public pays for: large-scale research centres, specialist science labs and testing facilities, radio observatories, proving grounds and subterranean spaces of various kinds. Many of the sites which make up this network are often considered as shining beacons of knowledge; while their presence is carefully mediated, the information available to the public did not necessarily throw any light on their unique situations. Many maintained their enclosures not simply through walls, barriers and perimeters, or careful landscaping, but through disinformation at different levels that allowed them to simply hide in plain sight.

This project started simply and with two aims: (1) to draw new maps, create new navigational tools and online resources that might start to describe a complex area of techno-scientific development; and (2) disseminate information to UK citizens who might develop new vistas into these knowledge spaces. As a group led by experienced media artists (including those in digital networked media and photography) who had undertaken former residencies on projects inside institutions and organised centres of research – the enclosures of the scientific establishment – we understood a perceived idea of the artist's role: one who can envision and communicate approved creative insights. However, our practice is based on a shared dialogue between academics and artists, and artistically has its roots in a number of long discourses surrounding a critical or radical art and technology practice in relation to the military-industrial complex, an approach that is quite often agonistic (Mouffe, 2007). Our work as artists and researchers draws on this and the traditions of socially engaged practice (such as Artist Placement Group, in 1966), which places the artist inside institutions of power. The challenge here, however, was how to remain neutral and objective, to address and examine the landscape as a laboratory, with each site a specimen node of a larger network, its presence measured in terms of location, information and critically, disinformation. As such our research had to be attuned to the play between the openness of our own approach and the required need for secrecy of the sites we encountered; the interplay of what is overt and covert needed to be addressed.

Our approach, Overt Research, has since been refined over a long period to become a distinct area of artistic inquiry and an area of artistic research practice. In this respect, as Henk Borgdorff (Bauer et al., 2012: 121) argues, the distinguishing dimensions of the artistic research method are not situated within academia but sit on the interdisciplinary border between a subject discipline and the domain of life itself. Thus our investigations began in 2007 with a rigorous process of networking among a group consisting of artists, academics and activists in order to gather information on our subject. In this case, we began our research after having attended an Economic and Social Research Council (ESRC) workshop series on spaces of secrecy and transparency, before homing in on the 'extraordinary spaces' that subsequently became our focus. Having made our selection of sites for visits, including Porton Down, Salisbury and the National Security Agency (NSA) site Echelon, Harrogate, our inquiries frequently resulted in poor or impartial information. This led to an initial methodological practice of classification by which we evaluated each site according to its 'level of transparency'. Having assessed the quality and openness of publicly available information as part of this process, we also ascertained the involvement of scientific funding councils in the activities carried out at each site in order to develop an understanding of the knowledge objectives of particular experimental facilities. Here, the work of Hans Jorg Rheinberger into 'epistemic things' was of great utility. Namely, in order to fully understand the connected and social nature of 'experimental systems' we need to also identify their "local, technical, instrumental, institutional, social and epistemic aspects" (Rheinberger, 1997: 38).

After this, we began mapping sites online using open source geolocative technologies. Such an approach is directly attributable to the work of artists such as Critical Art Ensemble (CAE), whose contestational tactics invert knowledge back into a complex from where it is developed (see e.g. Critical Art Ensemble, 1996). These contestational approaches are an important part of critical discourse in the recent history of tactical media and media arts, with artists very familiar with the origins of many technologies within the military industrial complex itself, including geospatial satellite systems we all use. So the geoimagings we used for online mapping were critically positioned amid such an enmeshed history of refined overhead and orbital surveillance systems developed by the very research establishments we are committed to exploring (for related approaches, see the projects of the Center for Land Use Interpretation, Trevor Paglen or Eyal Weizmann). We therefore turned our attention to a visible form of experimental fieldwork investigation: field reconnaissance, research excursions and exploratory documentation through photography. This fieldwork inverts the logic of surveillance per se as it relies both on observation and on remaining 'overtly' visible in the field. It entails approaching sites on public ground, working in plain sight, carrying identity cards as we start to photograph sites and facilities.

Overt field research means adhering to relevant laws and by-laws, and having legal advice with you, which is often useful both for the researcher and the security they might sometimes encounter. It also involves carefully analysing legal documents and case histories, including public statements from government officials pertaining to the use of photography and the Official Secrets Act, including the official UK government D-Notice committee that advises on documenting military installations. For example, in the discussion of the public access to online information on websites such as secretbases.co.uk, concerning the documentation of such sites at this time, Rear Admiral Nick Wilkinson (former Secretary (1999–2004) of Defence, Press and Broadcasting Advisory (D-Notice) Committee[1] states: "These sites should already not only be aware of what is public, but also have taken security measures accordingly." At all times the researcher must position themselves outside military sites but at the same time within the law and the public realm.

Once we have captured photographic and visual data, we sort, order, collate, edit and upload this information onto our website, dark-places.org.uk. Here, the data collected up to this point is added to by Internet researchers, complemented by discussions with dog walkers we might have encountered, or otherwise the curious and imaginative. Secrecy, we have come to understand, leads to rumour, myth and speculation; its operation always leaves traces. The steady bleeding of the oxygen of publicity leads to new stories whereby both parties (the researched and researcher) remain unaccounted for and unaccountable.

We therefore sense it is not enough for us to simply recount or remediate that which we have encountered online or as visual documentation. Therefore, where possible and plausible we take our audience with us on critical excursions into military-industrial landscapes where we enable others to subject sites to human presence. Here we explain how to observe, to identify patterns and small changes, to be attentive to shifts in perception. Circling perimeters, we point to and scrutinise the technological infrastructures of the military complex; buried cables, pipelines and checkpoints, the ownership of sites and the relationship between state secrecy and capital. In between sites, we present and discuss additional information including official information or conspiracy films, or recently declassified archive materials. The visible landscape thus shifts in our focus, remediated as we cruise through an enhanced vision of an overlapping research and intelligence complex that surrounds us.

The constant duality in our research method, performed overtly in plain sight, observing the covert spaces of knowledge, drawing on official and unofficial information, naturally leads us to places which are rife with speculations, sites which are massive in the social imaginary, notorious or celebrated. Taking Porton Down, Echelon or Bude as such examples, we make no judgements, but open up these spaces to dialogue, asking why, how and where the intelligence and development of research really takes place. We want to understand whose interests are being served by the public good' and how to defend or maintain the performance of secrecy in the open and transparent future of a postindustrial landscape. This is a journey across a landscape which is constantly (re)mediated and manipulated. Our presence as military researchers is a therefore a direct and situated intervention that embraces and shapes, leads and interprets both imaginary and physical encounters with militaristic sites of technological and scientific experimentation.

Note

1 The quote is taken from the Office of Experiments Field Researchers Tools Guide, pp. 18–19 (2009).

References

Bauer, U. M., Dombois, F., Maries, C. and Schwab, M. (Eds.) (2012) *Intellectual Birdhouse: Artistic Practice as Research*. London: Koenig Books.

Critical Art Ensemble (1996) *Electronic Civil Disobedience and other Unpopular Ideas*. Brooklyn, NY: Autonomedia.

Mouffe, C. (2007) Artistic Activism and Agonistic Spaces. *Journal of Art and Research* 1(2), 1–5.

Office of Experiments (2009) *Dark Places Catalogue*. Southampton: John Hansard Gallery.

Rheinberger, H.-J. (1997) *Toward a History of Epistemic Things: Synthesizing Proteins in the Test Tube*. Stanford: Stanford University Press.

Acknowledgements

Office of Experiments research would not be possible without Neal White, Steve Rowell and Lisa Haskell. With thanks to Nicola Triscott of Arts Catalyst and UCL's Department of Geography, Professor Gail Davies, to Mike Kenner and to the scores of others who watch and support us yet still fear to tread into the light.

Figure 29.1 "Porton and Winterbourne Gunnery Military Lands," September 2009. OS grid reference: SU 18415 34808.

Source: Office of Experiments' "Overt Research Project," 2008–10.

Figure 29.2   "Thorney Street entrance to MI-5 HQ at Thames House, London," May 2007. OS grid
reference: TQ 30209 78889.

Source: Office of Experiments' "Overt Research Project," 2008–10.

Figure 29.3   "Royal Ordnance Factory Burghfield warhead assembly complex," March 2009. OS grid
reference: SU 685 683.

Source: Office of Experiments' "Overt Research Project," 2008–10.

Figure 29.4 "Warning signs near the Broomway to Foulness Island," June 2008. OS grid reference: TQ 969 870.

Source: Office of Experiments' "Overt Research Project," 2008–10.

Figure 29.5 "On route to Foulness Island on the ancient Broomway at low tide, looking across the Maplin Sands to the North Sea," June 2008. OS grid reference: TQ 973 870.

Source: Office of Experiments' "Overt Research Project," 2008–10.

Figure 29.6 "Approach to Atomic Weapons Establishment Aldermaston," March 2009. OS grid reference: SU 598 636.

Source: Office of Experiments' "Overt Research Project," 2008–10.

Figure 29.7 "Recreational trail and disused US cruise missile bunkers at Greenham Common," March 2009. OS grid reference: SU 487 646.

Source: Office of Experiments' "Overt Research Project," 2008–10.

Figure 29.8 "Barricaded access road to MOD Corsham Computer Centre," March 2009. OS grid reference: ST 85489 68952.

Source: Office of Experiments' "Overt Research Project," 2008–10.

Figure 29.9 "Gate to Box Tunnel entrance near MOD Corsham Computer Centre," March 2009. OS grid reference: ST 85785 69450.

Source: Office of Experiments' "Overt Research Project," 2008–10.

Figure 29.10 "Orbiting the Government Communications Headquarters (housing estate perimeter)," March 2009. OS grid reference: SO 91926 22223.

Source: Office of Experiments' "Overt Research Project," 2008–10.

Figure 29.11 "GCHQ Hunter's Stones BACKBONE tower near RAF Menwith Hill," December 2007. OS grid reference: SE 21159 51462.

Source: Steve Rowell's "Ultimate High Ground" project, 2007–10.

Figure 29.12   "Testing radome and electronics laboratories at Government Communications Centre –
Hanslope Park," March 2009. OS grid reference: SP 81682 46112.

Source: Office of Experiments' "Overt Research Project," 2008–10.

Figure 29.13   "Climate-controlled USAF B-2 stealth bomber hanger at RAF Fairford," March 2009.
OS grid reference: SU 15516 97717.

Source: Office of Experiments' "Overt Research Project," 2008–10.

Figure 29.14 "US DOD/UK MOD antenna arrays at RAF Croughton," March 2009. OS grid reference: SP 56560 32029.

Source: Office of Experiments' "Overt Research Project," 2008–10.

Figure 29.15 "Brunton Airfield pastoral and camouflaged radome," June 2010. OS grid reference: NU 20882 25366.

Source: Office of Experiments' "Overt Research Project," 2008–10.

Figure 29.16 "Defence Science and Technology Laboratory – Porton Down, Salisbury," September 2009. OS grid reference: SU 20159 37190.

Source: Office of Experiments' "Overt Research Project," 2008–10.

Figure 29.17 "Along the Perimeter at DSTL Porton Down," June 2010. OS grid reference: SU 21022 37315.

Source: Office of Experiments' "Overt Research Project," 2008–10.

Figure 29.18   "International School for Security and Explosives Education, Chilmark, Salisbury," July
2009. OS grid reference: ST 98151 30133.

Source: Office of Experiments' "Overt Research Project," 2008–10.

Figure 29.19   "QinetiQ Land Magnetic Facilities, Portland Bill, Portland, Dorset," May 2010. OS grid
reference: SY 67584 68546.

Source: Office of Experiments' "Overt Research Project," 2008–10.

Figure 29.20 "Looking at RAF Spadeadam from Hadrian's Wall," June 2010. OS grid reference: NY 62510 66512.

Source: Steve Rowell's "Military Landscapes of the North-East" tour, Newcastle University, 2010.

Figure 29.21 "Armored Guard Post at MOD Otterburn Range," June 2010. OS grid reference: NY 86049 97752.

Source: Steve Rowell's "Military Landscapes of the North-East" tour, Newcastle University, 2010.

Figure 29.22 "Warning Signage at MOD Otterburn Range," June 2010. OS grid reference: NT 78257 03094.

Source: Steve Rowell's "Military Landscapes of the North-East" tour, Newcastle University, 2010.

Figure 29.23 "USA and RAF flags at main entrance to RAF Menwith Hill, North Yorkshire," November 2007. OS grid reference: SE 201 577.

Source: Steve Rowell's "Ultimate High Ground" project, 2007–10.

Figure 29.24   "MOD Perimeter Patrol Incident at RAF Menwith Hill," December 2007. OS grid reference SE 196 570.

Source: Steve Rowell's "Ultimate High Ground" project, 2007–10.

Figure 29.25   "Menwith Hill Pastoral," December 2007. OS grid reference: SE 197 555.

Source: Steve Rowell's "Ultimate High Ground" project, 2007–10.

Figure 29.26   "RAF Fylingdales Radar Station (Inner Perimeter Pasture)," May 2009. OS grid reference: SE 86515 97066.

Source: Steve Rowell's "Ultimate High Ground" project, 2007–10.

Figure 29.27   "Historic Mural in Officers' Mess at RAF Fylingdales Radar Station," May 2009. OS grid reference: SE 86515 97066.

Source: Steve Rowell's "Ultimate High Ground" project, 2007–10.

Figure 29.28 "Hurn Plant and Mechanical Handling Test Centre," March 2009. OS grid reference: SU 12984 03068.

Source: Office of Experiments' "Overt Research Project," 2008–10.

30

THE AUDIBLE COLD WAR

Louise K. Wilson

There are numerous iconic spaces and sites in the UK that haunt (and taunt) us with their inaccessibility – of right of entry, archived knowledge and legibility. Like the architectural equivalent of exotic fauna, these structures, specifically Cold War ones, exert a compelling fascination. There is an astounding affective power to these strange architectural structures that are familiar in material but otherwise utterly alien in form. They are enigmatic yet distinguished by their sheer scale and harsh geometries. The design is rigidly functional: the primary materials are often concrete, steel and earth – hard and soft materials combined to form reverberant monoliths. Yet gaining physical admittance to these abandoned defence sites can sometimes be as problematic as if they were still active. Additionally, access to detailed textual/visual information about the work undertaken there can be restricted or forbidden to the layperson, since many of the most important (nuclear) sites are still covered by the Official Secrets Act. Papers, photographs, films and other data are not yet, and may never be, in the public realm and former employees must remain silent about the work they performed there. Without knowledge to project onto these sites during the course of a visit, one must let the places speak for themselves. These architectural forms communicate more than their pure materiality, however, the difficulty of mediating the value of a monument is particularly acute for those monuments that Norbert Huse (1997)[1] described as being 'uncomfortable monuments':

> These are monuments that predominantly – because of their history – remind us of politically uncomfortable or painful times, or mark places of human suffering and fate, or cause uneasy emotions among visitors because of a clear attribution or use by predominantly authoritarian political systems.
>
> *(Schofield et al., 2006: 7–8)*

Monuments of the Cold War, defined as "structures built, or adapted, to carry out nuclear war between the end of the Second World War and 1989" (Cocroft and Thomas, 2003: 2) are profoundly paradoxical, moreover: often formally banal while radiating fearsome associations.

Many contemporary artists have sought to explore the 'uncomfortable monuments' extant from the hot and cold wars in Europe, the former Soviet Union, the United States and elsewhere. Generally artists have used (visual) time-based media to document the more dispersed nuclear landscapes and wastelands of spaces that are "literally filled with ideologies" (Lefebvre, 1976: 31). Peter Goin, Emmet Gowin, Stan Denniston and Richard Misrach are among those (North American) chroniclers who have forensically observed traces and repetitive forms on the ground and from aerial viewpoints. A brief survey of British-born or based artists would include Tacita Dean and Lisa Autogena, who have both made work about the enigmatic Acoustic or Sound Mirror structures (the early warning devices built shortly before the Second World War along the British coast). In addition, the moving-image work of Jane and Louise Wilson (no relation) has previously focused on summoning visible traces of the occupants in ideologically loaded sites such as the former Stasi prison Hohenschönhausen in East Berlin (*Stasi City*, 1997) and Greenham Common (*Gamma*, 1999).

This chapter will, however, discuss a project where the emphasis was primarily a sonic one. In 2005 I produced *A Record of Fear*, a series of sound-based artworks made for and about ex-AWRE (Atomic Weapons Research Establishment) Orford Ness in Suffolk. These works – temporarily installed on the shingle spit – were created in response to the site's unique history, architecture (and architectural acoustics), geography and ecology. The desire to explore these material remains through sound was important for numerous reasons. Briefly, these related to the prevalence of expanded notions of 'aurality' on the Ness – evident in, for example, the proliferation of unusual ambient sounds, the circulation of (cover) stories to create obfuscation and ideas of 'eavesdropping' constituent in some of the defence research carried out there. A listening post was hastily set up in 1957 to hear Sputnik, for example, and in the 1970s the top-secret Anglo-American Cobra Mist project was developing Over the Horizon backscatter radar to 'look down' into the Soviet Union. The project also acknowledged an ongoing fascination with the possibilities for audio-recording media to act as a conduit beyond tangible physical phenomena. This territory has been the subject of some fascinating and extensive media histories (and media archaeologies) in recent years – notably that of Friedrich Kittler, for whom "media always already provide the appearances of spectres" (Kittler, 1999: 12). Ultimately the decision to privilege listening as the principal mode for address was considered as prompting a more visceral and affective reading of these sites than may be offered by the primarily visual. Before examining this project and its rationale more closely, however, it is useful to ask what motivates artists to respond to peculiarly resonant (Cold War) sites of ruination – such as Orford Ness – and to cite two of the key contextual narrators whose writings frequently underpin this discourse.

The Lure of the Cold War

There are perhaps multiplicities of personal, political and psychoanalytical reasons why artists have wished to explore such sites. The Cold War certainly exerted a tangible effect as "one of the defining phenomena of the late twentieth century" (Slessor, 2004: n.p.) and touched lives in profound ways, not least with the prolonged persistence of fear. Joanna Bourke writes that in 1983 "over half of all teenagers in the United Kingdom believed that a nuclear war would occur in their lifetime"[2] (2005: 259). The fear the Cold War engendered was markedly different from that prompted by earlier conflicts. Previously unimaginable elements – driven by techno-scientific development – were pervasively present:

> courage, honour and hope had no place in the new warfare of the twentieth century. Everything was uncertain. Who was responsible for 'pushing the button'? Could people 'trust' machines and supercomputers not to malfunction? How real was the Soviet threat?
>
> *(Bourke, 2005: 285)*

Paradoxically, the fear of annihilation and the effects of radiation either galvanised (political) action or provoked inertia. Either way, the psychological effects have been long-lasting. For Adam Piette, "the traces of the gravitational pull generated by the alliance of power and secrecy have motivated a series of poets, writers and artists to reflect on the menace and mystery of the abandoned sites in the post–Cold War" (2009: 4). The attraction of secrecy is echoed in Angus Boulton's analysis that an artist's inquisitive nature plays a significant role, motivating the seeking out of access to many of the restricted areas that lay at the heart of this prolonged conflict. As he acknowledges,

> perhaps a crucial aspect of this period of history for artists derives from the fact that this war was played out principally in secret. Major decisions were taken behind closed doors, while training for war was routinely undertaken within the many large military complexes to be found throughout Europe and elsewhere.
>
> *(Boulton, 2006: 35)*

As he sees it, a desire to "unravel the world around them" and make "interpretations from a different, previously unforeseen perspective" (ibid.) underlies their practice.

This raises the question to what extent artists' motivations are quantifiably different from those of contemporary archaeologists, for example, who have explored these same sites. Arguably this shared interest *does* involve shared concerns: Angela Piccini writes:

> There is clearly something in the relationship between trauma, nostalgia, melancholy and loss that captures archaeological imaginations. The current interest in late industrial landscapes, the structures of the Cold War, the archaeologies of the motorway all seek to employ archaeology to understand the more painful sites of the twentieth and early twenty-first centuries.
>
> *(Holtorf and Piccini, 2009: 21–22)*

She acknowledges these places "hold fascination for a range of scholars across disciplines" (22). John Beck has noted how fortified bunkers in particular have visibly become "objects of troubled fascination for artists, architects and archaeologists" (Beck, 2011: 79). While the examples given clearly reveal nuanced sides to this enthralment, of particular interest for me are those photographic works by Erasmus Schroeter (2002) and Uta Kögelsberger (2004) that insist "that the bunkers participate in their own spectacle rather than be allowed to passively manage the effective environment merely by their presence" (Beck, 2011: 89). There are discernible traits that point to pressing contemporary concerns about the "accountability of . . . systems of authority and control" (80). In the public consciousness, bunkers – arguably the most iconic of Cold War structures – can stand in for other defence-related work environments, laboratories, manufacturing plants and launch sites to improvised shelters as well as the more psychological territory of imagined and metaphorical spaces (implied in the expression 'reds under the bed', for example).

Facilitating access inside Cold War sites can be time-consuming and complex, requiring varying permissions (or not) and conscious etiquettes of behaviour therein. As Angus Boulton hinted, the curiosity to explore nuclear test sites, ex-defence laboratories and so on may be coupled with the tantalising desire to see how far an artists' 'passport of admission' may be used to ease admittance. Bunkers and shelters have been noticeably alluring for physical intervention and durational engagement – both for sanctioned and uninvited guests. Stephen Felmingham's arts practice, for instance, necessitates illicit solo visits inside the (East Anglian) network of subterranean Royal Observer Corps (ROC) observation posts, confronting a sense "of being buried alive . . . of climbing down into an open grave, a potential grave."[3] In these charged spaces – breathing air that has been undisturbed for decades – Felmingham sketches the debris left behind. Later in the studio, these are developed into more detailed drawings. Archaic and symbolically laden materials such as ash, beeswax, rottenstone and soot are beautifully and precisely applied to produce meaning that move the imagery beyond straightforward documentation. These pieces quietly acknowledge the role of those ROC personnel who spent many hours occupying the posts – watching, monitoring and waiting.

Paul Virilio and W. G. Sebald: An Introduction to Orford Ness

Of particular interest when considering questions of (artistic) access and (architectural) legibility, are influential texts by Paul Virilio and the late W. G. Sebald, whose works are recurrently cited by artists, archaeologists, scholars and writers. It is the notion of the initial *encounter* with the militarised / Cold War structure that offers illuminating insights for this discussion. Paul Virilio's enthralment with the concrete fortifications that constituted the Atlantic Wall along the western coast of Europe, for example, says much about the strangely sublime attributes of such monoliths. They surely evoke David E. Nye's concept of the 'technological sublime', in which the heady cocktail of feelings of awe, wonder and terror conventionally accompanying encounters with spectacular natural sites has been reassigned to prodigious engineering and architectural feats. But while these structures date from the early 1940s, as John Beck notes, "Virilio's Atlantic Wall explorations are, I think, as much a response to the concealments of Cold War power as they are an historical excavation of the militarization of space during World War Two" (Beck, 2011: 98). In *Bunker Archaeology* (originally published in 1975), Virilio describes discovering "in the archaeological sense of the word" (1994: 10) one of the concrete masses for the first time. The phenomenological understanding that underlay his academic fascination with 'first contact' is palpable as he looked around the structure: "I was most impressed by a feeling, internal and external, of being immediately crushed" (11). On further reflection, he resolved that he "would hunt these gray forms until they would transmit to me a part of their mystery, a part of the secret a few phrases would sum up" (ibid.). In this encounter description (since the fortifications were chanced upon on holiday), Virilio's first impulse was to make sense of the structure in relation to his own physicality.

This quality of incomprehension, of not knowing but seeking to identify forms from other structures derived from memory, is also a striking feature in the writing of W. G. Sebald. In 1992 the German-born writer made a maiden visit to the former atomic weapon research establishment at Orford Ness in Suffolk. 'The island', as it is locally known, was, and in some ways still is, one of the most secret places in Britain. From 1913 until the early 1980s, it was a military testing site, and many of the activities undertaken there were of national and international significance, including the early invention of radar, atomic bomb environmental testing (where bombs and their components were subjected to the sorts of shocks, g-forces, vibration and extremes of temperature they might undergo before detonation) and Cold War surveillance. As Judith Palmer succinctly puts it "for seventy-two years, the army, navy and air force threw everything they had at Orford Ness and the waters which surround it" (2006: 99). Sebald's opportune visit offers a unique description of a place on the cusp of change (shortly before it was purchased and partially 'tidied up' by the National Trust). Sebald procured access by boat from Orford over the river Ore and then surveyed the Ness's scarred landscape. He observed:

> From a distance, the concrete shells, shored up with stones, in which for most of my lifetime hundreds of boffins had been at work devising new weapon systems, looked (probably because of their odd conical shape) like the tumuli in which the mighty and powerful were buried in prehistoric times with all their tools and utensils, silver and gold. My sense of being on ground intended for purposes transcending the profane was heightened by a number of buildings that resembled temples or pagodas, which seemed quite out of place in these military installations.
>
> *(1999: 235–236)*

Sebald was conducting a conscious and deliberate journey along the stretch of coastal East Anglia (for *The Rings of Saturn*, published in English in 1998). He allies the structures he sees from a distance with memories of equally 'powerful' structures. However, in the absence of any (textual) captioning, he remains mystified: "Where and in what time I truly was that day at Orfordness I cannot say, even now as I write these words" (237). For John Beck, Sebald's allusive 'first contact' with these monstrous presences

> does not ground historical understanding but is . . . radically destabilising . . . The buildings either seem to have been thrown to earth from some other place or else the visitor himself has been transported through time to glimpse the ruins of the future.
>
> *(Beck, 2011: 81)*

In short, "The past is both confirmed and denied by [the ruins'] obstinate presence" (ibid.). The paradoxical aspect of the Cold War–era structures – dating from such recent history yet resisting "assimilation" (ibid.) – is a repeated motif for visiting writers: Duncan McLaren observes that the buildings look like "ancient tumuli, Stone Age burial places . . . timeless and exotic" (McLaren, 2006: 34). For archaeologist John Schofield – in a text entitled *In the*

Figure 30.1   View of the 'pagoda' buildings from Black Beacon.

Source: Louise K. Wilson.

Other-World – the Ness's "material remains are mysterious and enigmatic" (Schofield, 2006: 75). But, as he pertinently reminds us, it "*is* part of the world we have ourselves created" (ibid., my emphasis). As Sebald noted, "during the Cold War, Orford inhabitants could only speculate about what went on at the Orfordness site, which, though, perfectly visible from the town, was effectively no easier to reach than the Nevada desert or an atoll in the South Seas" (233). But after the military finally left the Ness, the site lay empty for about eight years until the National Trust bought it in 1993, by which time it had been badly vandalized. After much discussion, the Trust guardians arrived at a policy of 'continued ruination' for the AWRE labs, some of which are considered iconic architectural landmarks. These were fenced off, and the natural world moved back in:

> When the [. . . .] AWRE relinquished the site they took their military menagerie with them: weapons like Blue Boar, Blue Bunny, Blue Peacock, and their colour-coded companions, Blue Danube, Red Beard, Yellow Sun Violet Club.[4] As the armaments moved out, the wildlife began to move back in: Brown Hare, Black-headed Gull, Red-legged Partridge.
>
> *(Palmer, 2006: 100)*

Animals, especially resident and migratory birds, occupy these buildings and broadcast their (seasonal) presence. Some downy white feathers on the grimy floor of a 'pagoda' building are the sole residue of a duck that became ensnared in the defensive razor wire. For Robert

Macfarlane, the Ness' "militarising influence" inflects vision, so that "everything I saw seemed bellicose, mechanized. A hare exploded from a shingle divot" (Macfarlane, 2008: 257). These bird and other animal tracks are mainly life affirming, however.

These vital presences are, additionally, supplanted by other more 'uncanny' presences that linger or are migratory. There is the persistent sweet, cloying smell carried on the wind from the armoury building, made all the more disturbing for not (apparently) having been identified. More potently here for this discussion, are the odd sounds that permeate buildings and combine with narratives to unsettle the visitor. The wind routinely animates the site, producing noises like oddly tuned musical or percussive instruments.

This charged and rich sensory environment site, then, is infused with multiple possibilities and readings. A cursory response should be mindful of the 'vibrant matter' that adheres to the bunkers and labs at Orford Ness, its "vital and lively properties become gradually more legible as the process of ruination takes its course" (DeSilvey and Edensor, 2012). The matter of the buildings starts to dissolve and merge with the natural. This entanglement provides rich material for aesthetic research, and numerous artists have travelled to the Ness to respond to this compelling and wounded landscape, whether propelled by their own originary interest or through National Trust or outside curatorial invitation. There isn't the space here to detail the artist and writer responses produced in recent times, save to highlight the role of (audio) field recording[5] which has latterly been a significant aspect in resulting time-based works. Emily Richardson's time-lapse film *Cobra Mist* (2008) is strongly enhanced by an evocative score designed by Benedict Drew using field recordings made by Chris Watson. *Untrue Island*, made four years later, featured the aforementioned Robert Macfarlane and Jane and Louise Wilson (whose sculptural interventions into the AWRE buildings included field recordings). Recently Dylan Ryan Byrne produced sound design for a video to contextualise Anya Gallacio's commissioned work referencing the early development of aerial photography and bombardment on the Ness (for SNAP, 2014). This nervy soundtrack of familiar Ness noises (wind and rain), musical elements and shutter clicks creates obvious synchresis as the camera repeatedly surveys Gallacio's scanning electron microscope images of shingle.

Essentially then – and to varying degrees – field recordings inform the mediation of place in these works. For artist and musician Peter Cusack, field recordings offer persuasive potential: effectively conveying "more than basic facts . . . Spectacular or not, they also transmit a powerful sense of spatiality, atmosphere and timing [. . .] They give a compelling impression of what it might actually be like to be there" (Cusack, 2013: 26). He reasons that this locative sound information ultimately "enables us to ask how we might feel and react in the circumstances" (ibid.). Arguably, then, as listeners we are implicated and positioned as active receivers; we are 'touched' by sound and align ourselves (psychologically and intellectually) accordingly. There is an element of novelty too about this foregrounded mediation of (disquieting) place and event. David Toop has written about the recent trend in the use of field recording towards exploration and 'unearthing' of the implicit and subtle resonances of politically unsettling spaces, along with their uncomfortable histories and associations. Toop acknowledges that field recording has varied origins – citing for example bird watching, the World Soundscape project, documentary filmmaking, radio drama and so on – and that these practices have accordingly been located "in looking as much as listening, and an uneasy relationship to landscape as visual spectacle" (Toop, 2007: 326). In the past, he argues, this has led to a tendency towards "the picturesque, benign and static" (ibid.) as sound artists have sought to align themselves with wider environmental issues and screen out or ignore more disturbing or seemingly unwanted aspects of urbanism and human intervention. In recent years, however, there has been a move to actively concentrate on such locations and spaces. Toop asks whether the current sites for (pre)occupation – including Cold War structures – "may have more potential to encourage *deep listening* than the paradise landscapes of the past" (ibid., my emphasis). He answers this by observing that reframing experience (through the engagement with "subliminal perceptions," for example) can prompt "heightened awareness" (ibid.). Field recording then can prompt the desire to attend to sites where layers of history and mythology are densely accreted.

A Record of Fear

The palpable sense of secrecy, strangeness and renewed 'occupation' on Orford Ness became the focus for *A Record of Fear*,[6] a series of audio works, video works, a self-guided walk and book that I produced a decade ago (2005–2006). It formed part of a body of work looking at the cultural and physical legacy of the Cold War. In 2000 I initiated a research project at RAF Spadeadam in Cumbria (where the Blue Streak engines were test-fired and where the test stand ruins still litter this now electronic warfare tactics range). This resulted in the production of a video, a soundwork and a participatory site tour around the concrete remains, in the company of a retired engineer. Following the project on the Ness, I travelled to the South Australian desert in 2006 to make (impulse response and ambient) field recordings and gather film and video material inside a radome at the disused US listening station at Nurrungar and on the rocket range at Woomera.

The research project at Spadeadam led to an increasing interest in using sound (recording and manipulation) as a methodology – specifically wishing to ask how far locations could be understood through sonic traces. This tallied with a desire to 'play' with beliefs around the apparent 'transportability' of the past and present through sound since, to paraphrase anthropologist David Tomas, sounds voyage through social spaces in a manner that makes them hard to define in terms of the subjects, objects and material processes that have 'produced' them and "ambient sonic spaces tend to give rise to ephemeral and fragmentary histories" (1990: 125–126). Tomas was analysing the collecting of audio recordings of 'vanishing sounds' (such as Industrial Revolution–era machinery) to ask what role anthologies play in forging and creating relationships with the past. For John Schofield, the acoustics of the sites themselves offers rich material for artists and archaeologists to consider change over time. He has spoken of his sonic observations across numerous militarized landscapes with "silence representing time before military occupation, and then these bizarre and totally unnatural sounds (some of which will reflect continuing use, and some decay and abandonment) that came after."[7] These two reflections were influential, and I subsequently became interested in the presence of reflected (reverberant) sound and the accumulation of 'archaeological' archive materials in the form of so-called impulse responses (typically produced either by making a loud rapid burst of sound or, more accurately, by generating a swept sine wave in a reverberant space). Reverberation, along with resonance, constitutes the acoustic signature of a room, playing a large part in the subjective qualities of its sound. It directly results from the architecture that produces it and although it is present in caves, rock overhangs and so on, is less common in the natural world. Reverberation reminds us that aural architecture, that is, the composite of multiple surfaces, objects and geometries, can also have a social meaning (Blesser and Salter, 2007: 3). This line of thought led to my wondering what we can specifically tell about a place from the existence of this acoustic 'fallout' that is generally such a feature in test laboratories, tunnels and control rooms. The acoustic phenomenon of reverberation is intrinsically connected with ideas of time and change. Certainly the removal of equipment, technology and people has inevitably meant an increased presence of reflected sound — audible in longer reverberation time. This phenomenon has meant that other iconic defence sites have been harnessed as visually and sonically dynamic temporary venues for live music. Aldeburgh Music used the abandoned Bentwaters Airbase in Suffolk for the *Faster Than Sound* music festival in 2006, and the decaying spy station Teufelsberg in Berlin regularly attracts sound artists and musicians. The latter's specific architectural design (constructed atop of a hill made from Second World War rubble) prompted acoustician Trevor Cox to ask for sonic preservation work to be done:

> There are three radomes on the roof of the disused listening station . . . but two of these have been very badly vandalised. Will someone capture the acoustic signature of the last radome before it, too, becomes damaged and the sound is lost forever?
>
> *(Cox, 2015: 275)*

This question of the extent to which field recordings can be considered valuable archaeological artefacts is apposite. I consider capturing the acoustic signature as analogous to collecting an archaeological artefact.

Practically, ideas for *A Record of Fear* evolved as a cumulative set of experiments during time spent listening and gathering field recordings on the Ness. The process of researching through walks, conversations, reading and so on unfolded over months. Talks with the National Trust guardians in particular yielded invaluable insights and knowledge about the site — its history, obviously, but also subtle observations about the strange sounds, seasonal changes, dangerous encounters but more positive, chance ones, too, with people once connected to the place.

The period of research was ultimately resolved as a one-event day in late September 2005, when visitors to the Ness had the unique opportunity to enter a number of the normally inaccessible test laboratories (closed because of their decrepitude) on a self-guided walk. In certain buildings they would hear specially installed sound works. It was intended that this series of pieces heard over time and foregrounding human and mechanical voices would prompt a moment of reflection on the past and the continuation of the military project. One central strategy here involved the installation of recordings of (female/male) voices — those of the Exmoor Singers. The London-based choir uses the Base Camp barracks at Orford Ness as a weekend rehearsal site once a year before playing concerts in local Suffolk churches and castles. For *A Record of Fear*, the Singers agreed to perform selected madrigals in some of the empty military buildings, principally testing out the acoustics of these spaces, in solo and choral arrangements. The vibrant human presence singing songs of love, yearning and an awareness of the passing of time would provide an affecting counterpoint to the stark and disturbing interiors. One building in particular provided a powerfully resonant visual and aural

Figure 30.2 Video still: Exmoor Singers singing around the WE177A nuclear weapon in the Control Room.

Source: Louise K. Wilson.

Figure 30.3 Exmoor Singers being filmed in Lab 5.

Source: Louise K. Wilson.

Figure 30.4   Video still from *A Record of Fear*, Lab 5.
Source: Louise K. Wilson.

location for recorded singing. Lab 6 was a combined centrifuge and vibration facility built in 1966 to subject electronic missile components, possibly from the Polaris missile, to gravity forces and (random) vibration tests. The 'whispering gallery' effect acoustics were found to be highly conducive for sung performance (recalling the aforementioned Bentwaters and Teufelsberg examples). According to the choir director James Jarvis:

> The most striking impression when singing John Bennet's madrigal 'Weep, O mine eyes' in the centrifuge pit was sonic. Though hardly designed as a performance space, the room's cylindrical shape and reflective walls formed an extraordinary acoustical environment when the singers arranged themselves in a coaxial circle facing into the centre: the sheer volume of sound that could be driven within this resonator by a few voices was overwhelming; the harmonic possibilities of the common chords could be exploited to the nth degree; and the dissonant suspensions had a searing effect.
>
> *(Wilson, 2006: 120)*

Other installations for *A Record of Fear* foregrounded field recordings over human voice such as those central to *Black Beacon Receiver* housed in the viewing gallery of a building (Black

Figure 30.5 Black Beacon Receiver.

Source: Richard Davies.

Beacon) once used to develop an experimental 'rotating loop' navigation device. This octagonal building provides an elevated viewing area over the site through seven viewing slots, akin to a bird hide or other camouflaged shelter constructed to observe wildlife. Looped soundscapes could be heard through a series of earpiece sculptures that looked both peculiar and oddly utilitarian, evoking the mysterious objects found washed up or half-buried on the Ness. Each earpiece 'broadcast' edited sounds appropriate to the view from that window, allowing the exploration of how a viewpoint might or might not appear to correspond with the likely source of sound. Visitors were invited to listen carefully to what could already be heard in the landscape as well as what was generally inaudible to the human ear. An array of contact microphones, hydrophones, ultrasonic recorders and directional and contact microphones had been used to capture the subtle ambient sounds of the site. Timeless natural processes were knitted together with those suggesting transitory human-made dereliction. These sounds, particularly those gathered from the natural world (bat sonar, the sound of a bird trapped in a ventilation duct, grasshoppers, waves hitting the shingle beach, the screech of an owl and so on) became more sinister and difficult to contextualize when recorded and played back in this fashion. The acknowledged uncanniness of the Ness was momentarily channelled and enhanced then through this participatory listening piece.

The sited soundwork in Lab 2 garnered the most response from visitors. This was the installation of a sound recording of an active centrifuge (previously in use in this same building). Lab 2 is now one of the most dilapidated and consequently one of the noisiest buildings on site, with the wind persistently activating loose metal sheeting. Before the building was cleaned out because of contamination (by hazardous materials such as asbestos), a cluster of elder bushes were valiantly occupying the centrifuge pit, giving the impression of "something you might see in a zoo which may have interesting tropical animals in it."[8] Lab 2's centrifuge machinery, "reputed to be the most powerful in the world at this time" (Heazell, 2010: 170), was transferred to AWE Aldermaston in 1971, where it is still in use. After much negotiation, I arranged to visit Aldermaston in order to record this machine in motion. It was a sound that had not been heard before (by those in current charge), there being no need for the technicians to hear its activity. It should be noted that visitors are not permitted to take cameras, phones and sketching materials into this contentious site, well known for maintaining the UK's nuclear deterrent. Ironically I was allowed in to attach boundary microphones to the centrifuge walls, to have the technicians run the machine at full tilt and so effect a situation that creates powerful vibrations that travel through the air and caress contiguous microphones. These recordings, made onto a hard disc recorder, were shared with the technicians and then taken away.

The mixed-down sound was amplified in the original centrifuge pit through paired speakers and a massive infrabass subwoofer: "There's an insistent heavy rhythm, like pterodactyl wingbeats, the fee-fi-fo-fum thump of some monstrous force hurtling faster, menacing, and unbearably loud" (Palmer, 2006: 113). The physical impact was palpable for the viewer. More than the other works, this rhythmic mechanized 'soundtrack' for the site prompted very intense reactions. A number of visitors commented on hearing it being a powerful and emotional experience. One visitor said that it was particularly relevant given what had been

Figure 30.6   Centrifuge (Lab 2).

Source: Louise K. Wilson.

happening in Iraq and another felt it to be "an exorcism of the past." As Judith Palmer noted, as an exercise "rich in paradox: the sound of a machine simultaneously present and absent, an echo of the past that is incontrovertibly current" (ibid.). In an interview with Bradley Garrett, Anja Kanngieser (writer/researcher on sonic geography), cited this centrifuge piece as an affective evocative of place:

> Such audio translations can be utterly compelling in a way that I often find visuals aren't. They can also speak to the politics of spaces and can express both subjective and meta critiques and affirmations of a particular place and its history, without reliance on linguistic and ideological discourses.[9]

This was affirmation for an experiment to find the means to articulate and expose a disregarded sound, one deemed of carrying no security risk. This raw, low-frequency recording (only minimal editing was applied) could be described as carrying possibly sinister associations since

> in many sound cultures, there is a connection between the low frequencies and danger, sadness, or melancholy. This is well illustrated by the western European knell, but also by many warning signals, for example, bells and foghorns, that require broad propagation and therefore must use low frequencies, thus inducing a feeling of fatality.
>
> *(Augoyard and Torgue, 2005: 42)*

This affect is difficult to quantify. Without any captioning, the viewer may experience a response of unease and foreboding then, but readings may be otherwise limited. The potency of being there, looking down at the ruination of the centrifuge pit, provided a stark context for the complex emotional, imaginative and even spiritual responses elicited.

Conclusion

Orford Ness, like other similar sites, can be considered as generative of narrative/rumor/myth, and technological modes of representation positioned as developing or extending these mythologies as opposed to simply capturing or disenchanting a place. In this sense, technology is employed in a transgressive fashion to some extent, perverting any purely scientific aims, to explore the poetic and haunting qualities of gathered/layered sound. Arguably, there is a recurring trope in sound art that is variously articulated as the desire to uncover "hidden resonances and meanings within the memory [of a building] and in particular the subtle traces that people and their actions leave behind" (Scanner quoted in Toop, 2004: 85). The notion of technological 'listening in' on empty spaces is a compelling process for the sound recordist. There is an oddly disturbing intimacy in listening to something through headphones that is not matched by looking at the same scene through a viewfinder. Tiny events and processes can become ominous through the act of recording. This forms part and inflects a wider process – of listening, recording, manipulating, playing back and so on – that has the potential to create an emotional analogue for these difficult, curiously handsome, sometimes banal Cold War sites. This can be viewed as constructive and even cathartic – assembling a transitory moment to explore and expose a painful past.

Notes

1 Huse, N. (1997) *Unbequeme Baudenkmale*. Munich: C. H. Beck. Quoted in Klausmeier et al., 2006, pp. 7–8.
2 In addition "nearly three-quarters believed it was inevitable at some point in the future" (Bourke, 2005: 259).
3 From the interview between Louise K. Wilson and Stephen Felmingham, Union 105, Leeds, April 2011.
4 Though as Palmer qualifies, "The testing of Blue Bunny, Blue Peacock and Violet Club on Orford Ness remains unconfirmed" (2006: 100).
5 "Broadly, field recording can be summarised as a diverse set of practices concerned with recording sound from atmospheric, hydrophonic, geophonic, electro-magnetic and other sources. It is a sprawling pursuit, but resolves toward an interest in creating and transmitting an impression of audition in time." http://www.factmag.com/2014/11/18/a-beginners-guide-to-field-recording/2/ (Accessed 1 June 2015).
6 For the project *Contemporary Art in Historic Places* in 2005, three artists were invited to create new work inspired by historic properties in the East of England. This shared initiative between the National Trust, English Heritage and Commissions East, was developed to 'attract new visitors to heritage sites or to engage existing visitors in new ways' (http://commseast.org.uk/cstudy/NT_EH.html). The properties were varied, with the seventeenth-century country house of Fellbrigg Hall (Norfolk), the linked locations of Dunstable Downs and De Grey Mausoleum (Bedfordshire) and the ex-military history site now nature reserve of Orford Ness (Suffolk) chosen for artists' intervention.
7 Email correspondence, 2003.
8 Interview with archaeologist Angus Wainwright (LKW, 2005, full date unknown).
9 http://www.placehacking.co.uk/2010/10/15/urbex-interview-anja-kanngieser/.

References

Augoyard, J.-F. and Torgue, H. (2005) *Sonic Experience: A Guide to Everyday Sounds*. Toronto: McGill-Queen's University Press.
Beck, J. (2011) Concrete Ambivalence: Inside the Bunker Complex. *Cultural Politics* 7(1), 79–102.
Blesser, B. and Salter, L.-R. (2007) *Spaces Speak, Are You Listening? Experiencing Aural Architecture*. Cambridge, MA: MIT Press.
Boulton, S. (2006) *Re-Mapping the Field: New Approaches in Conflict Archaeology*. Berlin: Westkreuz-Verlag.
Bourke, J. (2005) *Fear: A Cultural History*. London: Virago.
Cocroft, W. and Thomas, R.J.C. (2003) *Cold War: Building for Nuclear Confrontation, 1946–1989*. Swindon: English Heritage.
Cox, T. (2015) *Sonic Wonderland: A Scientific Odyssey of Sound*. London: Vintage.
Cusack, P. (2013) Field Recording as Sonic Journalism. In Carlyle A. and Lane, A. (Eds.), *On Listening*. Axminster: Uniform Books, pp. 20–30.
DeSilvey, C. and Edensor, T. (2012) Reckoning with Ruins. *Progress in Human Geography* 37(4), 465–485.
Heazell, P. (2010) *Most Secret: The Hidden History of Orford Ness*. Stroud: History Press.
Holtorf, C. and Piccini, A. (2009) *Contemporary Archaeologies: Excavating Now*. Frankfurt: Lang.
Huse, N. (1997) *Unbequeme Baudenkmale*. Munich: C. H. Beck.
Kittler, F. (1999) *Gramophone, Film, Typewriter*. Stanford: Stanford University Press.

Klausmeier, A. Purbrick, L. and Schofield, J. (2006) *Re-Mapping the Field: New Approaches in Conflict Archaeology*. Berlin: Westkreuz-Verlag.

Lefebvre, H. (1976) Reflections on the Politics of Space. *Antipode* 8(2), 30–37.

Macfarlane, R. (2008) *The Wild Places*. London: Granta.

McLaren, D. (2006) I Love the Soviet Union and the Soviet Union Loves Me. In Wilson, L. K. (Ed.), *A Record of Fear*. London: Commissions East and National Trust, pp. 30–40.

Palmer, J. (2006) Echoes of Destruction. In Wilson, L. K. (Ed.), *A Record of Fear*. London: Commissions East and National Trust, pp. 97–118.

Piette, A. (2009) *The Literary Cold War: 1945 to Vietnam*. Edinburgh: Edinburgh University Press.

Schofield, J. (2006) In the Other-World. In Wilson, L. K. (Ed.), *A Record of Fear*. London: Commissions East and National Trust, pp. 70–80.

Sebald, W. G. (1999) *Rings of Saturn*. London: Harvill Press.

Slessor, C. (2004) The Deactivated Landscape. In Watson, F. (Eds), *The Hush House: Cold War Sites in England*. Berkeley, CA: Hush House.

SNAP (2014) SNAP: Arts at the Aldeburgh Festival. Available at: http://www.snapaldeburgh. co.uk/2014/art_gallaccio.html (Accessed 2 February 2016).

Tomas, D. (1990) Collapsing Walls or Puffing, Smoking Sea Monsters? Ambient Sonic Spaces, Aural cultures, Marginal Histories. *Public 4–5*. Toronto: Public Access, pp. 125–126.

Toop, D. (2004) *Haunted Weather: Music, Silence and Memory*. London: Serpents Tail.

Toop, D. (2007) Soundscapes. *Uovo* 14 The Bookmakers Ed.: Italy, p. 326.

Toop, D. (2010) *Sinister Resonance: The Mediumship of the Listener*. New York: Continuum.

Virilio, P. (1994) *Bunker Archaeology*. Paris: Les editions du Demi-Cercle. Available at: http://www. snapaldeburgh.co.uk/2014/exhibition2014.html

Wilson, L. K. (2006) *A Record of Fear*. London: Commissions East and National Trust.

INDEX